Eclipse Bulletin:

Total Solar Eclipse of 2017 August 21

Fred Espenak and Jay Anderson

Edition 1.0
May 2015

Eclipse Bulletin: Total Solar Eclipse of 2017 August 21

Astropixels Publishing
P.O. Box 16197
Portal, AZ 85632

Astropixels Publishing Website: *astropixels.com/pubs*

This book may be ordered at: *astropixels.com/pubs/TSE2017.html*

Printed in the United States of America

ISBN 978-1-941983-04-1

Astropixels Publication: AP005 (Version 1.0e)

First Edition

Front Cover: An orthographic projection map illustrates the geographic regions within the path of the Moon's shadows during the total solar eclipse of 2017 August 21. A partial eclipse is visible from the large saddle-shaped zone outline with magenta and cyan curves. The narrow path of the total eclipse appears as the blue track. Map copyright © 2015 by Fred Espenak. More about the eclipse can be found at:
eclipsewise.com/solar/SEnews/TSE2017/TSE2017.html

Back Cover Photo of Fred Espenak: Copyright © 2015 by Patricia Espenak

Back Cover Photo of Jay Anderson: Copyright © 2015 by Fred Espenak

Preface

For over 40 years, the U. S. Naval Observatory produced a series of special publications known as the eclipse circulars. Each contained detailed predictions in the form of tables and maps for an upcoming total or annular solar eclipse and were of vital importance in the planning and execution of eclipse expeditions. Although intended for research scientists, the publications enjoyed an even greater circulation among amateur astronomers and eclipse chasers as the most authoritative guide and technical reference on each eclipse.

When the Naval Observatory terminated its eclipse circular series in 1992, Fred Espenak, then working at NASA's Goddard Space Flight Center, was contacted about producing a similar publication in support of the scientific community. Knowing that weather prospects along an eclipse track play a critical role in expedition planning, he invited Jay Anderson (meteorologist then working at Environment Canada) to join as coauthor, and the NASA eclipse bulletin series was born. From 1993 through 2008, they produced 13 eclipse bulletins in cooperation with the International Astronomical Union's Working Group on Eclipses. Modest funding allowed them to distribute each bulletin to both the professional and lay communities, including educators and the media.

When Espenak retired from NASA in 2009, funding ended for the eclipse bulletins. Various avenues were investigated for alternate funding without success. But with the recent development of print-on-demand publishing, another possibility presented itself. Using print-on-demand, it is no longer necessary to print thousands of books in advance, or handle inventory or shipping. To test it out, the print-on-demand process was used to produce the recent publication *Thousand Year Canon of Solar Eclipses* (Espenak, 2014).

Eclipse Bulletin: Total Solar Eclipse of 2017 August 21 is the first of a new series of publications. It contains tables presenting the umbral shadow path coordinates, a physical ephemeris of the umbra, local circumstances on the central line, topocentric data and path corrections due to the lunar limb profile, and mapping coordinates for the zones of grazing eclipse. High-resolution maps plot the total eclipse path across the USA. They show hundreds of cities and towns in the path, the location of major roads and highways, and the duration of totality with distance from the central line.

Local circumstance tables for more than 1000 cities within the USA provide times of each phase of the eclipse along with the eclipse magnitude and duration. Additional tables cover the eclipse circumstances for cities in Canada, Mexico, and Central and South America as well as Western Europe and North Africa. An exhaustive climatological investigation identifies locations along the eclipse path where the highest probability of favorable weather may be found. Finally, a site inspection report highlights interesting locations along the eclipse track from Oregon to South Carolina.

And now for the first time, the *Eclipse Bulletin* is available in both the standard black & white edition as well as a deluxe color edition. Color offers a significant advantage in revealing details especially in figures, maps, and photographs.

Readers unfamiliar with the *Eclipse Bulletins* should know these books assume an intermediate level or higher understanding of eclipses. There are plenty of introductory books covering eclipses for beginners (see the *Bibliography* section *Further Reading on Eclipses*) so the basics won't be covered here. Nevertheless, an effort has been made to make this new bulletin series more user friendly given the wide audience that has come to embrace these publications. Clear skies on 2017 August 21!

— *Fred Espenak & Jay Anderson*
May 2015

Acknowledgments

I am grateful to my wife, Patricia Totten Espenak, for tirelessly editing the manuscript at multiple stages of its development. My sincere thanks and appreciation go to Prof. Jay M. Pasachoff for his meticulous and thorough review of the manuscript and for his many helpful suggestions. Any remaining typographical, grammatical, or technical errors are solely my responsibility. I also want to thank Michael Zeiler for producing several of his beautiful eclipse maps exclusively for this publication.

A special thank you to Dr. B. Ralph Chou for a comprehensive discussion on solar eclipse eye safety (Sect. 4.1). Dr. Chou is Professor Emeritus, of the School of Optometry & Vision Science at the University of Waterloo with over 40 years of eclipse observing experience. As a leading authority on the subject, Dr. Chou's contribution should help dispel much of the fear and misinformation about safe eclipse viewing.

Finally, I want to thank my mentor, colleague, and friend Jean Meeus for a lifetime of inspiration, technical expertise and fascination in the calculation of solar system phenomena. His 1966 *Canon of Solar Eclipses* started me down the road of eclipse predictions shortly after viewing my first total solar eclipse in 1970.

— *Fred Espenak*
May 2015

I am a lucky person – lucky for the people who helped, mentors who gave advice and knowledge, opportunities that arose unexpectedly and led to new adventures, and parents who instilled a sense of daring and self-sufficiency. But mostly I want to acknowledge the lady to whom I'm married who said "do your own thing and take me along."

— *Jay Anderson*
May 2015

Dedication

To Patricia Totten Espenak, my wife and best friend. We first met in the shadow of the Moon twenty years ago and half a world away, and we've been chasing eclipses together ever since.

To Valerie Anne and Maggie Marie Delos-Reyes, my granddaughters who might see their very first total solar eclipse from their backyard in 2017.

And in loving memory of my parents Fred and Asie Espenak, and my sister Nancy J. Davies.

— *Fred Espenak*
May 2015

To Judy, who looks after me.

— *Jay Anderson*
May 2015

Table of Contents

The total solar eclipse of 2017 August 21 is the first to cross the continental United Stated since 1979. © 2015 Michael Zeiler

Section 1: Eclipse Predictions and Umbral Path

1.1 Introduction

On Monday, 2017 August 21, a total eclipse of the Sun is visible from within a narrow corridor that traverses the United States of America. The path of the Moon's umbral shadow begins in the northern Pacific and crosses the USA from west to east through parts of the following states: Oregon, Idaho, Wyoming, Nebraska, Kansas, Missouri, Illinois, Kentucky, Tennessee, Georgia, North Carolina, and South Carolina[1]. The Moon's penumbral shadow produces a partial eclipse visible from a much larger region covering all of North America.

This is the first total eclipse of the Sun visible from the contiguous United States since 1979. And that eclipse path covered only five states in the Pacific Northwest. The 2017 eclipse crosses a dozen states stretching across the country from the Pacific to Atlantic Oceans. To find an eclipse covering a comparable amount of real estate in the U.S., one must go back nearly a century to 1918. This rarity underscores the significance of the 2017 eclipse, which offers millions of Americans the opportunity to witness totality within 1,500 miles or less from home.

[1] A minuscule amount of real estate in the southwestern corners of Montana and Iowa also fall inside the path of totality.

Figure 1–1: Orthographic Projection Map of 2017 Eclipse Path

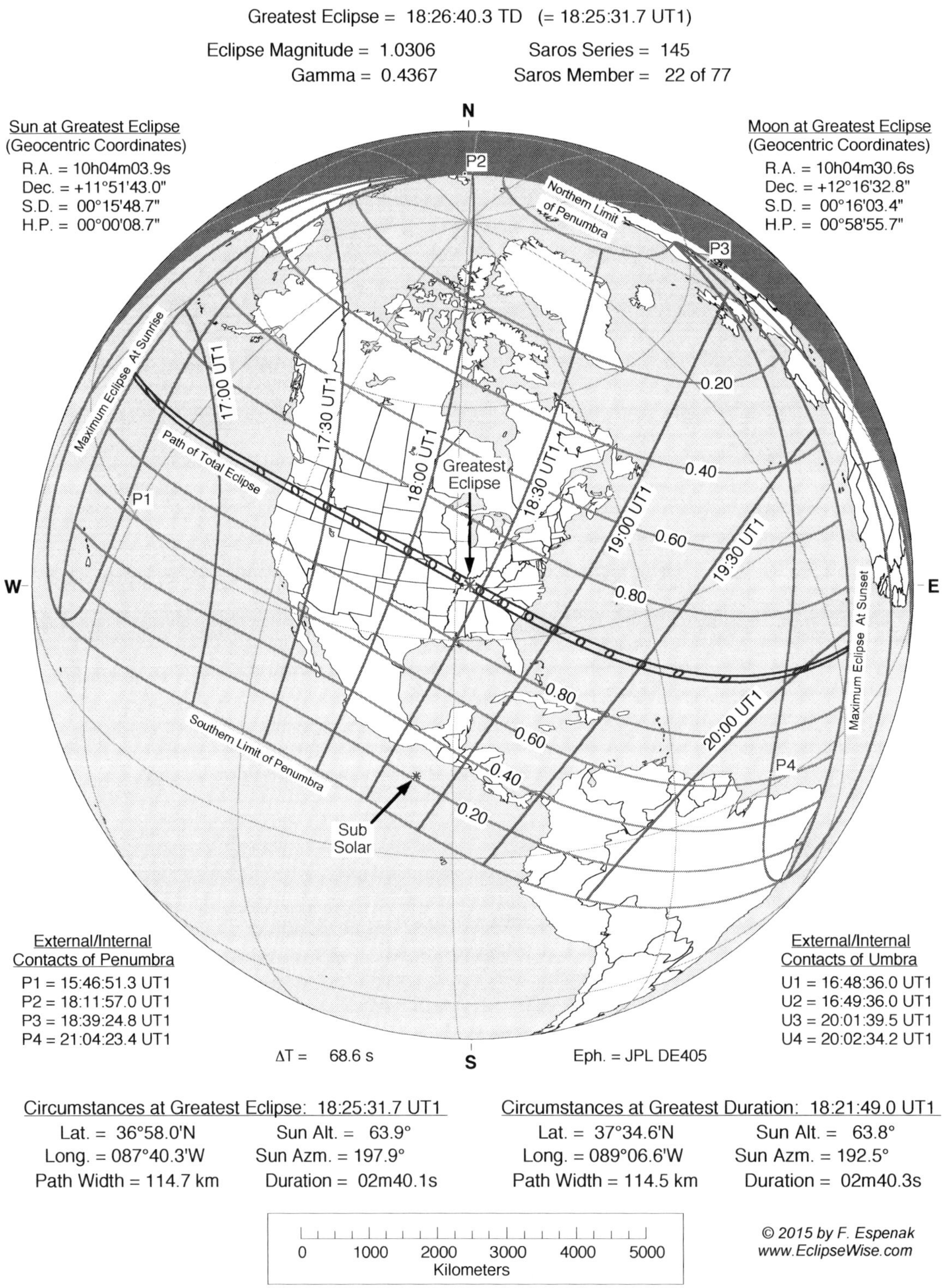

1.2 Orthographic Projection Map of the Eclipse Path

When the Moon's penumbral shadow strikes Earth, a partial eclipse of the Sun is visible from that region. The Moon's apparent motion with respect to the Sun is relatively slow — the partial phases can last about three hours. During this time, the Moon's dark limb slowly creeps across the Sun's disk. Figure 1–1 is an orthographic projection map of Earth (adapted from Espenak 1987) showing the path of penumbral (partial) and umbral (total) eclipse. The daylight terminator is plotted for the instant of greatest eclipse[2] with north at the top. The map is centered over the point of greatest eclipse and is indicated with an asterisk symbol. The sub-solar point[3] at that instant is also shown.

The limits of the Moon's penumbral shadow define the region of visibility of the partial eclipse. This saddle-shaped region often covers more than half of Earth's daylight hemisphere and consists of several distinct zones or limits. At the northern and southern boundaries lie the limits of the penumbra's path. Great loops at the western and eastern extremes of the penumbra's path identify the areas where the eclipse begins and ends at sunrise and sunset, respectively.

Bisecting the *eclipse begins and ends at sunrise and sunset* loops is the curve of maximum eclipse at sunrise (western loop) and sunset (eastern loop). The exterior tangency points *P1* and *P4* mark the coordinates where the penumbral shadow first contacts (partial eclipse begins) and last contacts (partial eclipse ends) Earth's surface. The path of the umbral shadow travels west to east within the penumbral path.

A curve of maximum eclipse is the locus of all points where the eclipse is at maximum at a given time. They are plotted at each half hour in Universal Time, and generally run in a north-south direction. The outline of the umbral shadow is plotted every 10 minutes in Universal Time. Curves of constant eclipse magnitude[4] delineate the locus of all points where the magnitude at maximum eclipse is constant. These curves run exclusively between the *curves of maximum eclipse at sunrise and sunset*. Furthermore, they are quasi-parallel to the northern and southern penumbral limits. These two limits may be thought of as curves of constant magnitude of 0.0, while the adjacent curves are for magnitudes of 0.2, 0.4, 0.6, and 0.8. The northern and southern limits of the path of total eclipse are curves of constant magnitude of 1.0.

Data pertinent to the eclipse also appear with the orthographic map in Figure 1–1. At the top is the instant of greatest eclipse, expressed in Terrestrial Dynamical Time and Universal Time. The eclipse magnitude is the geocentric ratio of diameters of the Moon and the Sun. Gamma is the minimum distance of the Moon's shadow axis from Earth's center in Earth radii at greatest eclipse. Finally, the Saros series of the eclipse is listed, followed by the Saros Member position. The first number identifies the sequence position of the eclipse in the Saros, while the second is the total number of eclipses in the series.

In the upper left and right corners are the geocentric coordinates of the Sun and the Moon, respectively, at the instant of greatest eclipse. They are:

R.A. – Right Ascension
Dec. – Declination
S.D. – Apparent Semi-Diameter
H.P. – Horizontal Parallax

---

[2] The instant of greatest eclipse occurs when the distance between the Moon's shadow axis and Earth's center reaches a minimum.

[3] The sub-solar point is the geographic location where the Sun appears directly overhead (zenith).

[4] Eclipse magnitude is defined as the fraction of the Sun's diameter occulted by the Moon. It is strictly a ratio of diameters and should not be confused with eclipse obscuration, which is a measure of the Sun's surface area occulted by the Moon. Eclipse magnitude is usually expressed as a decimal fraction (e.g., 0.50) but can also appear as a percentage (e.g., 50%).

To the lower left are exterior/interior contact times of the Moon's penumbral shadow with Earth, which are defined:

P1 - Instant of first exterior tangency of Penumbra with Earth's limb. (Partial Eclipse Begins)
P2 - Instant of first interior tangency of Penumbra with Earth's limb.
P3 - Instant of last interior tangency of Penumbra with Earth's limb.
P4 - Instant of last exterior tangency of Penumbra with Earth's limb. (Partial Eclipse Ends)

The lower right corner lists exterior/interior contact times of the Moon's umbral shadow with Earth's limb which are defined as follows:

U1 - Instant of first exterior tangency of Umbra with Earth's limb. (Umbral [Total] Eclipse Begins)
U2 - Instant of first interior tangency of Umbra with Earth's limb.
U3 - Instant of last interior tangency of Umbra with Earth's limb.
U4 - Instant of last exterior tangency of Umbra with Earth's limb. (Umbral [Total] Eclipse Ends)

At bottom left are the geographic coordinates of the position of greatest eclipse along with the local circumstances at that location (i.e., Sun altitude, Sun azimuth, path width and duration of totality).

At bottom right are the corresponding values for the position of greatest duration. Although the geographic locations of greatest duration and greatest eclipse differ by about 145 km (90 miles) and nearly 4 minutes along the eclipse track, the difference in the duration of totality is less than 0.2 second which is of no practical consequence.

1.3 Stereographic Projection Map of the Eclipse Path

The stereographic projection of Earth in Figure 1–2 depicts the paths of penumbral and umbral (total) eclipses. These paths cover a wide range of longitudes and latitudes, wrapping nearly half way around the globe. The stereographic projection is particularly useful for illustrating the sunrise/sunset regions of the path that are badly foreshortened in the orthographic projection used in Figure 1–1. Of course, the stereographic projection has its own shortcomings — it preserves neither distances nor the areas of continents, especially as the distance from the origin increases.

Figure 1–2 is oriented with north at the top. International political borders are shown and circles of latitude and longitude are plotted at 30° increments. The region of penumbral or partial eclipse is identified by its northern and southern limits, the curves of *eclipse begins/ends at sunrise/sunset*, and curves of *maximum eclipse at sunrise/sunset*, all of which are identified. Curves of *constant eclipse magnitude* are plotted for magnitudes 0.20, 0.40, 0.60, and 0.80, as are the limits of the path of total eclipse. Also included are curves of greatest eclipse at every half hour Universal Time.

1.4 Equidistant Conic Projection Map of the Eclipse Path

Figure 1–3 uses an equidistant conic projection chosen to minimize distortion, and that isolates the umbral path through the United States. A number of cities are plotted, as are state and province borders. Curves of maximum eclipse and constant eclipse magnitude are plotted and labeled at intervals of 30 min and 0.1 magnitudes, respectively. A linear scale is included for estimating approximate distances (in kilometers). Within the northern and southern limits of the path of totality, the outline of the umbral shadow is plotted at intervals of 10 minutes. The duration of totality (minutes and seconds) and the Sun's altitude correspond to the local circumstances on the central line at each shadow position.

Figures 1–1, 1–2 and 1–3 may be used to quickly determine the approximate time and magnitude of maximum eclipse at any location within the eclipse path.

Figure 1–2: Stereographic Projection Map of 2017 Eclipse Path

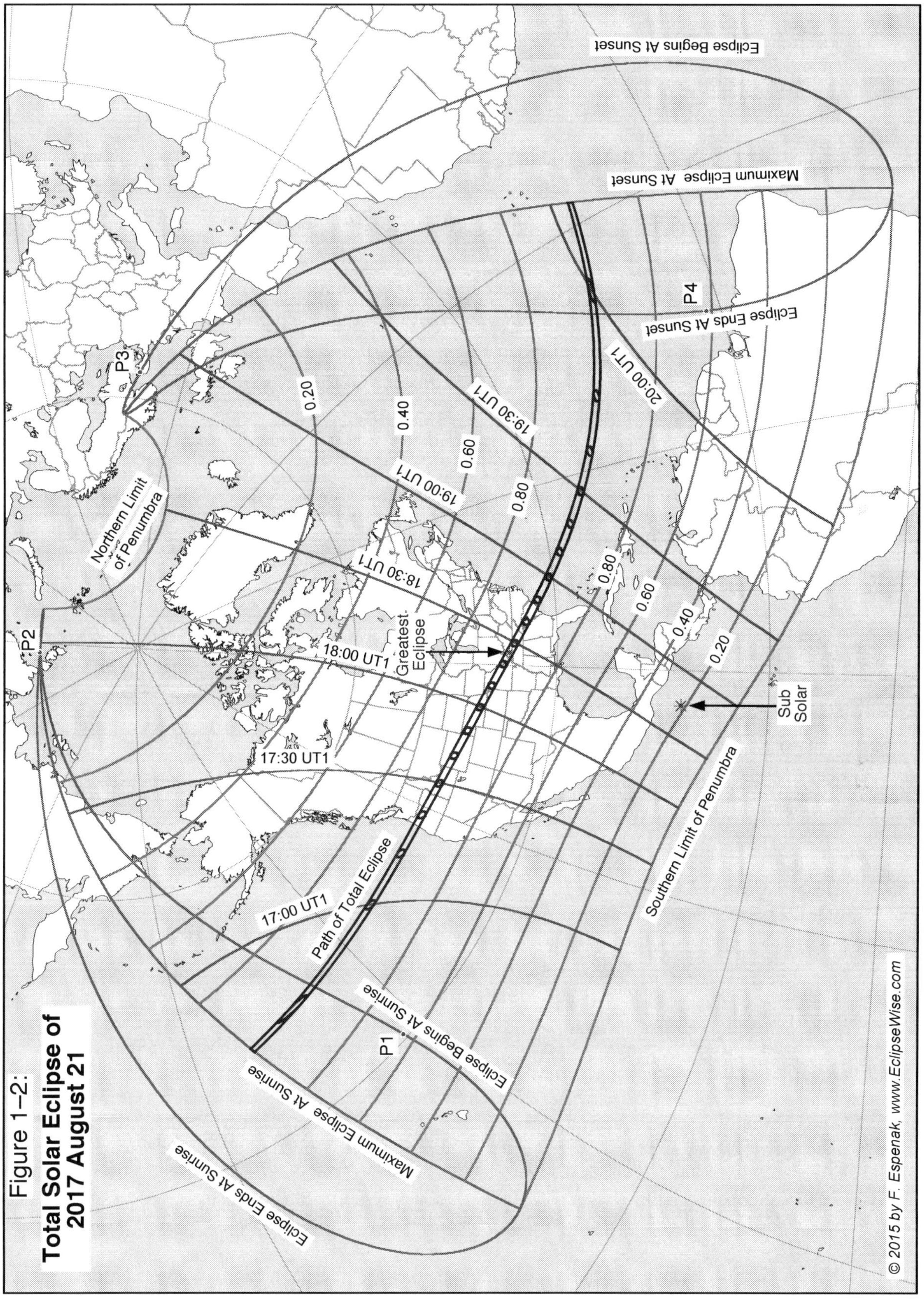

Figure 1–3: Equidistant Conic Projection Map of 2017 Eclipse Path

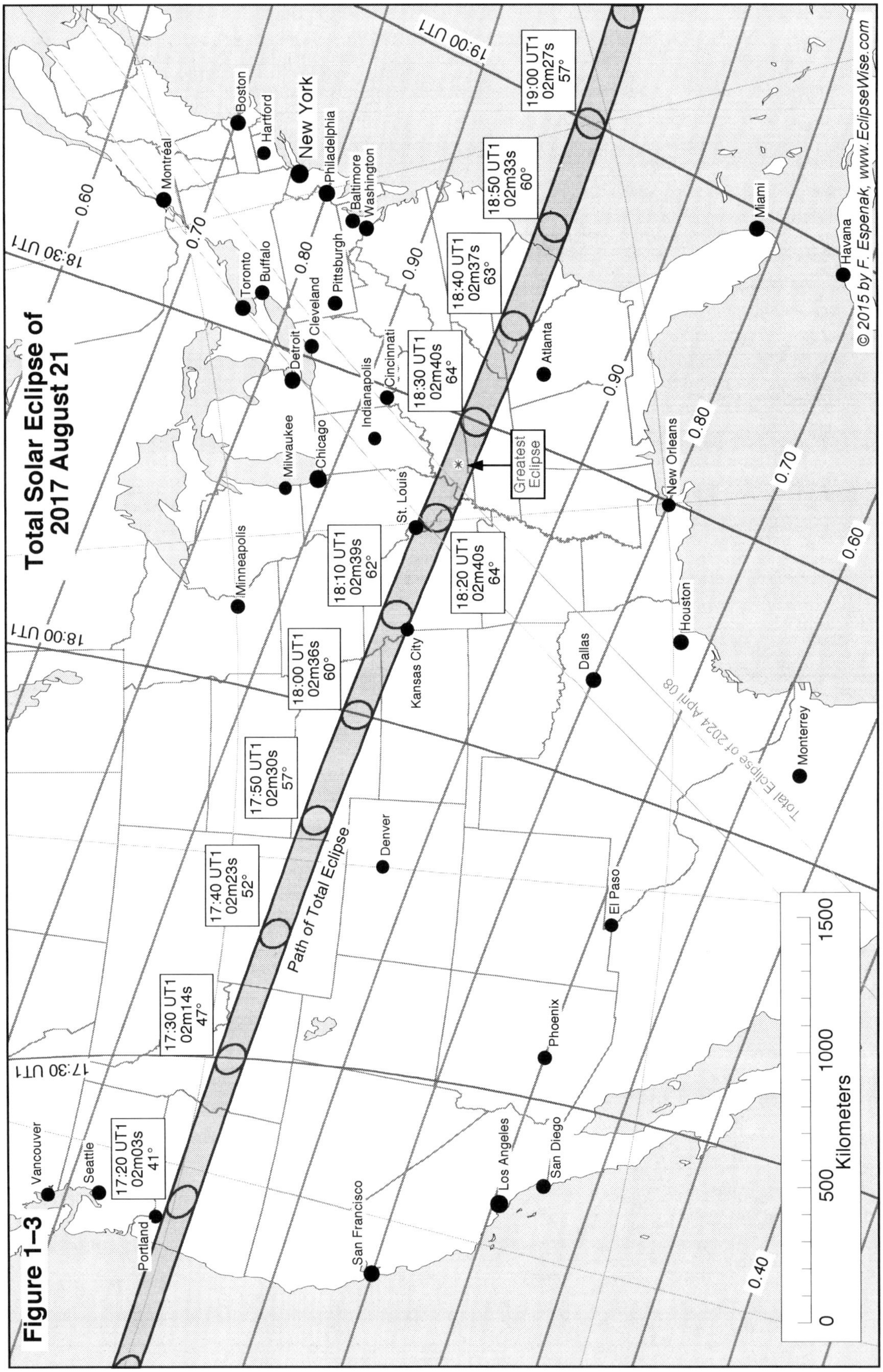

1.5 Description of the Umbral Path

The path of the Moon's umbral shadow begins at 16:49 UT1[5] in the North Pacific about 2400 km (1500 mi) northwest of the Hawaiian Islands. Along the sunrise terminator, the maximum duration is a mere 52 seconds as seen from the center of the narrow, 62 km (38 mi) wide path. No major landfall occurs during the first twenty-eight minutes of the central eclipse as the shadow sweeps across the North Pacific.

The umbra finally reaches terra firma along the western coast of Oregon at 17:16 UT1 (10:16 PDT). The central line passes through the south end of the community of Lincoln Beach. At this locale, it is already mid-morning with the Sun 39° above the eastern horizon. The duration of totality is just shy of 2 minutes (1 minute 59 seconds) and the path width has expanded to 99 km (62 mi) as the shadow pursues its eastward track with a ground velocity of 1.10 km/s (2400 mi/hr). This is still more than three times the speed of sound (0.34 km/s or 760 mi/hr) but the shadow's velocity is decreasing.

Racing over the Coast Range and into the Willamette Valley, the shadow engulfs the cities of Corvallis, Albany, and Salem. Although the central line passes through none of them, they still experience durations of from 1 minute 40 seconds, 1 minute 51 seconds, and 1 minute 55 seconds, respectively, with mid-totality occurring at approximately 17:18 UT1 (10:18 PDT).

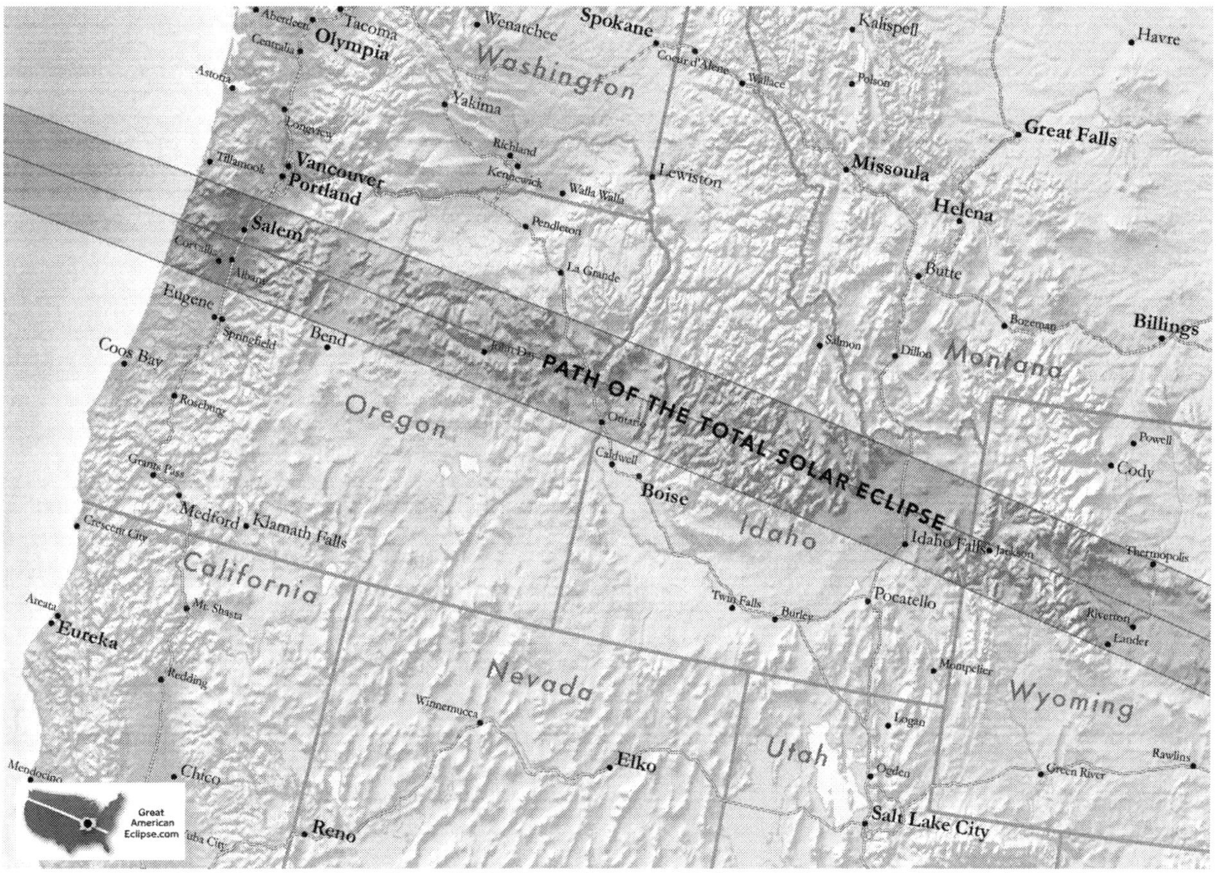

The path of the total solar eclipse of 2017 through Oregon, Idaho and Wyoming. © 2015 Michael Zeiler

The lunar shadow then sweeps across the 3,000-m (10,000-ft) peaks of the Cascade Range. Unfortunately, Mount Hood, one of the highest peaks in the range at 3,429 m (11,249 feet), lies 20 km (12 mi) outside the

---

[5] UT1 or Universal Time is the mean solar time on the Prime Meridian at Greenwich. Civil time signals are transmitted according to Coordinated Universal Time (UTC), which is based on International Atomic Time (TAI) with leap seconds added at irregular intervals to compensate for the slowing of Earth's rotation. The leap seconds keep UTC within 0.9 second of UT1.

path's northern limit to the disappointment of mountaineers hoping to witness totality from the summit. The eclipse here reaches a tantalizing magnitude of 0.995 — so close but still a miss.

The cities of Portland and Eugene are also outside the path and experience partial eclipses of magnitudes 0.991 and 0.992, respectively. While these magnitudes are very high, they won't produce the eerie twilight skies or exquisite views of the Sun's corona. These can only be seen during totality inside the track of the Moon's umbra.

Dropping down into "The Basin," east of the Cascades, the path passes through a region with the best weather prospects along the entire eclipse track. Just 8 km (5 mi) south of the central line lies the small city of Madras. Its 6,400 inhabitants experience a total phase lasting 2 minutes and 2 seconds at a Sun altitude of 42°.

Once past "The Basin", the shadow moves into the Ochoco Mountains, where the average elevation begins to climb. Eventually crossing the peaks of the Columbia Plateau, the path descends to the northwest corner of the Snake River Plain. The axis of the Moon's shadow crosses the Oregon/Idaho border formed by the Snake River at 17:26 UT1 (10:26 PDT). The duration of totality on the central line is 2 minutes and 10 seconds, the path width is 103 km (64 mi) and the shadow's velocity is 0.93 km/s (2080 mi/hr).

Boise, the state capitol and most populous city (pop. 214,200) in Idaho, stands 24 km (15 mi) outside the path's southern limit and receives a deep partial eclipse of magnitude 0.994. Nevertheless, the city makes a good staging point for the 137 km (85 mi) drive to the central line along Interstate 84.

After crossing the Sawtooth Range, the eclipse track descends into the lowlands formed by the Snake River Plain, a several hundred mile wide depression in the Rocky Mountains. Idaho Falls (pop. 58,300) is located inside the path but 33 km (21 mi) south of the central line. The duration of totality is 1 minute and 49 seconds — a 30 second penalty because of its distance from the path's center.

Leaving the Snake River Plain, the umbra enters Wyoming as it climbs over the Teton Range in Teton National Park. This is a popular eclipse-watching destination due in large part to the grandeur of the Tetons themselves. The central line conveniently passes directly though the Jackson Hole Airport, where totality lasts 2 minutes 20 seconds centered on 17:36 UT1 (11:36 MDT). As it continues east, the track crosses the Continental Divide in the Wind River Range and descends to the high plains of eastern Wyoming.

The central line passes through the south side of Casper (pop. 59,600) where the total eclipse lasts 2 minutes 26 seconds centered on 17:44 UT1 (11:44 MDT). This is 14 seconds shy of the longest possible duration that occurs 1600 km (980 mi) to the east in Illinois. The path width in Casper is 110 km (68 mi), the shadow velocity is 0.74 km/s (1655 mi/hr), and the Sun has an altitude of 54°. Because of its strategic location in the eclipse path, the Astronomical League has chosen Casper to host their national convention in the days preceding the eclipse.

As the eclipse track enters Nebraska, the prairie topography gradually descends from high plains to central lowlands. The small city of Alliance (pop. 8,500) lies just north of the central line and gets a 2 minute 30 second total eclipse beginning 17:49 UT1 (11:49 MDT). About 200 km (125 mi) to the southeast, the city of North Platte (pop. 24,500) is inside the path about 15 km (9 mi) north of the southern limit. The duration here is 1 minute 43 seconds — 51 seconds less than the central line. But North Platte's location along Interstate 80 makes it an easy destination to reach, especially if weather becomes an issue on eclipse day.

Strategically speaking, an even better location is Grand Island (pop. 50,500). Not only is it adjacent to I-80, but also the nearby central line offers a duration of 2 minutes 34 seconds. This is within 6 seconds of the maximum duration in southern Illinois. Mid-totality occurs at 18:00 UT1 (13:00 CDT) with the Sun's altitude at 60°, a path width of 112 km (69 mi), and shadow velocity of 0.69 km/s (1,550 mi/hr). Lincoln (pop. 268,700), Nebraska's capitol city, is located just inside the northern limit but larger Omaha (pop. 434,000) is well outside the track with an eclipse magnitude of 0.983.

The Moon's shadow straddles the border between Kansas and Missouri as it leaves Nebraska. Only the northeastern corner of Kansas falls within the umbra's path. Kansas City, the state's third largest city (pop. 148,400), straddles the eclipse track's southern limit. Consequently, the duration of totality ranges from zero (partial eclipse only) to over a minute depending on the observer's exact location. Note that the eastern half of the city is actually in the state of Missouri.

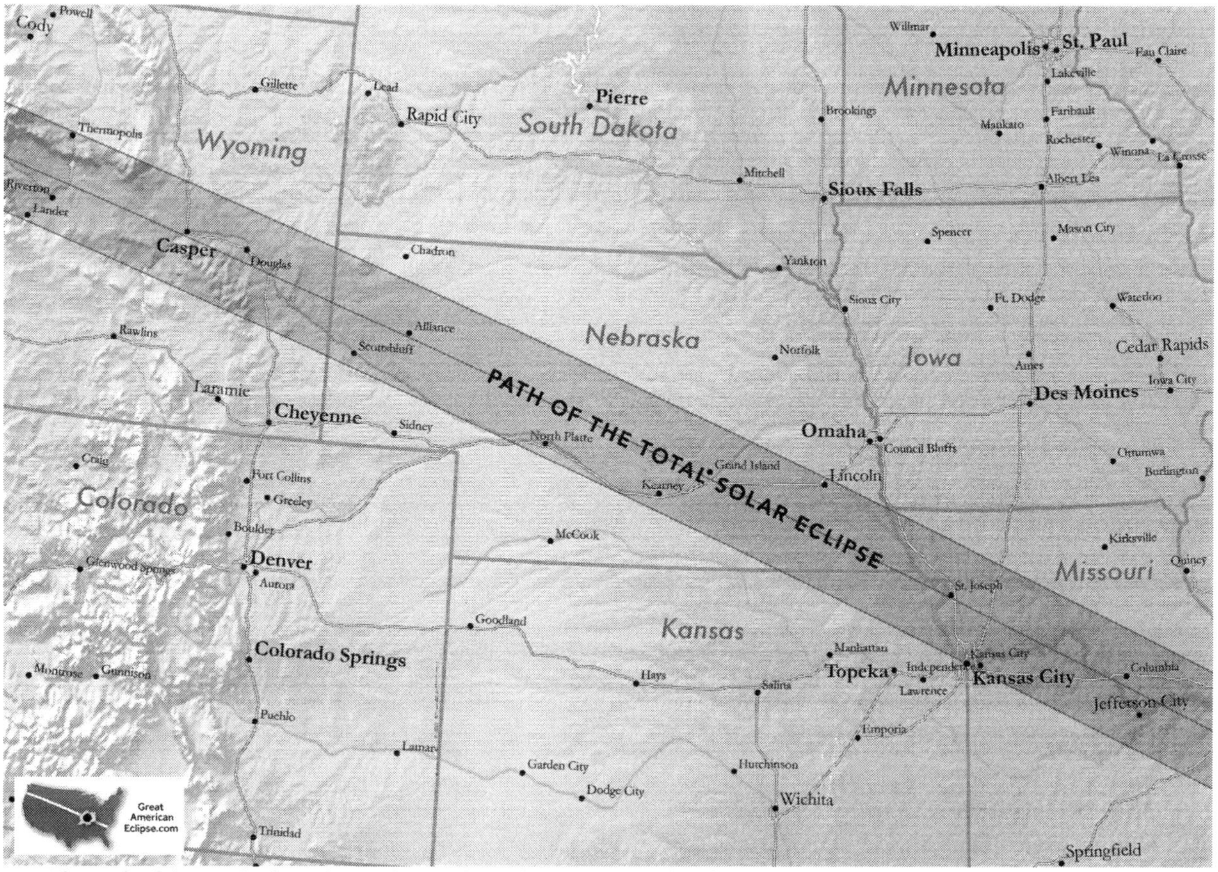

The path of the total solar eclipse of 2017 through Wyoming, Nebraska, Kansas and Missouri. © 2015 Michael Zeiler

In Missouri, the Missouri River crisscrosses the central line several times as the umbra engulfs Columbia (pop. 115,300) and Jefferson City (pop. 43,300). The central line passes between the two municipalities whose inhabitants witness a total eclipse lasting two and a half minutes or more. Further east, St. Louis (pop. 318,400) is bisected by the track's northern limit. As with Kansas City, the duration of totality depends on one's exact position with northern locations only experiencing a partial eclipse.

Crossing the Mississippi River, the Moon's shadow enters Illinois and encounters Carbondale, a small city (pop. 26,300) with a big future in eclipses. That's because the track of the total solar eclipse of 2024 April 08 also passes through the municipality. Furthermore, the greatest duration of totality along the entire track (2 minutes 40.3 seconds) occurs just 20 km (12 mi) southeast of Carbondale. The city itself must be content with a duration 4 seconds shorter due to its location 11 km (7 mi) north of the central line. It's no wonder that some people are calling Carbondale "Eclipse City."

A few minutes after entering Illinois, the shadow leaves the state as it crosses the Ohio River and arrives in Kentucky. It is here that the lunar shadow's axis makes its closest approach to Earth's center — the instant of greatest eclipse — at 18:25:32 UT1 (13:25:32 CDT). At this location (lat. = 36°58.0'N, long. = 087°40.3'W), totality lasts 2 minutes 40.1 seconds, the path width is 115 km (71 mi), the umbral velocity is 0.647 km/s (1447 mi/hr) and the Sun's altitude is 64°.

Hopkinsville (pop. 3,000) lies on the central line only 19 km (12 mi) southeast of greatest eclipse. The duration of totality here (the same as greatest eclipse) is within 0.1 second of the value of greatest duration near Carbondale, IL. Although some people are fixated on fractions of a second and the ultimate duration of totality in selecting an eclipse viewing site, these issues will fade into the noise as cloud avoidance and clear skies ascend to the rank of utmost importance on eclipse day. "A 30-second eclipse in a clear sky beats a 7-minute eclipse behind clouds."

Ever since its Oregon landfall 71 minutes earlier, the eclipse track has been gradually curving further and further south of east. Its azimuth is 115° as it crosses into Tennessee and the umbra begins the final 20 minutes of its trek through the USA.

Nashville (pop. 659,000), known as "Music City" around the world, is inside the path although 40 km (25 mi) south of the central line. Its location means a 44-second penalty but the duration is still a respectable 1 minute 56 seconds. But since Nashville covers a large geographic area, the actual duration is sensitive to an observer's exact position.

Knoxville, Chattanooga, and Atlanta are all near misses (eclipse magnitudes 0.997, 0.993 and 0.971) as the eclipse track crosses the Blue Ridge Mountains separating Tennessee from Georgia and North Carolina. However, only the northeast and southwest corners, respectively, of these two states lie within the path. The shadow quickly sweeps across them before entering South Carolina, the final state in the eclipse track.

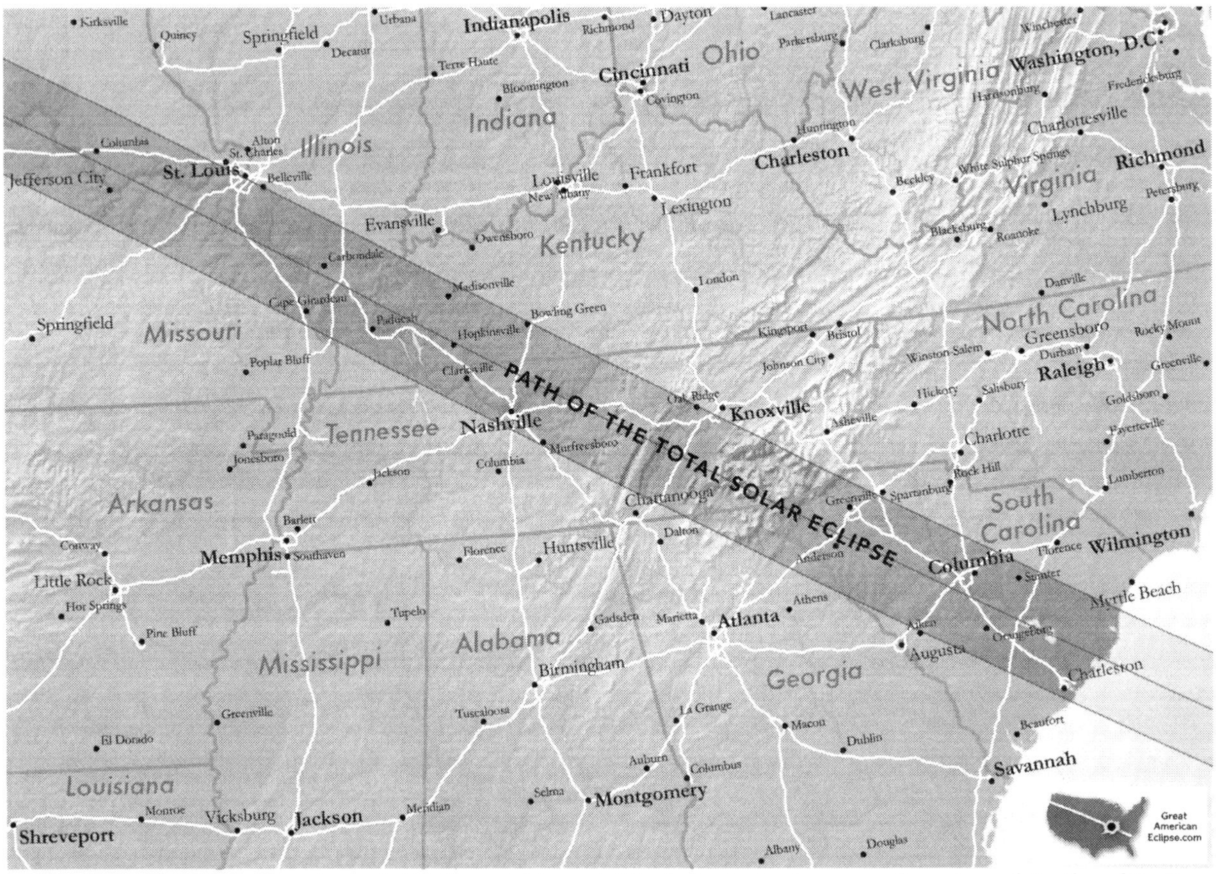

The path of the total solar eclipse of 2017 through Missouri, Illinois, Kentucky, Tennessee, Georgia, and North and South Carolina. © 2015 Michael Zeiler

Columbia (pop. 133,400), the capital and largest city in South Carolina, is 15 km (9 mi) north of the central line. Mid-totality occurs at 18:43 UT1 (14:43 EDT) with a duration of 2 minutes 30 seconds and a Sun altitude of 62°. After its 10-minute trip across the state, the umbra reaches the Atlantic coastline at Francis Marion National Forest. Historic Charleston (pop. 128,000) stands just inside the southern limit — totality there lasts 1 minute 29 seconds centered on 18:47 UT1 (14:47 EDT).

Leaving the American coastline, the lunar shadow begins a 75-minute-long journey covering 5800 km (3600 mi) across the Atlantic Ocean with no further landfall. At 20:02 UT1, the umbra reaches its terminus 630 km (390 mi) southwest of Cape Verde as it lifts off Earth's surface and returns to space. Over the course of 3 hours 13 minutes, the umbra travels along a 13,900 km (8640 mi) track that covers an area equal to 0.26% of Earth's surface.

1.6 Besselian Elements and Eclipse Path Tables

Section 1.6 contains a series of tables that characterize the eclipse and delineate the central umbral path of totality. The geocentric ephemeris for the Sun and Moon, various parameters, constants, and the Besselian elements (polynomial form) are given in Table 1–1. The coordinates of the Sun and the Moon were determined using the JPL DE405 (Jet Propulsion Laboratory Developmental Ephemeris 405). The DE405 is based upon the International Celestial Reference Frame (ICRF), the adopted reference frame of the International Astronomical Union (IAU). It includes the effects of both nutation and libration.

Unless otherwise stated, all predictions are based on center of mass positions for the Moon and Sun with no corrections made for center of figure, lunar limb profile, or atmospheric refraction. The predictions depart from normal International Astronomical Union (IAU) convention through the use of a smaller constant for the mean lunar radius k for all umbral contacts (see Sect. 1.9 "Lunar Limb Profile"). Times are expressed in either Terrestrial Dynamical Time (TDT) or in Universal Time, where the best value of ΔT (the difference between Terrestrial Dynamical Time and Universal Time) available at the time of preparation, is used.

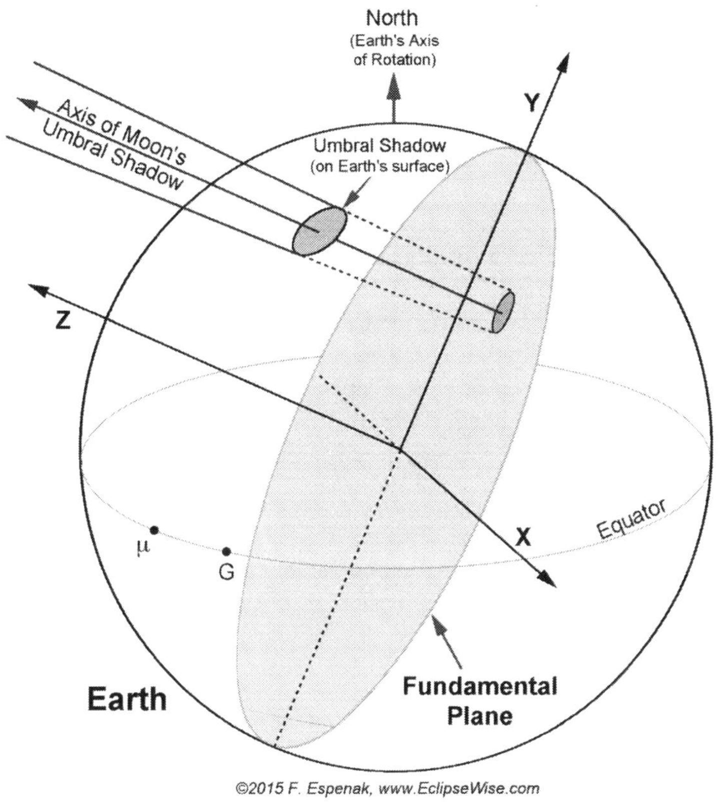

The Fundamental Plane passes through the center of Earth and is perpendicular to the Moon's shadow axis. The Besselian elements for the eclipse are defined with respect to the Fundamental Plane.

The Besselian elements are used to predict all aspects of a solar eclipse. The simplified geometry introduced by Bessel in 1824 transforms the orbital motions of the Sun and Moon into the position, motion, and size of the Moon's penumbral and umbral shadows with respect to a plane passing through Earth. This *fundamental plane* is constructed in an x–y rectangular coordinate system with its origin at Earth's center. The axes are oriented with north in the positive y direction and east in the positive x direction. The z-axis is perpendicular to the plane and parallel to the shadow axis.

The x and y coordinates of the shadow axis are expressed in units of the equatorial radius of Earth. The radii of the penumbral and umbral shadows on the fundamental plane are l_1 and l_2, respectively. The direction of the shadow axis on the celestial sphere is defined by its declination d and ephemeris hour angle μ. Finally, the angles that the penumbral and umbral shadow cones make with the shadow axis are expressed as f_1 and f_2, respectively. The details of actual eclipse calculations can be found in the *Explanatory Supplement* (Her Majesty's Nautical Almanac Office 1974) and *Elements of Solar Eclipses* (Meeus 1989).

From the polynomial form of the Besselian elements, any element can be evaluated for any time t_1 (in decimal hours) during the eclipse via the equation

$$a = a_0 + a_1 t + a_2 t^2 + a_3 t^3 \qquad (1\text{–}1)$$

$$(\text{or } a = \sum [a_n t^n];\ n = 0 \text{ to } 3),$$

where $a = x, y, d, l_1, l_2,$ or μ; and $t = t_1 - t_0$ (decimal hours) and $t_0 = 18.00$ TDT (2017 August 21).

The polynomial Besselian elements were derived from a least-squares fit to elements rigorously calculated at five separate times over a 6-hour period centered at t_0.

Table 1–2 lists all external and internal contacts of penumbral and umbral shadows with Earth. They include TDT and geodetic coordinates with and without corrections for ΔT.

The contacts are defined:

P1—Instant of first external tangency of penumbral shadow cone with Earth's limb (partial eclipse begins).
P2—Instant of first internal tangency of penumbral shadow cone with Earth's limb.
P3—Instant of last internal tangency of penumbral shadow cone with Earth's limb.
P4—Instant of last external tangency of penumbral shadow cone with Earth's limb (partial eclipse ends).

U1—Instant of first external tangency of umbral shadow cone with Earth's limb (total eclipse begins).
U2—Instant of first internal tangency of umbral shadow cone with Earth's limb.
U3—Instant of last internal tangency of umbral shadow cone with Earth's limb.
U4—Instant of last external tangency of umbral shadow cone with Earth's limb (total eclipse ends).

Similarly, the northern and southern extremes of the penumbral and umbral paths, and extreme limits of the central line are given. The IAU longitude convention is used throughout this publication (i.e., for longitude, east is positive and west is negative; for latitude, north is positive and south is negative).

The path of the umbral shadow is delineated at 5-minute intervals (in Universal Time) in Table 1–3. Coordinates of the northern limit, the southern limit, and the central line are listed to the nearest tenth of an arc minute (≈185 m at the Equator). The Sun's altitude, path width, and duration are calculated for the central line position. Table 1–4 presents a physical ephemeris for the umbral shadow at 5-minute intervals in Universal Time. The central line coordinates are followed by the topocentric ratio of the apparent diameters of the Moon and Sun, the eclipse obscuration (defined as the fraction of the Sun's surface area occulted by the Moon), and the Sun's altitude and azimuth at that instant. The umbral path width, the umbral shadow's major and minor axes, and its instantaneous velocity with respect to Earth's surface are included. Finally, the central line duration of the total phase is given.

Local circumstances for each central line position, listed in Table 1–3, are presented in Table 1–5. The first three columns give the Universal Time of maximum eclipse, the central line duration of totality, and the altitude of the Sun at that instant. The following columns list each of the four eclipse contact times followed by their related contact position angles and the corresponding altitude of the Sun. The four contacts identify significant stages in the progress of the eclipse. They are defined as follows:

First Contact: Instant of first external tangency between the Moon and Sun (partial eclipse begins).
Second Contact: Instant of first internal tangency between the Moon and Sun (total eclipse begins).
Third Contact: Instant of last internal tangency between the Moon and Sun (total eclipse ends).
Fourth Contact: Instant of last external tangency between the Moon and Sun (partial eclipse ends).

Major Contacts for a Total Solar Eclipse

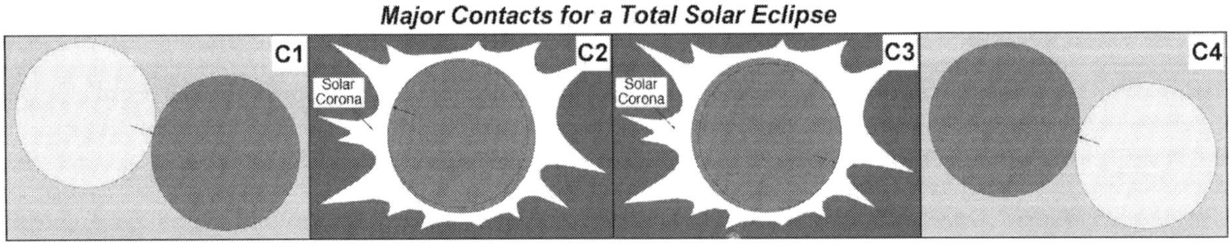

First Contact *(partial eclipse begins)* | *Second Contact* *(total eclipse begins)* | *Third Contact* *(total eclipse ends)* | *Fourth Contact* *(partial eclipse ends)*

©2015 F. Espenak, www.EclipseWise.com

The position angles P and V (P is defined as the contact angle measured counterclockwise from the equatorial north point of the Sun's disk, and V is defined as the contact angle measured counterclockwise from the local zenith point of the Sun's disk) identify the point along the Sun's disk where each contact occurs. Second and third contact altitudes are omitted because they are always within 1° of the altitude at maximum eclipse.

Table 1–6 presents topocentric values from the umbral path at maximum eclipse for the Moon's horizontal parallax, semi-diameter, relative angular velocity with respect to the Sun, and libration in longitude. The altitude and azimuth of the Sun are given along with the azimuth of the umbral path. The northern limit position angle identifies the point on the lunar disk defining the umbral path's northern limit. It is measured counterclockwise from the equatorial north point of the Moon. In addition, corrections to the path limits due to the lunar limb profile are listed (minutes of arc in latitude).

The irregular profile of the Moon results in a zone of "grazing eclipse" at each limit, which is delineated by interior and exterior contacts of lunar features with the Sun's limb. This geometry is described in greater detail in the Sect. 1.11 "Limb Corrections to the Path Limits: Graze Zones." Corrections to central line durations due to the lunar limb profile are also included. When added to the durations in Tables 1–3, 1–4, 1–5, and 1–7, a slightly longer central total phase is predicted along the path because of several high mountainous regions along the Moon's eastern and western limbs.

To aid and assist in the plotting of the umbral path on large scale maps, the path coordinates are also tabulated at 0.5° intervals in longitude in Tables 1–7a and 1–7b. The latitude of the northern limit, southern limit, and central line for each longitude is tabulated to the nearest hundredth of an arc minute (≈18.5 m at the Equator) along with the Universal Time of maximum eclipse at the central line position. Finally, local circumstances on the central line at maximum eclipse are listed and include the Sun's altitude and azimuth, the umbral path width, and the central duration of totality.

Tables 1–8a and 1–8b list the coordinates of the zones of grazing eclipse at 0.5° intervals in longitude. The time of maximum eclipse is given at both northern and southern limits, as well as the path's azimuth. The elevation and scale factors are also given (see Sect. 1.11 "Limb Corrections to the Path Limits: Graze Zones").

Expanded versions of Tables 1–7a & 1–7b and 1–8a & 1–8b using longitude steps of 7.5′ are available at:

eclipsewise.com/solar/SEnews/TSE2017/extra.html

Photo 1–1: The diamond ring effect heralds the end of the total solar eclipse of 1991 July 11.
© 1991 F. Espenak, www.MrEclipse.com

Table 1–1: Elements of the Total Solar Eclipse of 2017 August 21

Equatorial Conjunction: (Sun & Moon in R.A.)	18:14:22.83 TDT (=18:13:14.28 UT1)	J.D. = 2457987.259986
Ecliptic Conjunction: (Sun & Moon in Ec. Lo.)	18:31:19.58 TDT (=18:30:11.03 UT1)	J.D. = 2457987.271754
Instant of Greatest Eclipse:	18:26:40.26 TDT (=18:25:31.70 UT1)	J.D. = 2457987.268521

Geocentric Coordinates of Sun & Moon at Greatest Eclipse:

Sun:		Moon:	
R.A. =	10h04m03.939s	R.A. =	10h04m30.586s
Dec. =	+11°51'42.95"	Dec. =	+12°16'32.78"
Semi-Diameter =	15'48.68"	Semi-Diameter =	16'03.40"
Eq.Hor.Par. =	08.69"	Eq.Hor.Par. =	0°58'55.73"
Δ R.A. =	9.232s/h	Δ R.A. =	139.413s/h
Δ Dec. =	-50.27"/h	Δ Dec. =	-549.61"/h

Lunar Radius Constants: k1 = 0.2725076 (Penumbra), k2 = 0.2722810 (Umbra)

Shift in Lunar Position: Δb = 0.00", Δl = 0.00"

Geocentric Libration: (Optical + Physical) l = 4.6°, b = -0.6°, c = 21.9°

Brown Lun. No. = 1171

Saros Series = 145 (22/77)

nDot = -25.83 "/cy**2

Eclipse Magnitude = 1.03059   Gamma = 0.43671   ΔT = 68.6 s

Ephemeris = JPL DE405

Polynomial Besselian Elements for: 2017 August 21 18:00:00.0 TDT (=t0)

n	x	y	d	l_1	l_2	μ
0	-0.1295763	0.4854175	11.8669672	0.5421118	-0.0040253	89.245445
1	0.5406409	-0.1416394	-0.0136216	0.0001241	0.0001235	15.003937
2	-0.0000293	-0.0000905	-0.0000025	-0.0000118	-0.0000117	0.000001
3	-0.0000081	0.0000021	0.0000000	0.0000000	0.0000000	0.000000

Tan f_1 = 0.0046222   Tan f_2 = 0.0045992

At time t_1 (decimal hours), each Besselian element is evaluated by:

$$x = x_0 + x_1*t + x_2*t^2 + x_3*t^3 \quad (\text{or } x = \Sigma\ [x_n*t^n];\ n = 0 \text{ to } 3)$$

where: $t = t_1 - t_0$ (decimal hours) and $t_0 = 18.000$

The Besselian elements were derived from a least-squares fit to elements calculated at five uniformly spaced times over a six hour period centered at t0.
The Besselian elements are valid over the period 15.00 ≤ t_0 ≤ 21.00 TDT.

Note that all times are expressed in Terrestrial Dynamical Time (TDT).

Saros Series 145: Member 22 of 77 eclipses in series.

Table 1–2: Shadow Contacts and Circumstances

ΔT = 68.6 s
= 00°17'11.1"

		Terrestrial Dynamical Time (h m s)	Latitude	Ephemeris Longitude†	True Longitude*
External/Internal Contacts of Penumbra:	P1	15:47:59.9	30°32.7'N	153°22.7'W	153°05.5'W
	P2	18:13:05.6	78°02.9'N	094°30.8'E	094°48.0'E
	P3	18:40:33.4	57°00.0'N	009°28.6'E	009°45.8'E
	P4	21:05:31.9	01°42.4'N	045°17.0'W	044°59.8'W
Extreme North/South Limits of Penumbral Path:	N1	18:07:23.4	77°34.0'N	071°15.0'E	071°32.2'E
	S1	16:34:08.0	08°38.9'N	159°36.6'W	159°19.4'W
	N2	18:46:18.2	61°56.6'N	012°22.3'E	012°39.5'E
	S2	20:19:10.7	20°16.9'S	038°29.5'W	038°12.3'W
External/Internal Contacts of Umbra:	U1	16:49:44.5	39°37.5'N	171°42.6'W	171°25.4'W
	U2	16:50:44.6	39°51.0'N	172°02.5'W	171°45.3'W
	U3	20:02:48.0	11°07.5'N	027°35.5'W	027°18.3'W
	U4	20:03:42.8	10°55.0'N	027°51.9'W	027°34.7'W
Extreme North/South Limits of Umbral Path:	N1	16:50:30.0	39°59.7'N	172°02.0'W	171°44.8'W
	S1	16:49:59.2	39°28.8'N	171°43.2'W	171°26.0'W
	N2	20:03:01.3	11°15.6'N	027°37.0'W	027°19.8'W
	S2	20:03:29.3	10°46.9'N	027°50.3'W	027°33.1'W
Extreme Limits of Central Line:	C1	16:50:14.5	39°44.2'N	171°52.5'W	171°35.4'W
	C2	20:03:15.4	11°01.2'N	027°43.7'W	027°26.5'W
Instant of Greatest Eclipse:	G0	18:26:40.3	36°58.0'N	087°57.5'W	087°40.3'W
Instant of Greatest Duration:	GD	18:22:57.6	37°34.6'N	089°23.8'W	089°06.6'W

Circumstances at Greatest Eclipse:	Sun's Altitude = 63.9° Sun's Azimuth = 197.9°	Path Width = 114.7 km Central Duration = 02m40.12s
Circumstances at Greatest Duration:	Sun's Altitude = 63.8° Sun's Azimuth = 192.5°	Path Width = 114.5 km Central Duration = 02m40.25s

Distance (GD-GE) = 144.41 km   Δ Duration (GD-GE) = 0.131 s

---

† Ephemeris Longitude is the terrestrial dynamical longitude assuming a uniformly rotating Earth.

* True Longitude is calculated by correcting the Ephemeris Longitude for the non-uniform rotation of Earth.
(T.L. = E.L. + 1.002738*ΔT/240, where ΔT(seconds) = TDT - UT1)

Note: Longitude is measured positive to the East.

Because ΔT is not known in advance, the value used in the predictions is an extrapolation based on pre-2016 measurements. Nevertheless, the actual value is expected to fall within ±0.2 seconds of the estimated ΔT used here.

Table 1–3: Path of the Umbral Shadow

Universal Time (UT1)	Northern Limit Latitude	Northern Limit Longitude	Southern Limit Latitude	Southern Limit Longitude	Central Line Latitude	Central Line Longitude	Sun Alt °	Path Width km	Central Durat.
Limits	39°59.7'N	171°44.8'W	39°28.8'N	171°26.0'W	39°44.2'N	171°35.4'W	0	62	00m51.6s
16:50	41°29.9'N	164°29.4'W	41°25.3'N	161°23.7'W	41°29.3'N	162°50.3'W	7	70	01m01.7s
16:55	43°52.0'N	150°05.3'W	43°18.2'N	148°38.8'W	43°35.4'N	149°21.2'W	18	81	01m19.4s
17:00	44°45.1'N	142°02.6'W	44°02.9'N	140°57.5'W	44°24.2'N	141°29.4'W	24	87	01m30.9s
17:05	45°10.9'N	135°48.4'W	44°23.5'N	134°56.6'W	44°47.3'N	135°22.0'W	30	92	01m40.4s
17:10	45°21.0'N	130°33.4'W	44°30.0'N	129°52.1'W	44°55.6'N	130°12.3'W	34	95	01m48.6s
17:15	45°20.5'N	125°57.6'W	44°26.7'N	125°25.1'W	44°53.6'N	125°41.0'W	38	98	01m55.9s
17:20	45°11.9'N	121°50.5'W	44°16.0'N	121°25.9'W	44°44.0'N	121°37.8'W	41	101	02m02.6s
17:25	44°56.9'N	118°05.6'W	43°59.5'N	117°48.2'W	44°28.2'N	117°56.6'W	44	103	02m08.6s
17:30	44°36.7'N	114°38.9'W	43°38.2'N	114°28.0'W	44°07.4'N	114°33.2'W	47	105	02m14.0s
17:35	44°12.0'N	111°27.2'W	43°12.8'N	111°22.3'W	43°42.4'N	111°24.5'W	50	106	02m18.9s
17:40	43°43.6'N	108°28.2'W	42°43.9'N	108°28.8'W	43°13.7'N	108°28.4'W	52	108	02m23.3s
17:45	43°11.8'N	105°40.2'W	42°11.9'N	105°46.0'W	42°41.8'N	105°43.0'W	55	109	02m27.1s
17:50	42°37.1'N	103°01.8'W	41°37.2'N	103°12.2'W	42°07.1'N	103°06.9'W	57	110	02m30.5s
17:55	41°59.7'N	100°31.6'W	41°00.0'N	100°46.4'W	41°29.9'N	100°39.0'W	58	111	02m33.3s
18:00	41°20.0'N	098°08.8'W	40°20.6'N	098°27.6'W	40°50.3'N	098°18.2'W	60	112	02m35.7s
18:05	40°38.0'N	095°52.3'W	39°39.2'N	096°14.8'W	40°08.6'N	096°03.6'W	61	113	02m37.5s
18:10	39°54.0'N	093°41.5'W	38°55.8'N	094°07.2'W	39°24.9'N	093°54.4'W	62	113	02m38.9s
18:15	39°08.1'N	091°35.5'W	38°10.7'N	092°04.3'W	38°39.4'N	091°49.9'W	63	114	02m39.8s
18:20	38°20.5'N	089°33.6'W	37°23.8'N	090°05.2'W	37°52.2'N	089°49.5'W	64	114	02m40.2s
18:25	37°31.1'N	087°35.3'W	36°35.4'N	088°09.4'W	37°03.2'N	087°52.5'W	64	115	02m40.2s
18:30	36°40.1'N	085°39.9'W	35°45.3'N	086°16.3'W	36°12.7'N	085°58.2'W	64	115	02m39.6s
18:35	35°47.4'N	083°46.9'W	34°53.7'N	084°25.4'W	35°20.6'N	084°06.2'W	63	115	02m38.6s
18:40	34°53.2'N	081°55.5'W	34°00.6'N	082°35.9'W	34°26.9'N	082°15.8'W	63	115	02m37.2s
18:45	33°57.3'N	080°05.3'W	33°06.0'N	080°47.4'W	33°31.7'N	080°26.5'W	62	115	02m35.2s
18:50	32°59.9'N	078°15.5'W	32°09.7'N	078°59.2'W	32°34.8'N	078°37.5'W	60	115	02m32.9s
18:55	32°00.7'N	076°25.6'W	31°11.9'N	077°10.7'W	31°36.4'N	076°48.3'W	59	114	02m30.1s
19:00	30°59.9'N	074°34.8'W	30°12.4'N	075°21.2'W	30°36.2'N	074°58.1'W	57	114	02m26.8s
19:05	29°57.2'N	072°42.2'W	29°11.1'N	073°29.8'W	29°34.2'N	073°06.2'W	55	113	02m23.2s
19:10	28°52.5'N	070°47.0'W	28°07.9'N	071°35.7'W	28°30.3'N	071°11.5'W	53	112	02m19.0s
19:15	27°45.7'N	068°48.1'W	27°02.6'N	069°37.8'W	27°24.2'N	069°13.1'W	51	111	02m14.5s
19:20	26°36.4'N	066°44.1'W	25°54.9'N	067°34.8'W	26°15.7'N	067°09.6'W	48	109	02m09.5s
19:25	25°24.4'N	064°33.2'W	24°44.6'N	065°25.0'W	25°04.6'N	064°59.3'W	45	107	02m04.0s
19:30	24°09.1'N	062°13.4'W	23°31.3'N	063°06.3'W	23°50.3'N	062°40.0'W	42	105	01m58.1s
19:35	22°49.9'N	059°41.4'W	22°14.1'N	060°35.5'W	22°32.1'N	060°08.6'W	39	102	01m51.6s
19:40	21°25.8'N	056°52.7'W	20°52.4'N	057°48.4'W	21°09.2'N	057°20.7'W	35	99	01m44.5s
19:45	19°55.1'N	053°39.8'W	19°24.4'N	054°37.9'W	19°39.9'N	054°09.0'W	31	95	01m36.6s
19:50	18°14.8'N	049°49.5'W	17°47.6'N	050°51.2'W	18°01.3'N	050°20.6'W	26	89	01m27.7s
19:55	16°18.2'N	044°51.3'W	15°55.9'N	046°00.6'W	16°07.2'N	045°26.2'W	20	82	01m17.1s
20:00	13°39.5'N	036°48.2'W	13°28.6'N	038°27.6'W	13°34.5'N	037°39.1'W	11	71	01m02.5s
Limits	11°15.6'N	027°19.8'W	10°46.9'N	027°33.1'W	11°01.2'N	027°26.5'W	0	57	00m47.1s

Table 1–4: Physical Ephemeris of the Umbral Shadow

Universal Time (UT1)	Central Line Latitude	Central Line Longitude	Diameter Ratio	Eclipse Obscur.	Sun Alt °	Sun Azm °	Path Width km	Major Axis km	Minor Axis km	Umbra Veloc. km/s	Central Duration
16:49.1	39°44.2'N	171°35.4'W	1.0156	1.0314	0.0	74.5	62.2	-	53.4	-	00m51.6s
16:50	41°29.3'N	162°50.3'W	1.0177	1.0357	7.0	80.3	70.0	497.3	60.5	6.971	01m01.7s
16:55	43°35.4'N	149°21.2'W	1.0209	1.0422	18.0	90.6	81.4	231.0	71.3	2.548	01m19.4s
17:00	44°24.2'N	141°29.4'W	1.0227	1.0459	24.5	97.4	87.4	186.7	77.3	1.816	01m30.9s
17:05	44°47.3'N	135°22.0'W	1.0241	1.0487	29.6	103.3	91.9	165.8	81.8	1.472	01m40.4s
17:10	44°55.6'N	130°12.3'W	1.0251	1.0509	33.9	108.7	95.3	153.2	85.5	1.264	01m48.6s
17:15	44°53.6'N	125°41.0'W	1.0261	1.0528	37.7	113.8	98.2	144.6	88.5	1.123	01m55.9s
17:20	44°44.0'N	121°37.8'W	1.0268	1.0544	41.2	118.8	100.7	138.3	91.1	1.021	02m02.6s
17:25	44°28.2'N	117°56.6'W	1.0275	1.0558	44.3	123.7	102.8	133.5	93.3	0.942	02m08.6s
17:30	44°07.4'N	114°33.2'W	1.0281	1.0570	47.2	128.7	104.6	129.7	95.2	0.881	02m14.0s
17:35	43°42.4'N	111°24.5'W	1.0286	1.0580	49.8	133.8	106.2	126.7	96.9	0.832	02m18.9s
17:40	43°13.7'N	108°28.4'W	1.0290	1.0589	52.3	139.0	107.7	124.2	98.3	0.791	02m23.3s
17:45	42°41.8'N	105°43.0'W	1.0294	1.0597	54.5	144.3	108.9	122.2	99.5	0.759	02m27.1s
17:50	42°07.1'N	103°06.9'W	1.0297	1.0603	56.6	150.0	110.1	120.5	100.6	0.732	02m30.5s
17:55	41°29.9'N	100°39.0'W	1.0300	1.0609	58.4	155.9	111.1	119.1	101.4	0.709	02m33.3s
18:00	40°50.3'N	098°18.2'W	1.0302	1.0613	60.0	162.1	111.9	117.9	102.1	0.691	02m35.7s
18:05	40°08.6'N	096°03.6'W	1.0304	1.0617	61.3	168.6	112.7	117.0	102.7	0.677	02m37.5s
18:10	39°24.9'N	093°54.4'W	1.0305	1.0619	62.4	175.4	113.3	116.3	103.1	0.665	02m38.9s
18:15	38°39.4'N	091°49.9'W	1.0306	1.0621	63.2	182.5	113.9	115.7	103.3	0.657	02m39.8s
18:20	37°52.2'N	089°49.5'W	1.0306	1.0621	63.7	189.8	114.4	115.3	103.4	0.651	02m40.2s
18:25	37°03.2'N	087°52.5'W	1.0306	1.0621	63.9	197.2	114.7	115.1	103.4	0.648	02m40.2s
18:30	36°12.7'N	085°58.2'W	1.0305	1.0620	63.8	204.5	115.0	115.0	103.2	0.647	02m39.6s
18:35	35°20.6'N	084°06.2'W	1.0304	1.0618	63.4	211.6	115.1	115.1	102.9	0.649	02m38.6s
18:40	34°26.9'N	082°15.8'W	1.0303	1.0615	62.6	218.5	115.1	115.3	102.5	0.653	02m37.2s
18:45	33°31.7'N	080°26.5'W	1.0301	1.0612	61.6	224.9	115.0	115.7	101.9	0.660	02m35.2s
18:50	32°34.8'N	078°37.5'W	1.0299	1.0607	60.3	231.0	114.8	116.3	101.1	0.670	02m32.9s
18:55	31°36.4'N	076°48.3'W	1.0296	1.0601	58.8	236.6	114.4	117.1	100.2	0.684	02m30.1s
19:00	30°36.2'N	074°58.1'W	1.0293	1.0595	57.1	241.7	113.8	118.1	99.1	0.701	02m26.8s
19:05	29°34.2'N	073°06.2'W	1.0289	1.0587	55.1	246.4	113.1	119.4	97.9	0.722	02m23.2s
19:10	28°30.3'N	071°11.5'W	1.0285	1.0578	52.9	250.7	112.1	121.0	96.5	0.749	02m19.0s
19:15	27°24.2'N	069°13.1'W	1.0280	1.0568	50.5	254.6	110.8	122.9	94.9	0.782	02m14.5s
19:20	26°15.7'N	067°09.6'W	1.0274	1.0556	47.9	258.2	109.3	125.4	93.0	0.824	02m09.5s
19:25	25°04.6'N	064°59.3'W	1.0268	1.0543	45.1	261.5	107.4	128.4	90.9	0.877	02m04.0s
19:30	23°50.3'N	062°40.0'W	1.0261	1.0529	42.0	264.5	105.1	132.4	88.6	0.945	01m58.1s
19:35	22°32.1'N	060°08.6'W	1.0253	1.0512	38.6	267.3	102.3	137.5	85.8	1.034	01m51.6s
19:40	21°09.2'N	057°20.7'W	1.0243	1.0492	34.9	269.9	98.8	144.5	82.7	1.158	01m44.5s
19:45	19°39.9'N	054°09.0'W	1.0232	1.0469	30.7	272.4	94.5	154.7	79.0	1.338	01m36.6s
19:50	18°01.3'N	050°20.6'W	1.0219	1.0442	25.8	274.7	89.1	171.0	74.5	1.630	01m27.7s
19:55	16°07.2'N	045°26.2'W	1.0201	1.0407	19.8	277.1	81.9	203.1	68.7	2.209	01m17.1s
20:00	13°34.5'N	037°39.1'W	1.0174	1.0352	10.8	279.7	70.5	319.7	59.7	4.331	01m02.5s
20:02.1	11°01.2'N	027°26.5'W	1.0142	1.0286	0.0	282.1	56.7	-	48.8	-	00m47.1s

Table 1–5: Local Circumstances on the Central Line

Central Line Maximum Eclipse			First Contact				Second Contact			Third Contact			Fourth Contact			
UT1	Duration	Alt	UT1	P	V	Alt	UT1	P	V	UT1	P	V	UT1	P	V	Alt
		°		°	°	°		°	°		°	°		°	°	°
16:49	00m51.6s	1	-	-	-	-	16:48:40	100	149	16:49:32	280	329	17:47:08	100	152	11
16:50	01m01.7s	7	-	-	-	-	16:49:29	101	150	16:50:31	281	330	17:51:26	101	151	19
16:55	01m19.4s	18	15:53:50	281	328	7	16:54:20	102	150	16:55:40	282	330	18:02:17	103	149	30
17:00	01m30.9s	25	15:55:38	282	329	13	16:59:15	103	149	17:00:46	283	329	18:10:58	104	147	37
17:05	01m40.4s	30	15:57:58	283	329	18	17:04:10	104	149	17:05:50	284	329	18:18:53	105	144	42
17:10	01m48.6s	34	16:00:35	283	330	22	17:09:06	104	148	17:10:54	284	328	18:26:20	106	141	46
17:15	01m55.9s	38	16:03:24	284	330	26	17:14:02	105	147	17:15:58	285	327	18:33:26	107	138	49
17:20	02m02.6s	41	16:06:23	285	330	29	17:18:59	106	146	17:21:01	286	326	18:40:14	108	134	52
17:25	02m08.6s	44	16:09:30	285	330	32	17:23:56	107	144	17:26:04	287	324	18:46:46	108	129	54
17:30	02m14.0s	47	16:12:44	286	330	35	17:28:53	108	143	17:31:07	288	322	18:53:04	109	125	56
17:35	02m18.9s	50	16:16:05	287	330	38	17:33:51	108	141	17:36:09	288	320	18:59:08	110	120	58
17:40	02m23.3s	52	16:19:32	287	330	41	17:38:48	109	139	17:41:12	289	318	19:05:01	110	115	59
17:45	02m27.1s	55	16:23:05	288	329	43	17:43:46	110	136	17:46:14	290	315	19:10:43	111	109	59
17:50	02m30.5s	57	16:26:45	289	329	46	17:48:45	110	133	17:51:15	290	312	19:16:14	112	104	59
17:55	02m33.3s	58	16:30:32	289	328	48	17:53:43	111	130	17:56:17	291	309	19:21:36	112	99	59
18:00	02m35.7s	60	16:34:25	290	327	51	17:58:42	111	126	18:01:18	291	305	19:26:48	113	94	59
18:05	02m37.5s	61	16:38:24	290	326	53	18:03:41	112	121	18:06:19	292	300	19:31:51	113	89	58
18:10	02m38.9s	62	16:42:31	291	325	55	18:08:41	113	117	18:11:19	293	296	19:36:46	113	85	57
18:15	02m39.8s	63	16:46:45	292	323	57	18:13:40	113	112	18:16:20	293	290	19:41:34	114	80	56
18:20	02m40.2s	64	16:51:07	292	321	60	18:18:40	113	106	18:21:20	293	285	19:46:14	114	77	55
18:25	02m40.2s	64	16:55:36	293	318	62	18:23:40	114	100	18:26:20	294	279	19:50:47	114	73	54
18:30	02m39.6s	64	17:00:14	293	315	63	18:28:40	114	95	18:31:20	294	274	19:55:14	115	70	52
18:35	02m38.6s	63	17:05:00	294	311	65	18:33:41	115	89	18:36:19	295	268	19:59:35	115	67	51
18:40	02m37.2s	63	17:09:55	294	306	67	18:38:41	115	84	18:41:18	295	263	20:03:50	115	64	49
18:45	02m35.2s	62	17:14:58	295	300	68	18:43:42	115	78	18:46:17	295	258	20:08:00	115	62	47
18:50	02m32.9s	60	17:20:12	295	293	69	18:48:43	115	73	18:51:16	295	253	20:12:04	115	59	45
18:55	02m30.1s	59	17:25:34	295	286	70	18:53:45	115	69	18:56:15	295	248	20:16:03	115	57	43
19:00	02m26.8s	57	17:31:07	296	277	70	18:58:46	116	65	19:01:13	296	244	20:19:57	115	55	41
19:05	02m23.2s	55	17:36:50	296	269	70	19:03:48	116	61	19:06:11	296	241	20:23:46	115	53	39
19:10	02m19.0s	53	17:42:44	296	260	69	19:08:50	116	58	19:11:09	296	237	20:27:30	115	51	36
19:15	02m14.5s	51	17:48:48	296	252	68	19:13:53	116	54	19:16:07	296	234	20:31:09	115	50	34
19:20	02m09.5s	48	17:55:04	296	245	66	19:18:55	115	52	19:21:05	295	231	20:34:44	114	48	31
19:25	02m04.0s	45	18:01:32	296	239	63	19:23:58	115	49	19:26:02	295	229	20:38:12	114	46	29
19:30	01m58.1s	42	18:08:12	296	233	60	19:29:01	115	46	19:30:59	295	226	20:41:35	114	45	26
19:35	01m51.6s	39	18:15:06	296	229	57	19:34:04	115	44	19:35:56	295	224	20:44:51	114	43	23
19:40	01m44.5s	35	18:22:16	295	225	53	19:39:08	114	42	19:40:52	294	222	20:47:58	113	42	19
19:45	01m36.6s	31	18:29:44	295	221	48	19:44:12	114	40	19:45:48	294	220	20:50:55	113	40	15
19:50	01m27.7s	26	18:37:37	294	218	43	19:49:16	113	38	19:50:44	293	218	20:53:38	112	38	11
19:55	01m17.1s	20	18:46:06	294	215	36	19:54:21	112	35	19:55:38	292	215	20:55:56	111	36	5
20:00	01m02.5s	11	18:56:02	292	212	26	19:59:29	111	33	20:00:31	291	213	-	-	-	-
20:02	00m47.1s	1	19:03:27	291	209	14	20:01:43	109	31	20:02:30	289	211	-	-	-	-

Table 1–6: Topocentric Data and Path Corrections Due to Lunar Limb Profile

Universal Time (UT1)	Moon Topo H.P. "	Moon Topo S.D. "	Moon Rel. Ang.V "/s	Topo Lib. Long °	Sun Alt °	Sun Azm °	Path Azm °	North Limit P.A. °	Limit Int. ′	Limit Ext. ′	Limit Int. ′	Limit Ext. ′	Central Durat. Cor. s
16:50	3545.6	965.4	0.544	5.51	7.0	80.3	76.2	10.5	-0.1	0.8	0.5	-1.7	0.7
16:55	3556.8	968.5	0.499	5.46	18.0	90.6	80.3	11.8	-0.1	0.7	0.5	-0.7	0.8
17:00	3563.2	970.2	0.474	5.42	24.5	97.4	83.5	12.8	-0.1	0.3	0.5	-1.6	0.8
17:05	3567.9	971.5	0.455	5.38	29.6	103.3	86.5	13.6	-0.0	0.7	0.5	-2.2	0.8
17:10	3571.7	972.6	0.439	5.34	33.9	108.7	89.3	14.5	-0.0	0.9	0.5	-2.6	0.7
17:15	3575.0	973.4	0.427	5.30	37.7	113.8	91.9	15.3	-0.0	1.0	0.5	-2.7	0.7
17:20	3577.7	974.2	0.416	5.25	41.2	118.8	94.5	16.0	-0.0	0.8	0.5	-2.7	0.7
17:25	3580.1	974.8	0.406	5.21	44.3	123.7	97.0	16.8	-0.0	0.5	0.5	-2.4	0.6
17:30	3582.2	975.4	0.398	5.17	47.2	128.7	99.3	17.5	-0.0	0.4	0.4	-2.0	0.6
17:35	3583.9	975.8	0.391	5.13	49.8	133.8	101.6	18.2	-0.0	0.3	0.4	-1.5	0.6
17:40	3585.5	976.3	0.385	5.08	52.3	139.0	103.7	18.9	-0.0	0.4	0.4	-1.2	0.7
17:45	3586.8	976.6	0.379	5.04	54.5	144.3	105.8	19.6	-0.0	0.4	0.4	-1.4	0.7
17:50	3587.9	976.9	0.375	5.00	56.6	150.0	107.8	20.2	-0.0	0.5	0.4	-1.5	0.8
17:55	3588.9	977.2	0.371	4.96	58.4	155.9	109.6	20.9	-0.0	0.6	0.3	-1.5	0.8
18:00	3589.6	977.4	0.368	4.91	60.0	162.1	111.3	21.4	-0.0	0.7	0.3	-1.6	0.9
18:05	3590.2	977.5	0.366	4.87	61.3	168.6	113.0	22.0	0.0	0.7	0.3	-1.6	1.0
18:10	3590.7	977.6	0.364	4.83	62.4	175.4	114.5	22.5	0.0	0.8	0.3	-2.0	1.1
18:15	3590.9	977.7	0.363	4.79	63.2	182.5	115.9	23.0	0.1	0.9	0.2	-2.2	1.2
18:20	3591.1	977.7	0.362	4.74	63.7	189.8	117.2	23.5	0.1	0.9	0.2	-2.4	1.3
18:25	3591.0	977.7	0.362	4.70	63.9	197.2	118.3	23.9	0.1	0.9	0.2	-2.5	1.1
18:30	3590.8	977.7	0.363	4.66	63.8	204.5	119.4	24.3	0.1	0.9	0.2	-2.5	0.9
18:35	3590.5	977.6	0.364	4.61	63.4	211.6	120.3	24.6	0.2	0.9	0.1	-2.6	0.8
18:40	3590.0	977.5	0.366	4.57	62.6	218.5	121.0	24.9	0.2	0.9	0.1	-2.5	0.7
18:45	3589.4	977.3	0.368	4.53	61.6	224.9	121.7	25.1	0.2	0.9	0.1	-2.5	0.6
18:50	3588.6	977.1	0.371	4.49	60.3	231.0	122.1	25.3	0.2	0.8	0.1	-2.5	0.4
18:55	3587.6	976.8	0.374	4.44	58.8	236.6	122.4	25.5	0.2	0.9	0.1	-2.4	0.4
19:00	3586.5	976.5	0.379	4.40	57.1	241.7	122.6	25.6	0.2	1.0	0.1	-2.4	0.3
19:05	3585.2	976.2	0.383	4.36	55.1	246.4	122.6	25.6	0.2	1.0	0.1	-2.4	0.3
19:10	3583.6	975.8	0.389	4.32	52.9	250.7	122.4	25.6	0.2	1.0	0.1	-2.4	0.3
19:15	3581.9	975.3	0.395	4.27	50.5	254.6	122.1	25.6	0.2	1.0	0.1	-2.5	0.3
19:20	3579.9	974.8	0.402	4.23	47.9	258.2	121.6	25.5	0.2	0.9	0.1	-2.5	0.3
19:25	3577.7	974.2	0.410	4.19	45.1	261.5	120.8	25.3	0.2	0.9	0.1	-2.6	0.3
19:30	3575.2	973.5	0.419	4.15	42.0	264.5	119.9	25.0	0.2	0.9	0.1	-2.7	0.3
19:35	3572.3	972.7	0.430	4.10	38.6	267.3	118.7	24.7	0.1	0.9	0.2	-2.7	0.3
19:40	3568.9	971.8	0.442	4.06	34.9	269.9	117.3	24.3	0.1	0.9	0.2	-2.8	0.4
19:45	3565.0	970.7	0.456	4.02	30.7	272.4	115.6	23.8	0.1	0.9	0.3	-2.7	0.5
19:50	3560.3	969.4	0.473	3.98	25.8	274.7	113.4	23.2	0.0	1.0	0.4	-2.5	0.5
19:55	3554.2	967.8	0.495	3.94	19.8	277.1	110.7	22.4	0.0	0.8	0.5	-2.0	0.4
20:00	3544.9	965.3	0.529	3.89	10.8	279.7	106.7	21.1	-0.0	0.6	0.6	-1.7	0.4

Table 1–7a: Mapping Coordinates for the Path of Totality

Longitude	Latitude of:			Circumstances on Central Line				
	Northern Limit	Southern Limit	Central Line	Un. Time (UT1) h m s	Sun Alt °	Sun Az. °	Path Width km	Central Durat.
125°00.0'W	45°19.19'N	44°25.90'N	44°52.53'N	17:15:48	38.3	114.6	98.6	01m57.1s
124°30.0'W	45°18.34'N	44°24.87'N	44°51.60'N	17:16:24	38.7	115.2	98.9	01m57.9s
124°00.0'W	45°17.38'N	44°23.73'N	44°50.55'N	17:17:01	39.2	115.8	99.2	01m58.7s
123°30.0'W	45°16.31'N	44°22.48'N	44°49.38'N	17:17:38	39.6	116.4	99.6	01m59.5s
123°00.0'W	45°15.11'N	44°21.10'N	44°48.10'N	17:18:15	40.0	117.1	99.9	02m00.3s
122°30.0'W	45°13.80'N	44°19.61'N	44°46.69'N	17:18:53	40.4	117.7	100.1	02m01.2s
122°00.0'W	45°12.36'N	44°17.99'N	44°45.17'N	17:19:31	40.9	118.3	100.4	02m02.0s
121°30.0'W	45°10.81'N	44°16.25'N	44°43.52'N	17:20:10	41.3	119.0	100.7	02m02.8s
121°00.0'W	45°09.13'N	44°14.39'N	44°41.75'N	17:20:49	41.7	119.6	101.0	02m03.6s
120°30.0'W	45°07.33'N	44°12.41'N	44°39.86'N	17:21:29	42.1	120.3	101.3	02m04.4s
120°00.0'W	45°05.40'N	44°10.29'N	44°37.84'N	17:22:09	42.6	120.9	101.6	02m05.2s
119°30.0'W	45°03.35'N	44°08.06'N	44°35.70'N	17:22:50	43.0	121.6	101.9	02m06.1s
119°00.0'W	45°01.17'N	44°05.69'N	44°33.43'N	17:23:31	43.4	122.3	102.2	02m06.9s
118°30.0'W	44°58.87'N	44°03.19'N	44°31.03'N	17:24:13	43.8	123.0	102.5	02m07.7s
118°00.0'W	44°56.43'N	44°00.57'N	44°28.50'N	17:24:55	44.3	123.7	102.7	02m08.5s
117°30.0'W	44°53.87'N	43°57.81'N	44°25.84'N	17:25:38	44.7	124.4	103.0	02m09.3s
117°00.0'W	44°51.17'N	43°54.92'N	44°23.04'N	17:26:21	45.1	125.1	103.3	02m10.1s
116°30.0'W	44°48.34'N	43°51.89'N	44°20.12'N	17:27:05	45.5	125.8	103.6	02m10.9s
116°00.0'W	44°45.38'N	43°48.73'N	44°17.05'N	17:27:49	46.0	126.5	103.8	02m11.7s
115°30.0'W	44°42.28'N	43°45.43'N	44°13.86'N	17:28:34	46.4	127.3	104.1	02m12.5s
115°00.0'W	44°39.04'N	43°41.99'N	44°10.52'N	17:29:19	46.8	128.0	104.4	02m13.3s
114°30.0'W	44°35.67'N	43°38.42'N	44°07.05'N	17:30:05	47.2	128.8	104.6	02m14.1s
114°00.0'W	44°32.15'N	43°34.70'N	44°03.43'N	17:30:51	47.7	129.6	104.9	02m14.9s
113°30.0'W	44°28.50'N	43°30.84'N	43°59.67'N	17:31:38	48.1	130.3	105.2	02m15.7s
113°00.0'W	44°24.70'N	43°26.84'N	43°55.77'N	17:32:26	48.5	131.1	105.4	02m16.4s
112°30.0'W	44°20.76'N	43°22.69'N	43°51.73'N	17:33:13	48.9	132.0	105.7	02m17.2s
112°00.0'W	44°16.67'N	43°18.39'N	43°47.54'N	17:34:02	49.4	132.8	105.9	02m18.0s
111°30.0'W	44°12.44'N	43°13.95'N	43°43.20'N	17:34:51	49.8	133.6	106.2	02m18.8s
111°00.0'W	44°08.06'N	43°09.35'N	43°38.72'N	17:35:41	50.2	134.5	106.4	02m19.5s
110°30.0'W	44°03.54'N	43°04.61'N	43°34.08'N	17:36:31	50.6	135.3	106.7	02m20.3s
110°00.0'W	43°58.86'N	42°59.71'N	43°29.29'N	17:37:22	51.0	136.2	106.9	02m21.0s
109°30.0'W	43°54.03'N	42°54.66'N	43°24.35'N	17:38:13	51.4	137.1	107.2	02m21.8s
109°00.0'W	43°49.04'N	42°49.45'N	43°19.26'N	17:39:05	51.9	138.0	107.4	02m22.5s
108°30.0'W	43°43.90'N	42°44.08'N	43°14.01'N	17:39:57	52.3	138.9	107.7	02m23.2s
108°00.0'W	43°38.61'N	42°38.56'N	43°08.60'N	17:40:50	52.7	139.8	107.9	02m23.9s
107°30.0'W	43°33.15'N	42°32.87'N	43°03.03'N	17:41:44	53.1	140.8	108.1	02m24.6s
107°00.0'W	43°27.54'N	42°27.03'N	42°57.30'N	17:42:38	53.5	141.8	108.4	02m25.3s
106°30.0'W	43°21.76'N	42°21.02'N	42°51.40'N	17:43:33	53.9	142.8	108.6	02m26.0s
106°00.0'W	43°15.82'N	42°14.84'N	42°45.35'N	17:44:28	54.3	143.8	108.8	02m26.7s
105°30.0'W	43°09.72'N	42°08.50'N	42°39.13'N	17:45:24	54.7	144.8	109.0	02m27.4s
105°00.0'W	43°03.45'N	42°01.99'N	42°32.74'N	17:46:21	55.1	145.8	109.3	02m28.1s
104°30.0'W	42°57.01'N	41°55.31'N	42°26.18'N	17:47:18	55.5	146.9	109.5	02m28.7s
104°00.0'W	42°50.41'N	41°48.46'N	42°19.45'N	17:48:16	55.9	148.0	109.7	02m29.4s
103°30.0'W	42°43.63'N	41°41.43'N	42°12.55'N	17:49:15	56.3	149.1	109.9	02m30.0s
103°00.0'W	42°36.68'N	41°34.23'N	42°05.48'N	17:50:14	56.6	150.2	110.1	02m30.6s
102°30.0'W	42°29.56'N	41°26.86'N	41°58.23'N	17:51:13	57.0	151.4	110.3	02m31.2s
102°00.0'W	42°22.26'N	41°19.31'N	41°50.81'N	17:52:14	57.4	152.6	110.5	02m31.8s
101°30.0'W	42°14.78'N	41°11.58'N	41°43.20'N	17:53:15	57.8	153.8	110.7	02m32.4s

Table 1–7b: Mapping Coordinates for the Path of Totality

Longitude	Latitude of: Northern Limit	Southern Limit	Central Line	Circumstances on Central Line: Un. Time (UT1) h m s	Sun Alt °	Sun Az. °	Path Width km	Central Durat.
101°00.0'W	42°07.13'N	41°03.67'N	41°35.42'N	17:54:16	58.1	155.0	110.9	02m32.9s
100°30.0'W	41°59.30'N	40°55.57'N	41°27.46'N	17:55:19	58.5	156.2	111.1	02m33.5s
100°00.0'W	41°51.28'N	40°47.30'N	41°19.31'N	17:56:22	58.8	157.5	111.3	02m34.0s
099°30.0'W	41°43.08'N	40°38.84'N	41°10.98'N	17:57:25	59.2	158.8	111.5	02m34.5s
099°00.0'W	41°34.70'N	40°30.19'N	41°02.47'N	17:58:29	59.5	160.2	111.7	02m35.0s
098°30.0'W	41°26.14'N	40°21.36'N	40°53.77'N	17:59:34	59.8	161.5	111.9	02m35.5s
098°00.0'W	41°17.38'N	40°12.33'N	40°44.88'N	18:00:40	60.1	162.9	112.0	02m35.9s
097°30.0'W	41°08.44'N	40°03.12'N	40°35.81'N	18:01:46	60.5	164.3	112.2	02m36.4s
097°00.0'W	40°59.31'N	39°53.72'N	40°26.54'N	18:02:53	60.8	165.8	112.4	02m36.8s
096°30.0'W	40°49.98'N	39°44.13'N	40°17.08'N	18:04:00	61.1	167.3	112.5	02m37.2s
096°00.0'W	40°40.47'N	39°34.34'N	40°07.43'N	18:05:08	61.3	168.8	112.7	02m37.6s
095°30.0'W	40°30.76'N	39°24.36'N	39°57.59'N	18:06:17	61.6	170.3	112.9	02m37.9s
095°00.0'W	40°20.86'N	39°14.19'N	39°47.55'N	18:07:26	61.9	171.9	113.0	02m38.3s
094°30.0'W	40°10.77'N	39°03.82'N	39°37.32'N	18:08:36	62.1	173.5	113.2	02m38.6s
094°00.0'W	40°00.48'N	38°53.26'N	39°26.90'N	18:09:47	62.3	175.1	113.3	02m38.9s
093°30.0'W	39°49.99'N	38°42.50'N	39°16.27'N	18:10:58	62.6	176.8	113.5	02m39.1s
093°00.0'W	39°39.31'N	38°31.55'N	39°05.45'N	18:12:10	62.8	178.5	113.6	02m39.4s
092°30.0'W	39°28.43'N	38°20.40'N	38°54.44'N	18:13:22	63.0	180.2	113.7	02m39.6s
092°00.0'W	39°17.35'N	38°09.06'N	38°43.23'N	18:14:35	63.1	181.9	113.9	02m39.8s
091°30.0'W	39°06.08'N	37°57.51'N	38°31.82'N	18:15:49	63.3	183.7	114.0	02m39.9s
091°00.0'W	38°54.60'N	37°45.78'N	38°20.21'N	18:17:03	63.4	185.5	114.1	02m40.0s
090°30.0'W	38°42.94'N	37°33.85'N	38°08.41'N	18:18:18	63.6	187.3	114.2	02m40.1s
090°00.0'W	38°31.07'N	37°21.72'N	37°56.42'N	18:19:33	63.7	189.2	114.3	02m40.2s
089°30.0'W	38°19.00'N	37°09.40'N	37°44.22'N	18:20:49	63.7	191.0	114.4	02m40.2s
089°00.0'W	38°06.75'N	36°56.89'N	37°31.84'N	18:22:06	63.8	192.9	114.5	02m40.3s
088°30.0'W	37°54.29'N	36°44.19'N	37°19.26'N	18:23:23	63.9	194.8	114.6	02m40.2s
088°00.0'W	37°41.64'N	36°31.30'N	37°06.49'N	18:24:40	63.9	196.7	114.7	02m40.2s
087°30.0'W	37°28.80'N	36°18.22'N	36°53.52'N	18:25:58	63.9	198.6	114.8	02m40.1s
087°00.0'W	37°15.77'N	36°04.96'N	36°40.37'N	18:27:17	63.9	200.5	114.8	02m40.0s
086°30.0'W	37°02.54'N	35°51.51'N	36°27.04'N	18:28:36	63.8	202.4	114.9	02m39.8s
086°00.0'W	36°49.13'N	35°37.88'N	36°13.52'N	18:29:55	63.8	204.4	115.0	02m39.6s
085°30.0'W	36°35.54'N	35°24.07'N	35°59.81'N	18:31:15	63.7	206.3	115.0	02m39.4s
085°00.0'W	36°21.76'N	35°10.09'N	35°45.93'N	18:32:35	63.6	208.2	115.0	02m39.2s
084°30.0'W	36°07.80'N	34°55.94'N	35°31.87'N	18:33:56	63.5	210.1	115.1	02m38.9s
084°00.0'W	35°53.66'N	34°41.62'N	35°17.64'N	18:35:17	63.3	212.0	115.1	02m38.5s
083°30.0'W	35°39.35'N	34°27.13'N	35°03.24'N	18:36:38	63.2	213.9	115.1	02m38.2s
083°00.0'W	35°24.86'N	34°12.49'N	34°48.67'N	18:38:00	63.0	215.7	115.1	02m37.8s
082°30.0'W	35°10.21'N	33°57.69'N	34°33.94'N	18:39:21	62.7	217.6	115.1	02m37.4s
082°00.0'W	34°55.40'N	33°42.73'N	34°19.05'N	18:40:43	62.5	219.4	115.1	02m36.9s
081°30.0'W	34°40.43'N	33°27.63'N	34°04.02'N	18:42:05	62.2	221.2	115.1	02m36.4s
081°00.0'W	34°25.30'N	33°12.39'N	33°48.83'N	18:43:28	62.0	223.0	115.1	02m35.9s
080°30.0'W	34°10.03'N	32°57.02'N	33°33.50'N	18:44:50	61.7	224.7	115.0	02m35.3s
080°00.0'W	33°54.61'N	32°41.51'N	33°18.04'N	18:46:13	61.3	226.4	115.0	02m34.7s
079°30.0'W	33°39.05'N	32°25.88'N	33°02.44'N	18:47:35	61.0	228.1	114.9	02m34.1s
079°00.0'W	33°23.36'N	32°10.14'N	32°46.72'N	18:48:58	60.6	229.8	114.8	02m33.4s

Table 1–8a: Mapping Coordinates for the Zones of Grazing Eclipse

Longitude	North Graze Zone Latitudes Exterior Limit	North Graze Zone Latitudes Interior Limit	Northern Limit Un. Time UT1 h m s	South Graze Zone Latitudes Interior Limit	South Graze Zone Latitudes Exterior Limit	Southern Limit Un. Time (UT1) h m s	Path Azm °	Elev Fact	Scale Fact km/"
125°00.0'W	45°20.13'N	45°19.17'N	17:16:07	44°26.41'N	44°23.18'N	17:15:30	92.4	-0.48	2.00
124°30.0'W	45°19.28'N	45°18.32'N	17:16:43	44°25.38'N	44°22.15'N	17:16:06	92.7	-0.48	2.00
124°00.0'W	45°18.30'N	45°17.36'N	17:17:19	44°24.24'N	44°21.02'N	17:16:43	93.0	-0.48	2.00
123°30.0'W	45°17.20'N	45°16.28'N	17:17:56	44°22.98'N	44°19.77'N	17:17:21	93.3	-0.48	2.00
123°00.0'W	45°15.98'N	45°15.09'N	17:18:33	44°21.60'N	44°18.40'N	17:17:58	93.6	-0.47	2.00
122°30.0'W	45°14.64'N	45°13.77'N	17:19:10	44°20.11'N	44°16.92'N	17:18:37	93.9	-0.47	1.99
122°00.0'W	45°13.18'N	45°12.34'N	17:19:48	44°18.49'N	44°15.31'N	17:19:15	94.3	-0.47	1.99
121°30.0'W	45°11.59'N	45°10.78'N	17:20:26	44°16.75'N	44°13.59'N	17:19:55	94.6	-0.47	1.99
121°00.0'W	45°09.88'N	45°09.10'N	17:21:05	44°14.88'N	44°11.75'N	17:20:34	94.9	-0.47	1.99
120°30.0'W	45°08.04'N	45°07.30'N	17:21:44	44°12.89'N	44°09.79'N	17:21:15	95.2	-0.47	1.99
120°00.0'W	45°06.07'N	45°05.37'N	17:22:24	44°10.78'N	44°07.71'N	17:21:55	95.6	-0.47	1.99
119°30.0'W	45°03.98'N	45°03.32'N	17:23:04	44°08.54'N	44°05.50'N	17:22:36	95.9	-0.47	1.99
119°00.0'W	45°01.75'N	45°01.14'N	17:23:45	44°06.17'N	44°03.17'N	17:23:18	96.2	-0.46	1.99
118°30.0'W	44°59.42'N	44°58.83'N	17:24:26	44°03.67'N	44°00.71'N	17:24:00	96.6	-0.46	1.99
118°00.0'W	44°56.96'N	44°56.39'N	17:25:08	44°01.04'N	43°58.13'N	17:24:43	96.9	-0.46	1.99
117°30.0'W	44°54.36'N	44°53.83'N	17:25:50	43°58.28'N	43°55.42'N	17:25:26	97.3	-0.46	1.99
117°00.0'W	44°51.63'N	44°51.13'N	17:26:33	43°55.38'N	43°52.57'N	17:26:10	97.6	-0.46	1.98
116°30.0'W	44°48.80'N	44°48.30'N	17:27:16	43°52.35'N	43°49.60'N	17:26:54	97.9	-0.46	1.98
116°00.0'W	44°45.83'N	44°45.33'N	17:27:59	43°49.19'N	43°46.50'N	17:27:39	98.3	-0.46	1.98
115°30.0'W	44°42.73'N	44°42.23'N	17:28:44	43°45.88'N	43°43.27'N	17:28:25	98.6	-0.46	1.98
115°00.0'W	44°39.49'N	44°38.99'N	17:29:28	43°42.44'N	43°39.90'N	17:29:10	99.0	-0.46	1.98
114°30.0'W	44°36.11'N	44°35.62'N	17:30:13	43°38.86'N	43°36.39'N	17:29:57	99.3	-0.45	1.98
114°00.0'W	44°32.58'N	44°32.10'N	17:30:59	43°35.14'N	43°32.75'N	17:30:44	99.7	-0.45	1.98
113°30.0'W	44°28.91'N	44°28.45'N	17:31:45	43°31.28'N	43°28.97'N	17:31:31	100.1	-0.45	1.98
113°00.0'W	44°25.09'N	44°24.65'N	17:32:32	43°27.27'N	43°25.06'N	17:32:19	100.4	-0.45	1.98
112°30.0'W	44°21.12'N	44°20.71'N	17:33:19	43°23.11'N	43°21.00'N	17:33:08	100.8	-0.45	1.98
112°00.0'W	44°17.01'N	44°16.63'N	17:34:07	43°18.81'N	43°16.80'N	17:33:57	101.1	-0.45	1.98
111°30.0'W	44°12.74'N	44°12.39'N	17:34:55	43°14.36'N	43°12.46'N	17:34:47	101.5	-0.45	1.98
111°00.0'W	44°08.35'N	44°08.02'N	17:35:44	43°09.76'N	43°07.97'N	17:35:37	101.9	-0.45	1.98
110°30.0'W	44°03.84'N	44°03.49'N	17:36:34	43°05.01'N	43°03.34'N	17:36:28	102.2	-0.45	1.98
110°00.0'W	43°59.18'N	43°58.81'N	17:37:24	43°00.10'N	42°58.52'N	17:37:20	102.6	-0.45	1.98
109°30.0'W	43°54.38'N	43°53.98'N	17:38:14	42°55.11'N	42°53.49'N	17:38:12	103.0	-0.45	1.98
109°00.0'W	43°49.42'N	43°48.99'N	17:39:05	42°49.89'N	42°48.33'N	17:39:05	103.3	-0.45	1.97
108°30.0'W	43°44.30'N	43°43.85'N	17:39:57	42°44.52'N	42°42.91'N	17:39:58	103.7	-0.45	1.97
108°00.0'W	43°39.02'N	43°38.56'N	17:40:49	42°38.99'N	42°37.34'N	17:40:52	104.1	-0.45	1.97
107°30.0'W	43°33.58'N	43°33.10'N	17:41:42	42°33.29'N	42°31.61'N	17:41:46	104.4	-0.45	1.97
107°00.0'W	43°27.98'N	43°27.49'N	17:42:35	42°27.44'N	42°25.72'N	17:42:41	104.8	-0.44	1.97
106°30.0'W	43°22.20'N	43°21.71'N	17:43:29	42°21.42'N	42°19.67'N	17:43:37	105.2	-0.44	1.97
106°00.0'W	43°16.26'N	43°15.78'N	17:44:24	42°15.24'N	42°13.46'N	17:44:33	105.6	-0.44	1.97
105°30.0'W	43°10.16'N	43°09.67'N	17:45:19	42°08.89'N	42°07.09'N	17:45:30	105.9	-0.44	1.97
105°00.0'W	43°03.88'N	43°03.41'N	17:46:15	42°02.37'N	42°00.56'N	17:46:28	106.3	-0.44	1.97
104°30.0'W	42°57.48'N	42°56.97'N	17:47:11	41°55.68'N	41°53.86'N	17:47:26	106.7	-0.44	1.97
104°00.0'W	42°50.90'N	42°50.37'N	17:48:08	41°48.82'N	41°46.99'N	17:48:25	107.1	-0.44	1.97
103°30.0'W	42°44.15'N	42°43.59'N	17:49:05	41°41.79'N	41°39.95'N	17:49:24	107.4	-0.44	1.97
103°00.0'W	42°37.22'N	42°36.65'N	17:50:03	41°34.58'N	41°32.74'N	17:50:25	107.8	-0.44	1.97
102°30.0'W	42°30.11'N	42°29.52'N	17:51:02	41°27.20'N	41°25.36'N	17:51:25	108.2	-0.44	1.97
102°00.0'W	42°22.82'N	42°22.23'N	17:52:02	41°19.64'N	41°17.80'N	17:52:27	108.6	-0.44	1.97
101°30.0'W	42°15.35'N	42°14.76'N	17:53:02	41°11.90'N	41°10.08'N	17:53:29	109.0	-0.44	1.97
101°00.0'W	42°07.70'N	42°07.11'N	17:54:02	41°03.98'N	41°02.14'N	17:54:31	109.3	-0.44	1.97
100°30.0'W	41°59.87'N	41°59.28'N	17:55:03	40°55.88'N	40°54.02'N	17:55:35	109.7	-0.45	1.97
100°00.0'W	41°51.90'N	41°51.26'N	17:56:05	40°47.59'N	40°45.72'N	17:56:39	110.1	-0.45	1.97

Table 1–8b: Mapping Coordinates for the Zones of Grazing Eclipse

Longitude	North Graze Zone Latitudes Exterior Limit	North Graze Zone Latitudes Interior Limit	Northern Limit Un. Time UT1 h m s	South Graze Zone Latitudes Interior Limit	South Graze Zone Latitudes Exterior Limit	Southern Limit Un. Time (UT1) h m s	Path Azm °	Elev Fact	Scale Fact km/"
099°30.0'W	41°43.73'N	41°43.07'N	17:57:08	40°39.12'N	40°37.24'N	17:57:43	110.5	-0.45	1.97
099°00.0'W	41°35.38'N	41°34.69'N	17:58:11	40°30.47'N	40°28.58'N	17:58:49	110.8	-0.45	1.98
098°30.0'W	41°26.84'N	41°26.13'N	17:59:15	40°21.64'N	40°19.74'N	17:59:55	111.2	-0.45	1.98
098°00.0'W	41°18.10'N	41°17.38'N	18:00:19	40°12.61'N	40°10.72'N	18:01:01	111.6	-0.45	1.98
097°30.0'W	41°09.17'N	41°08.44'N	18:01:24	40°03.41'N	40°01.51'N	18:02:08	111.9	-0.45	1.98
097°00.0'W	41°00.05'N	40°59.31'N	18:02:30	39°54.00'N	39°52.12'N	18:03:16	112.3	-0.45	1.98
096°30.0'W	40°50.73'N	40°49.99'N	18:03:36	39°44.41'N	39°42.53'N	18:04:25	112.7	-0.45	1.98
096°00.0'W	40°41.22'N	40°40.48'N	18:04:43	39°34.62'N	39°32.66'N	18:05:34	113.0	-0.45	1.98
095°30.0'W	40°31.50'N	40°30.78'N	18:05:50	39°24.63'N	39°22.59'N	18:06:44	113.4	-0.45	1.98
095°00.0'W	40°21.61'N	40°20.89'N	18:06:59	39°14.46'N	39°12.34'N	18:07:54	113.7	-0.45	1.98
094°30.0'W	40°11.55'N	40°10.80'N	18:08:07	39°04.09'N	39°01.89'N	18:09:06	114.1	-0.46	1.98
094°00.0'W	40°01.30'N	40°00.51'N	18:09:17	38°53.52'N	38°51.26'N	18:10:17	114.4	-0.46	1.98
093°30.0'W	39°50.85'N	39°50.03'N	18:10:27	38°42.76'N	38°40.43'N	18:11:30	114.8	-0.46	1.98
093°00.0'W	39°40.19'N	39°39.35'N	18:11:38	38°31.80'N	38°29.42'N	18:12:43	115.1	-0.46	1.99
092°30.0'W	39°29.33'N	39°28.48'N	18:12:49	38°20.65'N	38°18.21'N	18:13:56	115.5	-0.46	1.99
092°00.0'W	39°18.28'N	39°17.41'N	18:14:01	38°09.30'N	38°06.82'N	18:15:11	115.8	-0.46	1.99
091°30.0'W	39°07.01'N	39°06.14'N	18:15:13	37°57.75'N	37°55.23'N	18:16:25	116.1	-0.47	1.99
091°00.0'W	38°55.55'N	38°54.67'N	18:16:26	37°46.01'N	37°43.45'N	18:17:41	116.4	-0.47	1.99
090°30.0'W	38°43.88'N	38°43.01'N	18:17:40	37°34.07'N	37°31.48'N	18:18:57	116.8	-0.47	1.99
090°00.0'W	38°32.02'N	38°31.15'N	18:18:54	37°21.94'N	37°19.32'N	18:20:13	117.1	-0.47	1.99
089°30.0'W	38°19.95'N	38°19.09'N	18:20:09	37°09.62'N	37°06.98'N	18:21:30	117.4	-0.47	2.00
089°00.0'W	38°07.68'N	38°06.84'N	18:21:24	36°57.10'N	36°54.44'N	18:22:48	117.7	-0.48	2.00
088°30.0'W	37°55.21'N	37°54.39'N	18:22:40	36°44.40'N	36°41.72'N	18:24:06	118.0	-0.48	2.00
088°00.0'W	37°42.55'N	37°41.75'N	18:23:57	36°31.49'N	36°28.80'N	18:25:25	118.3	-0.48	2.00
087°30.0'W	37°29.69'N	37°28.91'N	18:25:14	36°18.40'N	36°15.71'N	18:26:44	118.5	-0.48	2.00
087°00.0'W	37°16.63'N	37°15.89'N	18:26:31	36°05.13'N	36°02.43'N	18:28:03	118.8	-0.49	2.00
086°30.0'W	37°03.40'N	37°02.67'N	18:27:49	35°51.68'N	35°48.97'N	18:29:23	119.1	-0.49	2.01
086°00.0'W	36°50.00'N	36°49.27'N	18:29:07	35°38.05'N	35°35.33'N	18:30:44	119.4	-0.49	2.01
085°30.0'W	36°36.42'N	36°35.68'N	18:30:26	35°24.23'N	35°21.52'N	18:32:05	119.6	-0.49	2.01
085°00.0'W	36°22.64'N	36°21.90'N	18:31:45	35°10.25'N	35°07.54'N	18:33:26	119.9	-0.50	2.01
084°30.0'W	36°08.69'N	36°07.95'N	18:33:05	34°56.09'N	34°53.39'N	18:34:47	120.1	-0.50	2.02
084°00.0'W	35°54.55'N	35°53.82'N	18:34:25	34°41.76'N	34°39.07'N	18:36:09	120.3	-0.50	2.02
083°30.0'W	35°40.23'N	35°39.51'N	18:35:45	34°27.27'N	34°24.59'N	18:37:31	120.5	-0.51	2.02
083°00.0'W	35°25.74'N	35°25.03'N	18:37:06	34°12.62'N	34°09.95'N	18:38:54	120.8	-0.51	2.02
082°30.0'W	35°11.09'N	35°10.39'N	18:38:27	33°57.82'N	33°55.16'N	18:40:16	120.9	-0.51	2.03
082°00.0'W	34°56.27'N	34°55.58'N	18:39:48	33°42.86'N	33°40.22'N	18:41:39	121.1	-0.52	2.03
081°30.0'W	34°41.29'N	34°40.62'N	18:41:09	33°27.76'N	33°25.13'N	18:43:02	121.3	-0.52	2.03
081°00.0'W	34°26.17'N	34°25.50'N	18:42:31	33°12.51'N	33°09.91'N	18:44:25	121.5	-0.52	2.03
080°30.0'W	34°10.89'N	34°10.23'N	18:43:53	32°57.13'N	32°54.54'N	18:45:48	121.6	-0.53	2.04
080°00.0'W	33°55.47'N	33°54.81'N	18:45:14	32°41.62'N	32°39.05'N	18:47:11	121.8	-0.53	2.04
079°30.0'W	33°39.90'N	33°39.26'N	18:46:36	32°25.98'N	32°23.41'N	18:48:35	121.9	-0.53	2.04
079°00.0'W	33°24.21'N	33°23.57'N	18:47:58	32°10.23'N	32°07.68'N	18:49:58	122.0	-0.54	2.05
078°30.0'W	33°08.38'N	33°07.76'N	18:49:20	31°54.37'N	31°51.84'N	18:51:21	122.1	-0.54	2.05
078°00.0'W	32°52.45'N	32°51.83'N	18:50:42	31°38.41'N	31°35.89'N	18:52:44	122.2	-0.54	2.05

1.7 Mean Lunar Radius

A fundamental parameter used in the prediction of solar eclipses is the Moon's mean radius k, expressed in units of Earth's equatorial radius. The actual radius of the Moon varies as a function of position angle and libration due to the irregularity of the lunar limb profile. From 1968 through 1980, the Nautical Almanac Office used two separate values for k in their eclipse predictions. The larger value (k=0.2724880), representing a mean over lunar topographic features, was used for all penumbral (i.e., exterior) contacts and for annular eclipses. A smaller value (k=0.272281), representing a mean minimum radius, was reserved exclusively for umbral (i.e., interior) contact calculations of total eclipses (*Explanatory Supplement*, 1974). Unfortunately, the use of two different values of k for central eclipses introduces a discontinuity in the case of hybrid or annular-total eclipses.

In August 1982, the IAU General Assembly adopted a value of k=0.2725076 for the mean lunar radius. This value is currently used by the Nautical Almanac Office for all solar eclipse predictions (Fiala and Lukac, 1983) and is believed to be the best mean radius, averaging mountain peaks and low valleys along the Moon's rugged limb. The adoption of one single value for k eliminates the discontinuity in the case of annular-total eclipses and ends confusion arising from the use of two different values. However, the use of even the best 'mean' value for the Moon's radius introduces a problem in predicting the character and duration of central eclipses, particularly total eclipses. A total eclipse can be defined as an eclipse in which the Sun's disk is completely occulted by the Moon. This cannot occur so long as any photospheric rays are visible through deep valleys along the Moon's limb (Meeus, Grosjean and Vanderleen, 1966). But the use of the IAU's mean k guarantees that some annular or annular-total eclipses will be misidentified as total. For example, the *Astronomical Almanac for 1986* identified eclipse of 1986 Oct 03 as total with a 3 seconds duration when in it was in fact a beaded annular eclipse. Clearly, a smaller value of k is needed since it is more representative of the deepest lunar valley floors, hence the minimum solid disk radius, and ensures that an eclipse is truly total.

Of primary interest to most observers are the times when the central eclipse begins and ends (second and third contacts, respectively) and the duration of the central phase. When the IAU's mean value for k is used to calculate these times, they must be corrected to accommodate low valleys (total) or high mountains (annular) along the Moon's limb. The calculation of these corrections are essential, especially if one plans to observe near the path limits (Herald, 1983). For observers near the central line of a total eclipse, the limb corrections can be closely approximated by using a smaller value of k, which accounts for valleys along the profile.

This work uses the IAU's accepted value of k (k=0.2725076) for all penumbral (exterior) contacts. In order to avoid eclipse type misidentification and to predict central durations, which are closer to the actual durations observed at total eclipses, we depart from the IAU convention by adopting the smaller value for k (k=0.272281) for all central (interior) contacts. This is consistent with predictions published in *Five Millennium Canon of Solar Eclipses* (Espenak and Meeus, 2006) and *Thousand Year Canon of Solar Eclipses* (Espenak, 2014). Consequently, the smaller k produces shorter central durations and narrower paths for total eclipses when compared with calculations using the IAU value for k. Similarly, the smaller k predicts longer central durations and wider paths for annular eclipses.

1.8 Central Line and Duration of Totality

The axis of the Moon's shadow determines the central line of the path of totality (and annularity). For the purposes of this discussion, it is assumed that the central line lies in the middle of the eclipse path.[6]

The duration of totality is longest on the central line, so great effort is made to get as close as possible to it. However, the duration actually drops off quite slowly with distance from the central line. For example, a location 20% from the center to the edge of the path still has a totality lasting 98% of the central line duration. Even if one travels half way to the path limit, the duration is still 87% of the central line value.

---

[6] The position of the central line is actually offset from the center. The difference is greatest in cases where the Sun's altitude is low. Curvature of Earth's surface is responsible for this shift.

Figure 1–4: Duration Vs. Distance from the Central Line

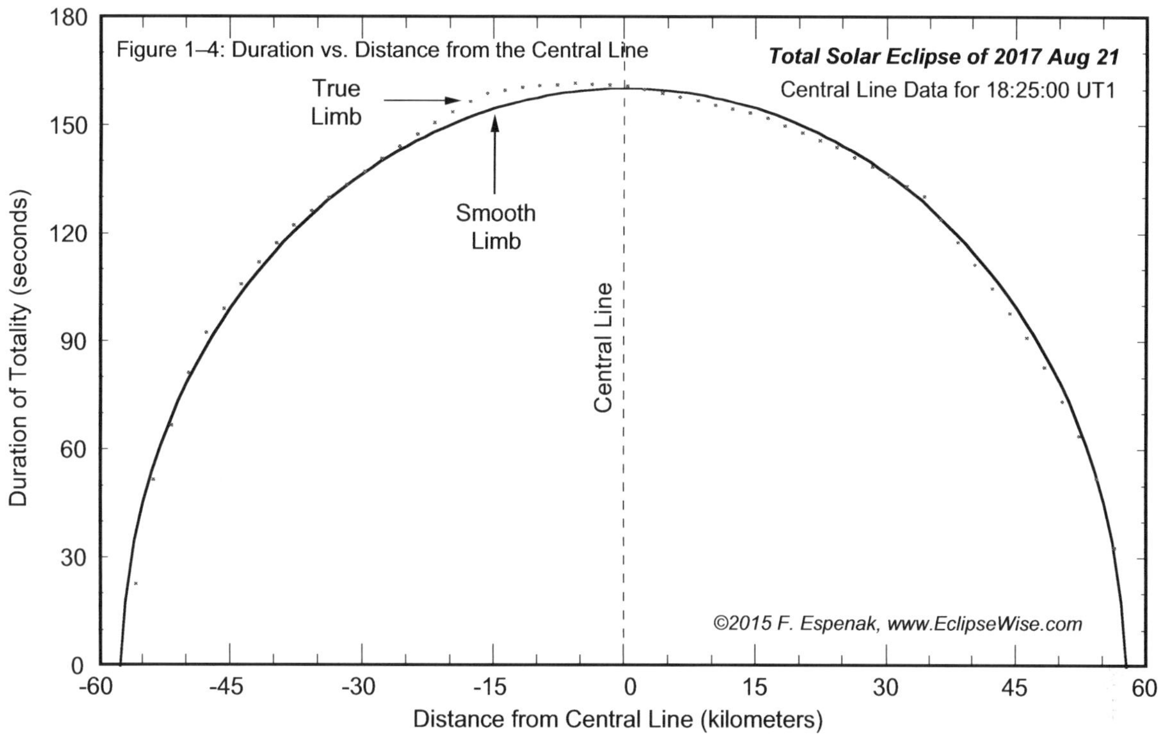

Figure 1–4 shows how the duration of totality changes with distance from the central line. The solid curve represents the duration using the Moon's mean limb. At a distance ~87% from the central line to the path limit, the duration drops to 50%. It is only when the distance from the central line to the path limit reaches ~90% that the duration takes a precipitous drop. But even at a distance of 96%, the duration is still 28% of the central line value. The dotted curve uses the Moon's true limb to calculate the duration using Kaguya/Herald data.

Although figure 1–4 plots data for the 2017 total solar eclipse, the same relationship holds for all total eclipse paths. The solid curve is calculated assuming a smooth lunar limb. In reality, the Moon's limb profile has mountains that extend the duration and valleys that shorten it. These topographic features typically change the duration by a second or two for most of the path — their effects become significantly larger near the umbral path limits. The dotted line in figure 1–4 has been calculated using the Moon's true profile.

The following expression is useful for calculating the duration of totality for any location at a perpendicular distance of δ kilometers from the central line. It assumes a smooth profile for the Moon's limb.

$$d = D \times [\, 1 - (\, \delta/Z\,)^2\,]^{1/2} \text{ seconds} \qquad (1\text{–}2)$$

Where: d = duration of totality at point of interest (seconds)
D = duration of totality on the central line (seconds)
δ = perpendicular distance from the central line to position of interest (kilometers)
Z = perpendicular distance from the central line to the path edge (kilometers)

The Moon's limb profile varies with the lunar libration, which changes from eclipse to eclipse. Consequently, the effects of the limb profile are different at each eclipse.

1.9 Greatest Eclipse and Greatest Duration

Greatest Eclipse is defined as the instant when the axis of the Moon's shadow passes closest to the center of Earth. The computation of the duration of the total (or annular) phase at this point is typically done using a smooth Moon that ignores the effects of mountains and valleys along the lunar limb. For total eclipses, the instant of *Greatest Eclipse* offers a very good approximation (typically within 0.1 seconds) to the *Greatest Duration* of totality along the entire eclipse path.

Because of its rigorous geometric definition, the instant of *Greatest Eclipse* is easily calculated for total, annular and partial eclipses, and is the standard time used for comparing different eclipses with each other. For example, the time and date of each eclipse in the *Thousand Year Canon of Solar Eclipses: 1501 – 2500* (Espenak 2014) corresponds to the instant of *Greatest Eclipse*.

Greatest Duration is defined as the instant when the length of the total (or annular) phase reaches a maximum along the central eclipse path. The computation of the eclipse duration is again done using a smooth Moon that ignores the effects of mountains and valleys along the lunar limb. Although the location of *Greatest Duration* may be relatively close to the point of *Greatest Eclipse*, it differs slightly because of several factors including Earth's oblateness, the relative motion of the Moon's shadow with respect to Earth's Equator, and the latitude of the shadow axis.

The length of the total eclipse calculated at *Greatest Duration* is typically 0.1 seconds longer compared with *Greatest Eclipse*, and the geographic location typically differs by ~75 kilometers. These values are based on the 67 central total solar eclipses occurring during the 21st Century. The statistics for *Greatest Eclipse (GE)* vs. *Greatest Duration (GD)* for these eclipses are as follows.

Greatest Eclipse and Greatest Duration for Total Solar Eclipses of the 21st Century

Statistic	Duration (GD-GE)	Distance (GD-GE)
Average	0.084 sec	74.2 km
Std. Dev.	0.138 sec	66.2 km
Minimum	0.000 sec	2.5 km
Maximum	0.608 sec	251.6 km

Unlike *Greatest Eclipse*, there is no explicit or analytical solution for the determination of *Greatest Duration*. It may be calculated though an iterative series of approximations. When the highest accuracy is needed, a *Corrected Greatest Duration* must be calculated that includes the effects of the Moon's limb profile, which may differ by a couple seconds from the uncorrected value. (see next section)

The table below summarizes the values for *Greatest Eclipse* and *Greatest Duration* for the total solar eclipse of 2017 August 21. The central line duration of totality is 0.13 seconds longer at the point of *Greatest Duration*, which lies 144.4 km west of the point of *Greatest Eclipse*.

Greatest Eclipse and Greatest Duration for 2017 August 21

Event	Time (UT1)	Latitude	Longitude	Duration
Greatest Eclipse:	18:25:31.7	36°58.0'N	087°40.3'W	02m40.12s
Greatest Duration:	18:21:49.0	37°34.6'N	089°06.6'W	02m40.25s

1.10 Lunar Limb Profile

Eclipse contact times, magnitude, and duration of totality all depend on the angular diameters and relative velocities of the Moon and Sun. Unfortunately, these calculations are limited in accuracy by the departure of the Moon's limb from a perfectly circular figure. The Moon's surface exhibits a dramatic topography, which manifests itself as an irregular limb when seen in profile. Most eclipse calculations assume a mean radius that averages high mountain peaks and low valleys along the Moon's limb. Such an approximation is acceptable for many applications, but when higher accuracy is needed the Moon's actual limb profile must be considered.

For many years, Watts's limb charts (Watts 1963) constituted the most detailed maps of lunar mountain profiles at all libration phases. These data were the product of a photographic survey of the marginal zone of the Moon and give limb profile heights with respect to an adopted smooth reference surface (or datum).

In 2009, a new lunar profile was published based on the laser altimeter (LALT) onboard the Japan Aerospace Exploration Agency's lunar orbiter spacecraft SELENE (Selenological and Engineering Explorer), better known by its nickname Kaguya. Australian grazing occultation observer David Herald (private communication) reduced the Kaguya laser-altimeter time-series dataset and converted it into a Watts-like format including the difference between the center of mass and center of figure. It provides the angular height of the actual lunar limb above a mean lunar limb (k=0.2725076 in Earth radii of 6378.137 km) as a function of position angle (0.2° increments) and libration (0.2° increments in L and B) relative to the Moon's center of mass, with the Moon being at its mean distance (384398.55 km).

Along the entire 2017 total eclipse path, the Moon's topocentric libration (physical plus optical) in longitude ranges from l = +5.5° to +3.9° (l = +5.3° to +4.5° across the section of the path through the USA). Thus, a limb profile with the appropriate libration is required in any detailed analysis of contact times, central durations, etc. However, a profile with an intermediate value is useful for planning purposes and may even be adequate for many applications.

The lunar limb profile in Figure 1–5 is derived from the Kaguya/Herald dataset with respect to the Moon's center of mass. It is generated for 18:00:00 UT1, which corresponds to the integral hour nearest the instant of greatest eclipse. The umbral shadow is then located in eastern Nebraska at latitude 40° 50.3'N and longitude 98° 18.1'W. The Moon's topocentric libration is l = +4.91°, and the topocentric semi-diameters of the Sun and Moon are 948.7 and 977.4 arc-seconds, respectively. The Moon's angular velocity with respect to the Sun is 0.368 arc-seconds/second.

Lunar limb profiles for two additional positions along the central line (for 17:30:00 UT1 and 18:30:00 UT1) are available at: *eclipsewise.com/solar/SEnews/TSE2017/extra.html*

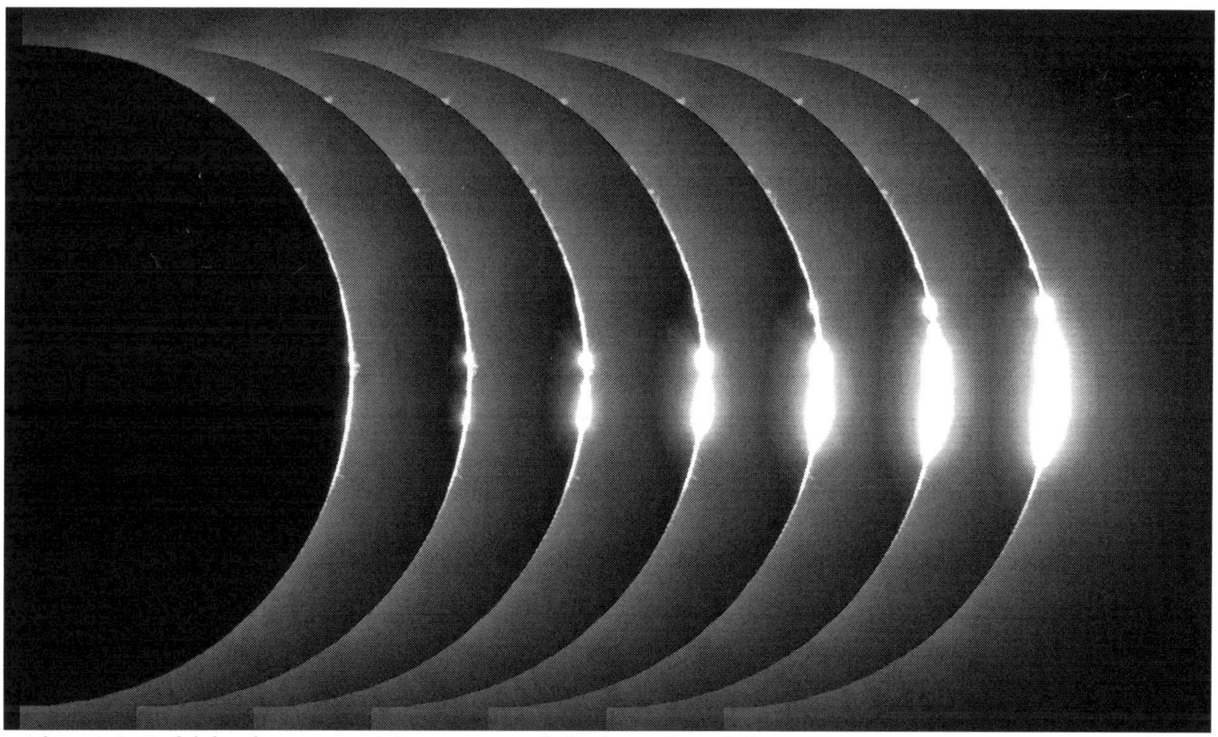

Photo 1–2: Baily's beads appear at 3rd contact as sunlight shines through deep valleys along the Moon's limb, This image sequence was shot at the total solar eclipse of 2006 Mar 29 from Jalu, Libya. © 2006 by F. Espenak, www.MrEclipse.com

Figure 1–5: Lunar Limb Profile for 2017 August 21 at 18:00:00 UT1

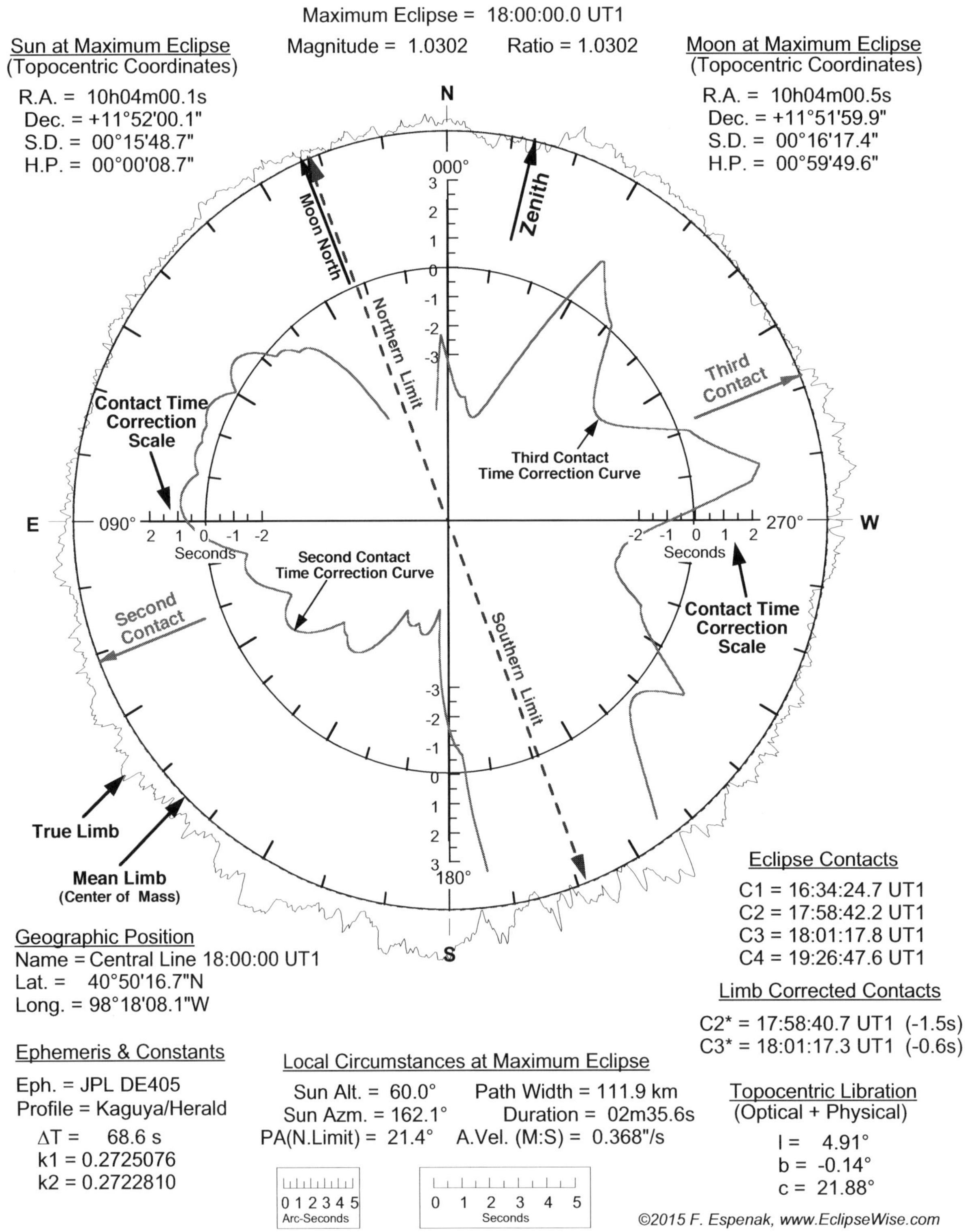

The radial scale of the limb profile in Figure 1–5 is greatly exaggerated so that the true limb's departure from the mean limb is readily apparent. Note that most of the predictions presented in this publication are calculated with respect to the mean limb unless otherwise noted. Position angles of various lunar features can be read using the protractor marks along the Moon's mean limb. The position angles of second and third contact are clearly marked, as are the north pole of the Moon's axis of rotation and the observer's zenith at mid-totality. The dashed line with arrows at either end identifies the contact points on the limb corresponding to the northern and southern limits of the path. To the upper left of the profile, are the Sun's topocentric coordinates at maximum eclipse. They include the right ascension *(R.A.)*, declination *(Dec.)*, semi-diameter *(S.D.)*, and horizontal parallax *(H.P.)*. The corresponding topocentric coordinates for the Moon appear to the upper right. Below and left of the profile are the geographic coordinates of the central line at selected location, while the times of the four eclipse contacts at that location appear to the lower right. The limb-corrected times of second and third contacts are listed with the applied correction to the center of mass smooth limb prediction.

Directly below the limb profile are the local circumstances at maximum eclipse. They include the Sun's altitude and azimuth, the path width, and umbral duration. The position angle of the path's northern-to-southern limit axis is *PA (N.Limit)* and the angular velocity of the Moon with respect to the Sun is *A.Vel. (M:S)*. At the bottom left are a number of parameters used in the predictions, and the topocentric lunar librations appear at the lower right.

As mentioned earlier, the four eclipse contact times for the central line position at 18:00 UT1 are listed to the lower right of the limb profile (Figure 1–5). These are calculated using a smooth limb. When the Kaguya/Herald limb profile is applied to C2 and C3 (start and end of totality), the resulting values appear under *Limb Corrected Contacts*. The numbers in the brackets, (–1.5s) and (–0.6s), are the corrections added to smooth limb C2 and C3 times to compensate for the actual lunar limb profile. In this case, the duration of totality is 0.9 seconds longer (2 min 36.6 sec) than the smooth limb duration (2 min 35.7 sec).

The extended duration of totality at this central line position is not unique. The large, mountainous terrain in the southeast quadrant of the lunar limb profile in Figure 1–5 is largely responsible for extending the smooth limb durations (from 0.4 to 1.3 seconds) along the entire eclipse track across the USA (Table 1–9).

Table 1–9: Correction to Central Line Duration of Totality Due To Lunar Limb Profile

Universal Time UT1	Central Line Longitude	Topo. Lib. Long.	Central Line Duration (smooth)	Lunar Limb Correct. (sec)	Central Line Duration (corrected)	Maximum Duration Shift (km)	Maximum Duration (corrected)
17:15	125° 41.0'W	5.30	01m55.9s	0.7	01m56.6s	-3.60	01m57.36s
17:20	121° 37.8'W	5.25	02m02.6s	0.7	02m03.3s	-4.40	02m04.09s
17:25	117° 56.6'W	5.21	02m08.6s	0.6	02m09.2s	-5.20	02m10.20s
17:30	114° 33.2'W	5.17	02m14.0s	0.6	02m14.6s	-5.80	02m15.77s
17:35	111° 24.5'W	5.13	02m18.9s	0.6	02m19.5s	-6.60	02m20.82s
17:40	108° 28.4'W	5.08	02m23.3s	0.7	02m24.0s	-5.60	02m25.08s
17:45	105° 43.0'W	5.04	02m27.1s	0.7	02m27.8s	-5.00	02m28.84s
17:50	103° 06.9'W	5.00	02m30.5s	0.8	02m31.3s	-4.40	02m32.11s
17:55	100° 39.0'W	4.96	02m33.3s	0.8	02m34.1s	-3.80	02m34.90s
18:00	*098° 18.2'W*	*4.91*	*02m35.7s*	*0.9*	*02m36.6s*	*-3.20*	*02m37.21s*
18:05	096° 03.6'W	4.87	02m37.5s	1.0	02m38.5s	-2.00	02m38.99s
18:10	093° 54.4'W	4.83	02m38.9s	1.1	02m40.0s	-1.60	02m40.34s
18:15	091° 49.9'W	4.79	02m39.8s	1.2	02m41.0s	-1.00	02m41.25s
18:20	***089° 49.5'W***	***4.74***	***02m40.2s***	***1.3***	***02m41.5s***	***-0.60***	***02m41.65s***
18:25	087° 52.5'W	4.70	02m40.2s	1.1	02m41.2s	-0.60	02m41.23s
18:30	085° 58.2'W	4.66	02m39.6s	0.9	02m40.5s	-1.00	02m40.56s
18:35	084° 06.2'W	4.61	02m38.6s	0.8	02m39.4s	-1.40	02m39.44s
18:40	082° 15.8'W	4.57	02m37.2s	0.7	02m37.9s	-1.80	02m37.89s
18:45	080° 26.5'W	4.53	02m35.2s	0.6	02m35.8s	-2.00	02m35.89s
18:50	078° 37.5'W	4.49	02m32.9s	0.4	02m33.3s	-2.60	02m33.36s

Figure 1–6: Limb Profile Effects on the Duration of Totality

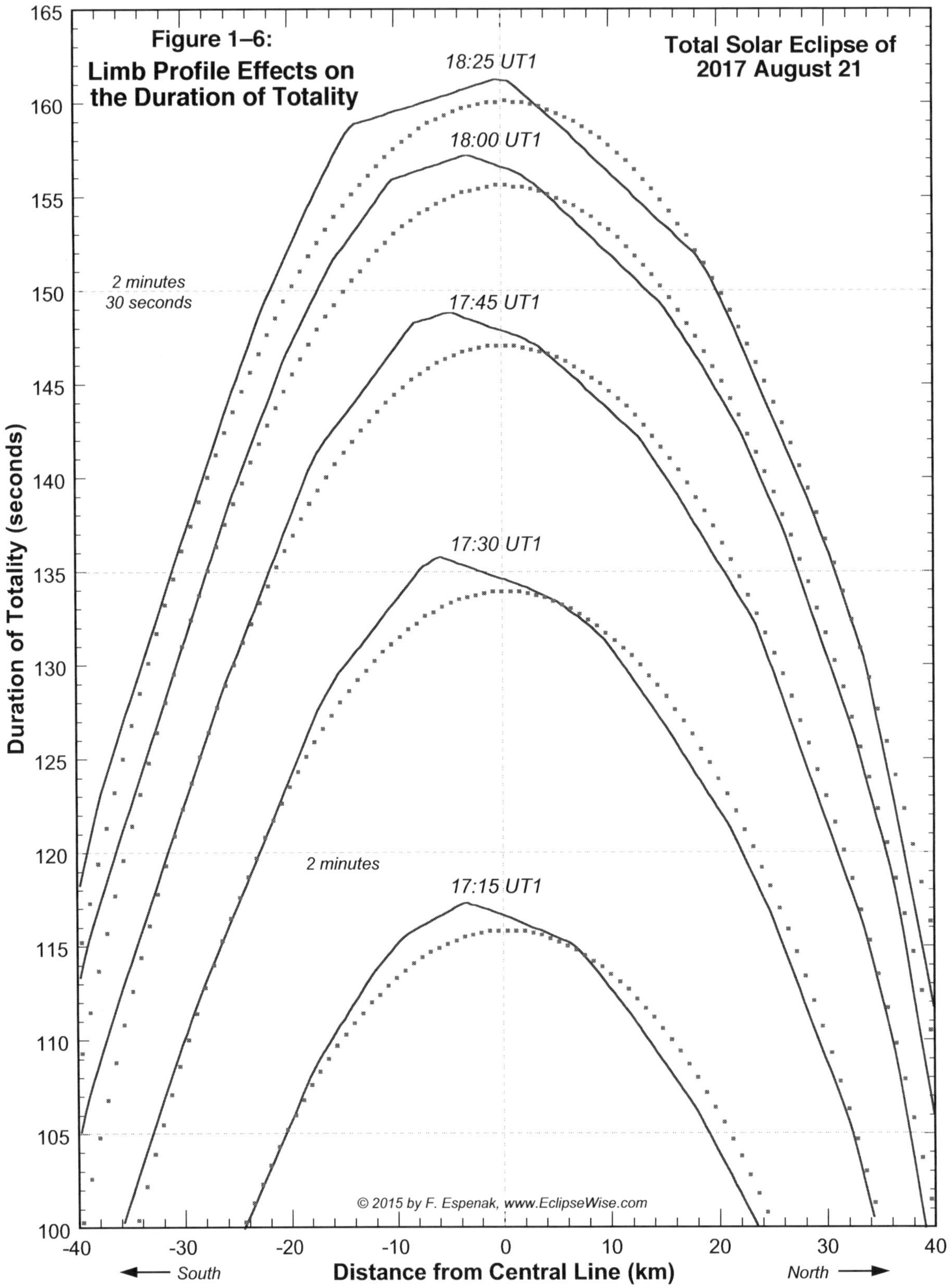

Table 1–9 consists, in part, of data extracted from Tables 1–3 and 1–6. It lists the central line time of mid-eclipse and longitude of the umbral path across the USA, the Moon's topocentric libration in longitude and the central line duration calculated using a smooth lunar limb. The computed limb correction due to the Kaguya/Herald profile, and the resulting central line duration are given next (the last two columns will be discussed shortly). Of particular interest is the data for the 18:20 UT1 central line position. It shows the longest duration of totality along the entire path of 2 minutes 41.5 seconds, an increase of 1.3 seconds over the smooth limb duration.

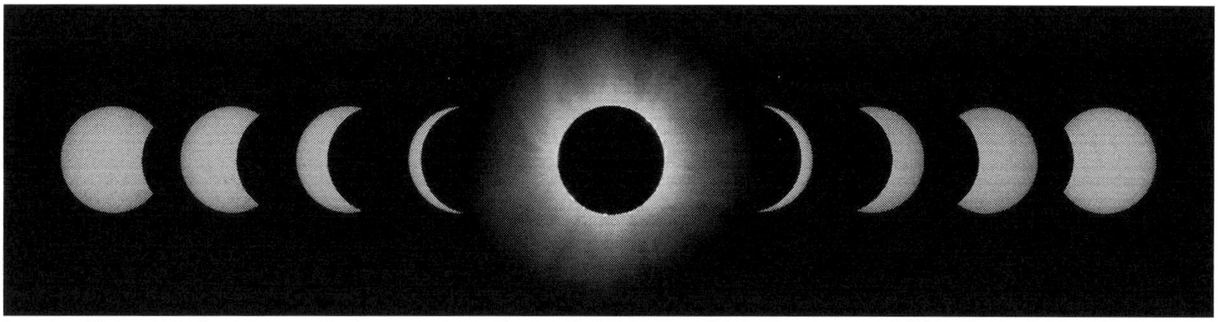

Photo 1–3: A sequence of images captures the partial phases and the solar corona visible only during totality. The total solar eclipse of 1999 August 11 from Lake Hazar, Turkey. © 1999 by F. Espenak, www.MrEclipse.com

1.11 Limb Profile Effects on the Maximum Duration of Totality

In the previous section, we discussed how the irregularities in the Moon's limb profile affect the duration of totality on the central line. For some eclipses, deep valleys along the limb can shorten the length of totality. But in the case of the 2017 eclipse, the duration is lengthened as much as 1.3 seconds because of a mountainous region in the southeast quadrant of the Moon.

Figure 1–6 shows a series of calculations for the duration of totality within ±40 kilometers of the central line for several locations spaced at 15-minute intervals along the path through the USA. For a given time, the duration of totality is calculated at 0.2 kilometer intervals perpendicular to the path within a 80 kilometer zone centered on the central line. Predictions using the Moon's center of mass and mean (smooth) limb are represented by the dotted curves. Predictions using the Kaguya/Herald limb profile to calculate corrected contact times and the resulting durations of totality are plotted as solid curves.

What is immediately apparent in Figure 1–6, is the skewing of the limb-corrected duration curves. This is a consequence of the complex Sun/Moon limb geometry, which changes significantly with the observer's position along the path. While the mean limb duration curves (dotted) appear symmetric about the central line, the limb-corrected duration curves are decidedly asymmetric and peak several kilometers to the south of the central line. Furthermore, the durations at these shifted positions actually exceed the durations on the central line. In other words, when the lunar limb profile is taken into account, the maximum duration does *NOT* occur on the central line as conventional wisdom would suggest, but rather at several kilometers to the south of the central line (at least, in the case of the 2017 total eclipse).

The shifts south of the central line to maximize the duration, as well as the corresponding duration values appear in the last two columns of Table 1–9. The shifts to maximum duration range from 0.6 to 6.6 km south of the central line with gains in duration from 0.0 to 1.3 seconds. The smallest gains are east of the Mississippi River (>90°W), while those exceeding 1 second are from Idaho through Wyoming (118° W to 105° W). Observers wishing to witness the maximum possible duration of totality from a given section of the path can use Table 1–9 to optimize their location with respect to the central line.

Figure 1–7 illustrates this information with two graphs covering longitudes from 105°W to 78°W. The top graph shows the shift south from the central line needed to reach the maximum duration. The lower graph plots two curves – the duration on the central line and the duration at the point of maximum duration south of the central line.

Figure 1–7: Duration on the Central Line and the Point of Maximum Duration

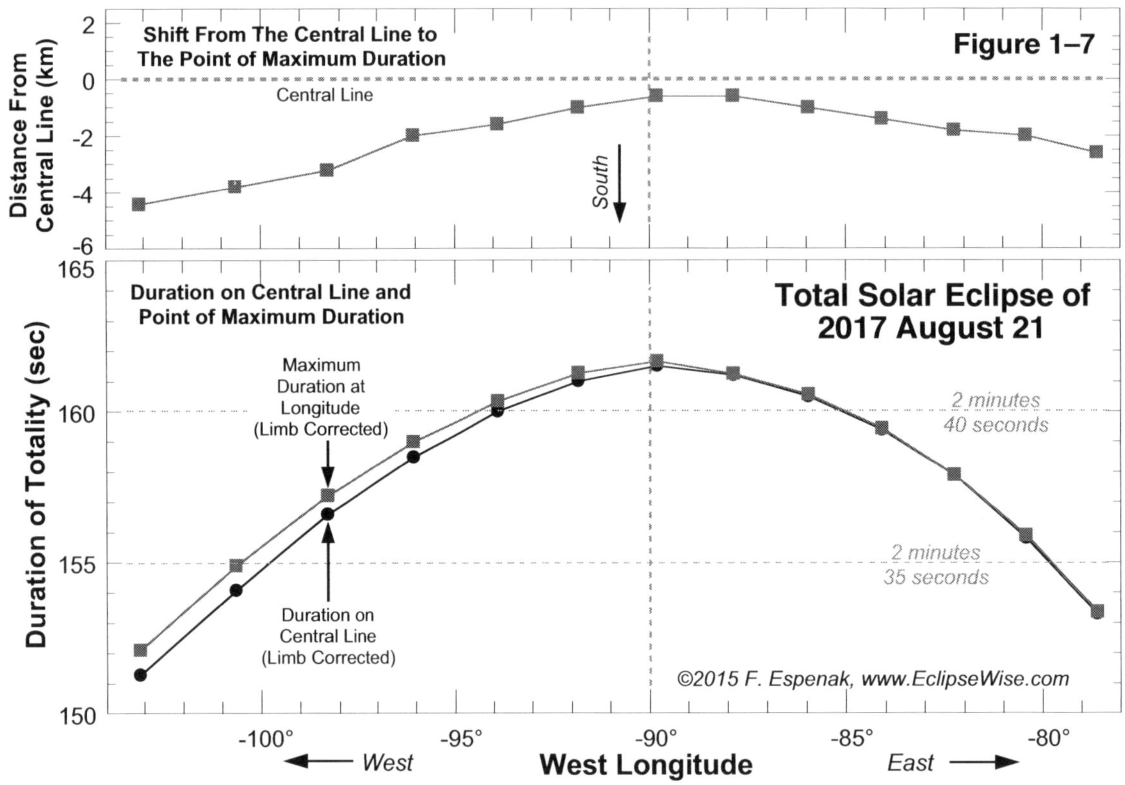

1.12 Limb Corrections to the Path Limits: Graze Zones

The northern and southern umbral limits provided in this publication were derived using the Moon's center of mass and a mean lunar radius. They have not been corrected for the effects of the lunar limb profile. In applications where precise limits are required, The Kaguya/Herald limb data must be used to correct the nominal or mean path. Unfortunately, a single correction at each limit is not possible because the Moon's libration in longitude and the contact points of the limits along the Moon's limb each vary as a function of time and position along the umbral path. This makes it necessary to calculate a unique correction to the limits at each point along the path. Furthermore, the northern and southern limits of the umbral path are actually paralleled by a relatively narrow zone where the eclipse is neither penumbral nor umbral. An observer positioned here will witness a slender solar crescent that is fragmented into a series of bright beads and short segments whose morphology changes quickly with the rapidly varying geometry between the limbs of the Moon and the Sun. These beading phenomena are caused by the appearance of photospheric rays that alternately pass through deep lunar valleys and hide behind high mountain peaks, as the Moon's irregular limb grazes the edge of the Sun's disk. The geometry is directly analogous to the case of grazing occultations of stars by the Moon. The graze zone is typically 5–10 km wide and its interior and exterior boundaries can be predicted using the lunar limb profile. The interior boundaries define the actual limits of the umbral eclipse (both total and annular) while the exterior boundaries set the outer limits of the grazing eclipse zone.

Table 1–6 provides topocentric data and corrections to the path limits due to the true lunar limb profile. At 5-minute intervals, the table lists the Moon's topocentric horizontal parallax, semi-diameter, relative angular velocity with respect to the Sun, and lunar libration in longitude. The Sun's central line altitude and azimuth is given, followed by the azimuth of the umbral path. The position angle of the point on the Moon's limb, which defines the northern limit of the path, is measured counterclockwise (i.e., eastward) from the equatorial north point on the limb. The path corrections to the northern and southern limits are listed as interior and exterior components in order to define the graze zone. Positive corrections are in the northern sense, while negative shifts are in the southern sense. These corrections (minutes of arc in latitude) may be added directly to the path coordinates listed in Table 1–3. Corrections to the central line umbral durations due to the lunar limb profile are also included and they are all positive; thus, when added to the central durations given in Tables 1–3, 1–4, 1–5, and 1–7, a slightly longer central total phase is predicted. This effect is due to a mountainous region in the southeast quadrant of the Moon for the predicted librations during the 2017 eclipse.

Detailed coordinates for the zones of grazing eclipse at each limit along the path are presented in Tables 1–8a and 1–8b. Given the uncertainties in the limb data, these predictions should be accurate to ±0.3 arc-seconds. (The interior graze coordinates take into account the deepest valleys along the Moon's limb, which produce the simultaneous second and third contacts at the path limits; thus, the interior coordinates that define the true edge of the path of totality.) They are calculated from an algorithm that searches the path limits for the extreme positions where no photospheric beads are visible along a ±30° segment of the Moon's limb, symmetric about the extreme contact points at the instant of maximum eclipse. The exterior graze coordinates are arbitrarily defined and calculated for the geodetic positions where an unbroken photospheric crescent of 60° in angular extent is visible at maximum eclipse.

Photo 1–4 : Baily's beads and solar prominences are captured at second contact during the total solar eclipse of 2006 March 29 from Jalu, Libya. © 2006 by F. Espenak, www.MrEclipse.com

In Tables 1–8a and 1–8b, the graze zone latitudes are listed every 0.5° in longitude (at sea level) and include the time of maximum eclipse at the northern and southern limits, as well as the path's azimuth. To correct the path for locations above sea level, *Elev Fact* (elevation factor) is a multiplicative factor by which the path must be shifted north or south perpendicular to itself, i.e., perpendicular to path azimuth, for each unit of elevation (height) above sea level.

The elevation factor is the product, tan(90–A) × sin(D), where A is the altitude of the Sun, and D is the difference between the azimuth of the Sun and the azimuth of the limit line, with the sign selected to be positive if the path should be shifted north with positive elevations above sea level. To calculate the shift, a location's elevation is multiplied by the elevation factor value. Negative values (usually the case for eclipses in the Northern Hemisphere) indicate that the path must be shifted south. For instance, if one's elevation is 1000 m above sea level and the elevation factor value is –0.50, then the shift is –500 m (= 1000 m × –0.50); thus, the observer must shift the path coordinates 500 m in a direction perpendicular to the path and in a negative or southerly sense.

The final column of Tables 1–8a and 1–8b lists the *Scale Fact* (in kilometers per arc second). This parameter provides an indication of the width of the zone of grazing phenomena, due to the topocentric distance of the Moon and the projection geometry of the Moon's shadow on Earth's surface. Because the solar chromosphere has an apparent thickness of about 3 arc-seconds, and assuming a scaling factor value of 1.75 km/arc-second, then the chromosphere should be visible continuously during totality for any observer in the path who is within 5.2 km (=1.75 × 3) of each interior limit. The most dynamic beading phenomena, however, occurs within 1.5 arc-seconds of the Moon's limb. Using the above scaling factor, this translates to the first 2.6 km inside the interior limits, but observers should position themselves at least 1 km inside the interior limits (south of the northern interior limit or north of the southern interior limit) in order to ensure that they are inside the path because of small uncertainties in the Kaguya/Herald data and the actual path limits.

For applications where the zones of grazing eclipse are needed at a higher frequency of longitude interval, tables of coordinates every 7.5′ in longitude are available at:

eclipsewise.com/solar/SEnews/TSE2017/extra.html

1.13 Saros History

The periodicity and recurrence of solar (and lunar) eclipses is governed by the Saros cycle, a period of approximately 6,585.3 days (= 18 years 11 days 8 hours). When two eclipses are separated by a period of one Saros, they share a very similar geometry. The eclipses occur at the same node with the Moon at nearly the same distance from Earth and at the same time of year. Thus, the Saros is useful for organizing eclipses into families or series. Each series typically lasts 12 or 13 centuries and contains 70 or more eclipses.

The total eclipse of 2017 is the 22nd member of Saros series 145 (Table 1–10), as defined by van den Bergh (1955). All eclipses in an odd numbered Saros series occur at the Moon's ascending node and the Moon moves southward (i.e., gamma decreases) with each succeeding member in the family.

The series is a young one which began with a minuscule partial eclipse at high northern hemisphere latitudes on 1639 January 04. After 14 partial eclipses each of increasing magnitude, the first central eclipse occurred on 1891 June 06. The event was a 6-second annular eclipse with a path sweeping through eastern Siberia and the Arctic Ocean. Although the vertex of the umbral shadow fell just short of Earth's surface, the Moon's distance was gradually decreasing with each subsequent eclipse in the series. In fact, the very next eclipse was a hybrid or annular/total eclipse on 1909 June 17. Greatest eclipse occurred in the Arctic Ocean near the North Pole where totality lasted 24 seconds.

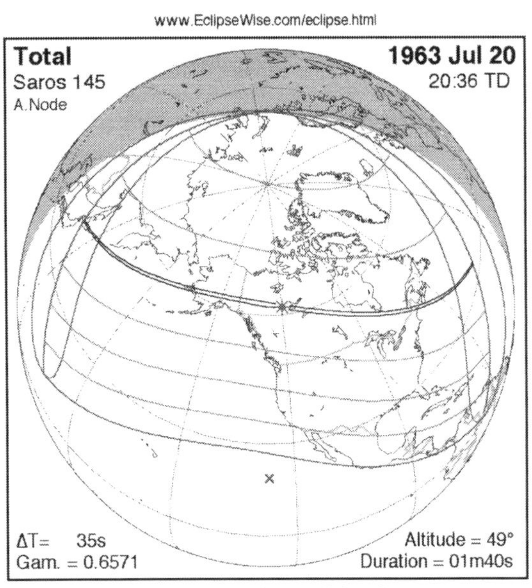

Thousand Year Canon of Solar Eclipses
©2014 by Fred Espenak

The 3rd central eclipse of Saros 145 occurred on 1927 June 29. It was the first total eclipse of the family and passed through England in addition to Scandinavia and Siberia. The duration of totality was now 50 seconds. On 1945 July 09, the path of totality began in Idaho and quickly swept northeast through Montana, Saskatchewan and Manitoba. After crossing Hudson Bay, Greenland, and the North Atlantic, the umbra returned to Scandinavia and Siberia.

The 5th central eclipse occurred on 1963 July 20 and is well known to many eclipse observers. Its path began in Japan, crossed the North Pacific and swept through Alaska, central and eastern Canada and Maine. The event drew a great deal of media attention and a beautiful article about the eclipse appeared months later in the pages of NATIONAL GEOGRAPHIC (November 1963). In fact, one of the authors (Espenak) has fond memories of watching the partial phases of his first eclipse as a boy from his grandparents' home in New York.

The track of the following eclipse of 1981 July 31 crossed central Siberia, Sakhalin Island, and the Pacific Ocean where it ended north of Hawaii. At the point of greatest eclipse the duration just broke the 2-minute mark.

The most recent eclipse of the series took place on 1999 August 11. The shadow track began in the Atlantic and swept through a dozen countries in central Europe (England, France, Belgium, Luxembourg, Germany, Austria, Slovenia, Croatia, Hungary, Yugoslavia, Romania and Bulgaria). The umbra then crossed the Black Sea, Turkey, Iran, Pakistan, and India where the path ended in the Bay of Bengal. Interestingly, this was the first total eclipse visible in England since the 1927 eclipse of the same Saros series 72 years earlier.

www.EclipseWise.com/eclipse.html

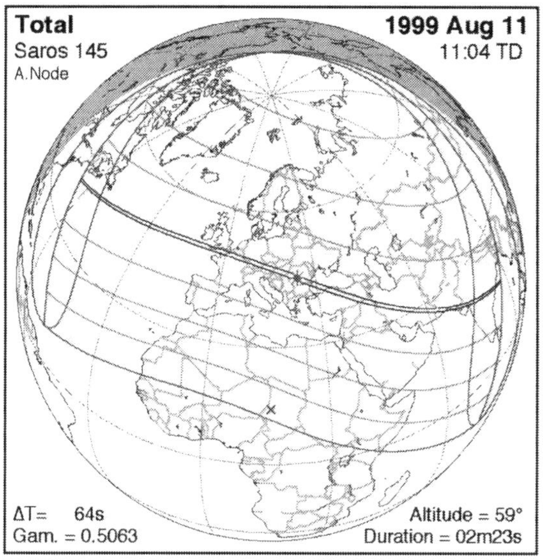

Thousand Year Canon of Solar Eclipses
©2014 by Fred Espenak

European TV stations carried substantial coverage of the eclipse. The umbral shadow was also observed from the Russian Mir space station as it raced across Europe. Both authors of this publication observed the eclipse from two clear locations in eastern Turkey.

The next eclipse of Saros 145 occurs on 2035 September 02. Its central path begins in western China, crosses Japan, and ends in the Pacific Ocean 2100 km (1300 mi) southeast of the Hawaiian Islands. The duration at greatest eclipse will fall 6 seconds short of 3 minutes.

During the 21st through 24th centuries, Saros 145 continues to produce total solar eclipses of increasing duration as the path of each event shifts southward. By the time the midpoint of the series is reached (2324 February 25), the duration of totality exceeds 4 minutes. The duration continues to increase into the 25th and 26th centuries. The maximum duration of totality peaks at 7 minutes 12 seconds on 2522 June 25 when the path crosses southern Africa and the Indian Ocean.

In spite of the long duration of 2522, only 6 total eclipses remain in the series – the duration of each rapidly dropping. The final total eclipse occurs on 2648 September 09. It belongs to the rare case of central eclipses, which have no southern limit. Still, the umbral path offers a duration of a respectable 2 minutes 48 seconds that encompasses the southern part of the Antarctic Peninsula.

For the next 3½ centuries, a long series of 20 partial eclipses of progressively decreasing magnitude take place. The final event occurs on 3009 April 17 from the polar regions of the Southern Hemisphere. A detailed list of eclipses in Saros series 145 appears in Table 1–10.

In summary, Saros 145 includes 77 eclipses with the following order: 14 partial, 1 annular, 1 hybrid, 41 total, and 20 partial. Complete details and links to maps for each eclipse in the series can be found at:

eclipsewise.com/solar/SEsaros/SEsaros145.html

Table 1–10: Solar Eclipses of Saros Series 145

First Eclipse: 1639 Jan 04 Duration of Series: 1370.3 yrs.
Last Eclipse: 3009 Apr 17 Number of Eclipses: 77

Saros Summary: Partial: 34 Annular: 1 Total: 41 Hybrid: 1

Eclipse Date	Type	Gamma	Mag./ Width	Center Durat.
1639 Jan 04	Pb	1.565	0.001	
1657 Jan 14	P	1.554	0.017	
1675 Jan 25	P	1.543	0.034	
1693 Feb 05	P	1.527	0.059	
1711 Feb 17	P	1.507	0.091	
1729 Feb 27	P	1.481	0.134	
1747 Mar 11	P	1.450	0.186	
1765 Mar 21	P	1.412	0.251	
1783 Apr 01	P	1.367	0.329	
1801 Apr 13	P	1.315	0.420	
1819 Apr 24	P	1.258	0.522	
1837 May 04	P	1.193	0.638	
1855 May 16	P	1.125	0.762	
1873 May 26	P	1.051	0.897	
1891 Jun 06	A	0.975	33	00m06s
1909 Jun 17	H	0.896	51	00m24s
1927 Jun 29	T	0.816	77	00m50s
1945 Jul 09	T	0.736	92	01m15s
1963 Jul 20	T	0.657	101	01m40s
1981 Jul 31	T	0.579	108	02m02s
1999 Aug 11	T	0.506	112	02m23s
2017 Aug 21	T	0.437	115	02m40s
2035 Sep 02	T	0.373	116	02m54s
2053 Sep 12	T	0.314	116	03m04s
2071 Sep 23	T	0.262	116	03m11s
2089 Oct 04	T	0.217	115	03m14s
2107 Oct 16	T	0.178	114	03m16s
2125 Oct 26	T	0.146	112	03m15s
2143 Nov 07	T	0.121	111	03m14s
2161 Nov 17	T	0.101	110	03m13s
2179 Nov 28	T	0.087	110	03m12s
2197 Dec 09	T	0.077	111	03m13s
2215 Dec 21	T	0.070	114	03m14s
2233 Dec 31	T	0.065	117	03m18s
2252 Jan 12	T	0.061	123	03m23s
2270 Jan 22	T	0.056	129	03m29s
2288 Feb 02	T	0.050	138	03m38s
2306 Feb 14	T	0.040	147	03m49s
2324 Feb 25	Tm	0.026	158	04m02s
2342 Mar 08	T	0.008	169	04m16s

Eclipse Date	Type	Gamma	Mag./ Width	Center Durat.
2360 Mar 18	T	-0.017	181	04m33s
2378 Mar 29	T	-0.048	193	04m51s
2396 Apr 09	T	-0.085	206	05m12s
2414 Apr 20	T	-0.128	217	05m33s
2432 Apr 30	T	-0.178	229	05m56s
2450 May 12	T	-0.233	241	06m19s
2468 May 22	T	-0.293	252	06m41s
2486 Jun 02	T	-0.358	263	06m59s
2504 Jun 14	T	-0.427	275	07m10s
2522 Jun 25	T	-0.499	287	07m12s
2540 Jul 05	T	-0.572	300	07m04s
2558 Jul 16	T	-0.646	315	06m43s
2576 Jul 27	T	-0.720	334	06m12s
2594 Aug 07	T	-0.792	361	05m32s
2612 Aug 18	T	-0.862	406	04m45s
2630 Aug 30	T	-0.930	512	03m54s
2648 Sep 09	Ts	-0.992	-	02m49s
2666 Sep 20	P	-1.050	0.919	
2684 Oct 01	P	-1.103	0.817	
2702 Oct 13	P	-1.150	0.727	
2720 Oct 23	P	-1.191	0.648	
2738 Nov 04	P	-1.225	0.584	
2756 Nov 14	P	-1.255	0.528	
2774 Nov 25	P	-1.278	0.486	
2792 Dec 06	P	-1.297	0.451	
2810 Dec 17	P	-1.311	0.426	
2828 Dec 27	P	-1.322	0.405	
2847 Jan 08	P	-1.331	0.389	
2865 Jan 18	P	-1.339	0.374	
2883 Jan 30	P	-1.348	0.359	
2901 Feb 10	P	-1.359	0.339	
2919 Feb 21	P	-1.371	0.316	
2937 Mar 04	P	-1.389	0.285	
2955 Mar 15	P	-1.411	0.245	
2973 Mar 25	P	-1.439	0.194	
2991 Apr 06	P	-1.472	0.134	
3009 Apr 17	Pe	-1.514	0.059	

Eclipse Type:
- P - Partial
- A - Annular
- T - Total
- H - Hybrid (Annular/Total)
- Pb - Partial Eclipse (Saros Series Begins)
- Pe - Partial Eclipse (Saros Series Ends)
- Ts - Total Eclipse (no southern limit)

Note: Mag./Width column gives either the eclipse magnitude (for partial eclipses) or the umbral path width in kilometers (for total and annular eclipses).

Section 2: Local Circumstances for the Eclipse

2.1 Introduction

As the Moon's umbral and penumbral shadows sweep across Earth's surface, different geographic locations are in different parts of either of the two shadows at any given time. The penumbral shadow is quite large and has a diameter of about 6900 km (4300 mi) during the 2017 eclipse. Because of this, the entire North American continent, the Atlantic Ocean, and the northern third of South America fall within its path, and a partial solar eclipse will be visible from the region.

In contrast, the umbral shadow is quite narrow and an observer must be somewhere within its path in order to see the total phase of the eclipse. The width of the umbral path ranges from 57 to 115 kilometers (35 to 71 miles) during the 2017 eclipse.

Figure 2–1: Moon's Shadows and Local Circumstances

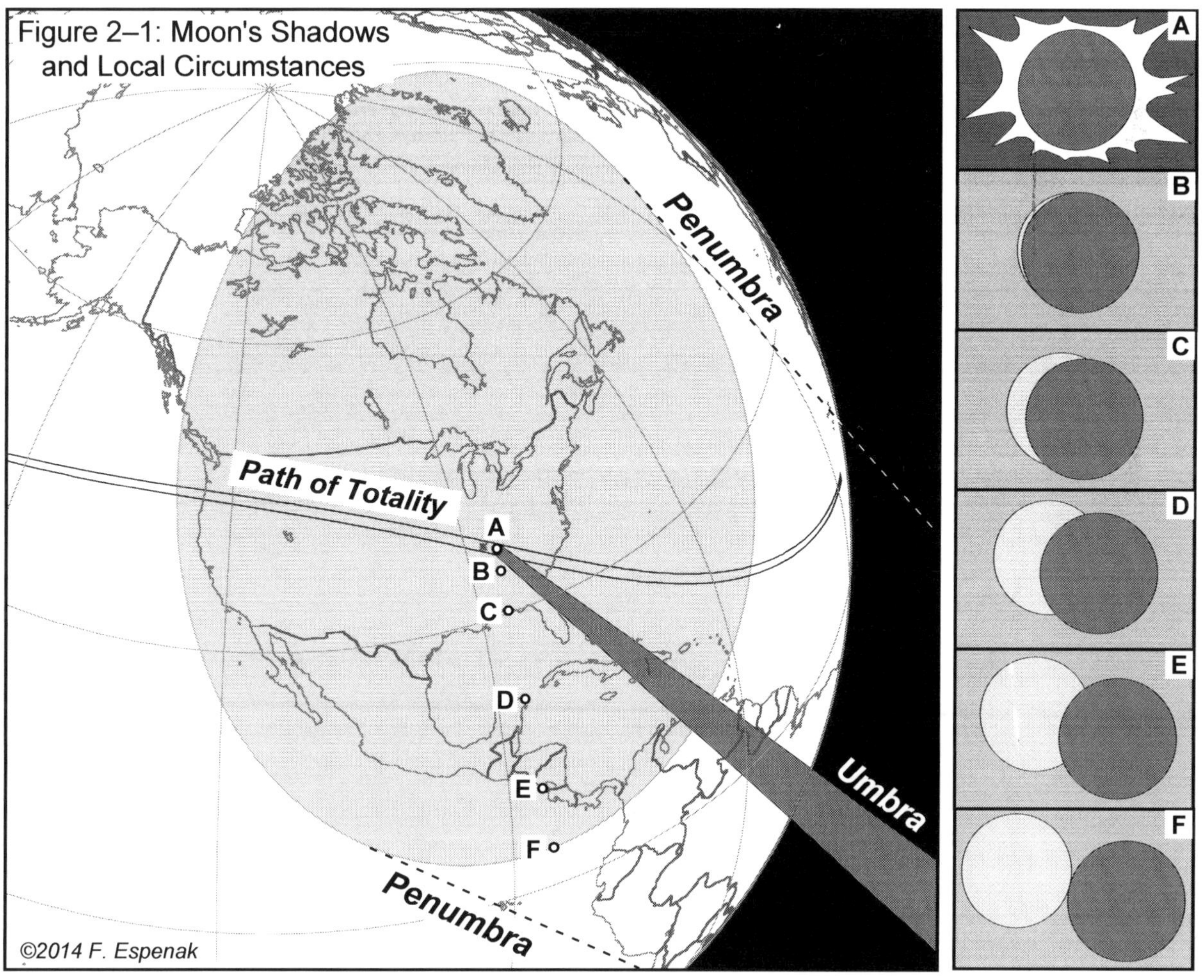

Figure 2–1 illustrates the path of totality for the total solar eclipse of 2017 August 21. The appearance of maximum eclipse from six different geographic locations on the map (labeled A through F) is shown in the side bar. The total eclipse is visible only from position A because it lies in the path of totality.

Figure 2-1 illustrates how observers at different locations can observe different phases of the eclipse (or even no eclipse) at the same instant. Because of this, the beginning, middle and end times of the eclipse, its magnitude at maximum, and the Sun's altitude and azimuth at each phase of the eclipse (i.e., the local circumstances of the eclipse), will be different for any two locations.

2.2 Solar Eclipse Contacts

During the course of a solar eclipse, the instants when the Moon's disk becomes tangent to the Sun's disk are known as eclipse contacts. They mark various stages or phases of a solar eclipse.

Partial solar eclipses have two primary contacts.

First Contact (C1) — Instant of first exterior tangency of the Moon with the Sun (Partial Eclipse Begins)
Fourth Contact (C4) — Instant of last exterior tangency of the Moon with the Sun (Partial Eclipse Ends)

Total solar eclipses have four primary contacts. Contacts C2 and C3 mark the instants when the Moon's disk is first and last internally tangent to the Sun. These are the times when the annular or total phase of the eclipse begins and ends, respectively.

First Contact (C1) — Instant of first exterior tangency of the Moon with the Sun (Partial Eclipse Begins)
Second Contact (C2) — Instant of first interior tangency of the Moon with the Sun (Total Eclipse Begins)
Third Contact (C3) — Instant of last interior tangency of the Moon with the Sun (Total Eclipse Ends)
Fourth Contact (C4) — Instant of last exterior tangency of the Moon with the Sun (Partial Eclipse Ends)

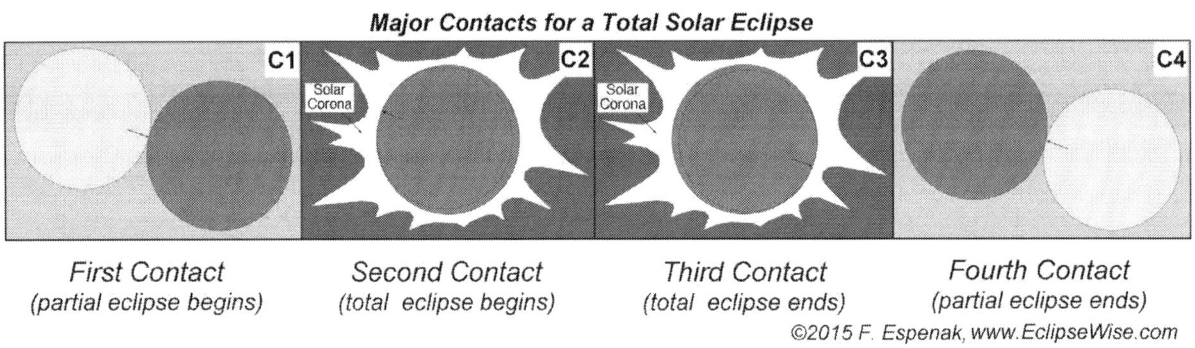

Figure 2-1 illustrates the four contacts for a total solar eclipse. The arrows indicate the contact point of the Sun's limb in each diagram.

2.3 Local Circumstances Tables

Local circumstances for the 2017 eclipse as seen from over 1,000 cities throughout the United States are found in Tables 2-1 through 2-23. The eclipse circumstances for 120 cities in Canada appear in Tables 2-24 through 2-26. Local circumstances for cities in other regions include Mexico and Central America (Table 2-27), South America (Table 2-28), Europe (Table 2-29), and Africa and the North Atlantic (Table 2-30).

The tables give the local circumstances at each contact and at maximum eclipse for every location. For locations where the eclipse is partial, maximum eclipse is the instant when the greatest fraction of the Sun's diameter is occulted. For locations where the eclipse is total, maximum eclipse is the instant of mid-totality.

The times listed for the United States and Canada (Tables 2-1 through 2-26) are in *local time* for each city. The eclipse occurs during mid-summer when most states and provinces observe *daylight saving time or DST* (Arizona, Hawaii, and Saskatchewan are exceptions). Where applicable, DST has been included in calculating the local time.

For cities in Mexico, Central and South America, Europe, Africa, and the North Atlantic (Tables 2–27 through 2–30), the times listed are in *Universal Time (UT1)*.

The time of each contact is given to 1 second, along with position angles **P** and **V** and the altitude of the Sun. The position angles identify the point along the Sun's disk where each contact occurs and are measured counterclockwise (i.e., eastward) from the north and zenith points, respectively. Locations outside the umbral path miss the total eclipse and only witness first and fourth contacts. The time of maximum eclipse (either partial or total) is also listed to the nearest second. Next, the position angles **P** and **V** of the Moon's disk with respect to the Sun are given followed by the altitude and azimuth of the Sun at maximum eclipse. Finally, the corresponding eclipse magnitude[7] and obscuration[8] are listed. For total eclipses, the eclipse magnitude given is actually the topocentric ratio of the Moon and Sun's apparent diameters, rather than the fraction of the Sun's diameter occulted.

Two additional columns are included if the location lies within the path of the Moon's umbral shadow. The *Umbral Depth* is a relative measure of a location's position with respect to the central line and umbral path limits. It is a unitless parameter, which is defined as

$$u = 1 - (2\,x/W), \tag{2–1}$$

where:

u is the *Umbral Depth*,
x is the perpendicular distance from the central line in kilometers, and
W is the width of the path in kilometers.

The *Umbral Depth* for a location in the path of totality varies from 0.0 to 1.0. A position at the umbral path limits corresponds to a value of 0.0, while a position on the central line has a value of 1.0. The parameter can be used to quickly determine the corresponding central line duration. Thus, it is a useful tool for evaluating the trade-off in duration of a location's position relative to the central line. Using the location's duration and *Umbral Depth*, the central line duration is calculated as

$$D = d \ / \ [1 - (1 - u)^2]^{1/2}, \tag{2–2}$$

where:

D is the duration of totality on the central line (seconds),
d is the duration of totality at location (seconds), and
u is the *Umbral Depth*.

The final column *Umbral Durat.* gives the duration of totality. The effects of refraction have not been included in these calculations, nor have there been any corrections for the lunar limb profile.

Locations were chosen based on general geographic distribution, population, and proximity to the path. The primary source for geographic coordinates is *The New International Atlas* (Rand McNally 1991). In this rapidly changing political world, it is often difficult to ascertain the correct name or spelling for a given location; therefore, the information presented here is for location purposes only and is not meant to be authoritative. Corrections to names, spellings, and coordinates, should be forwarded to the authors in order to update the geographic database for future eclipse predictions (*eclipsewise.com/main/contact.html*).

2.4 Lunar Limb Corrections to Contact Times

In investigations where accurate contact times are needed, the lunar limb profile can be used to correct the nominal or mean limb predictions. For any given position angle, there will be a high mountain (annular eclipses) or a low valley (total eclipses) in the vicinity that ultimately determines the true instant of contact. The difference, in time, between the Sun's position when tangent to the contact point on the mean limb and tangent to the highest mountain (annular) or lowest valley (total) at actual contact is the desired correction to the predicted contact time. On the exaggerated radial scale of Figure 1–5, the Sun's limb can be represented as

[7] Eclipse magnitude is defined as the fraction of the Sun's *diameter* occulted by the Moon.
[8] Eclipse obscuration is defined as the fraction of the Sun's *area* occulted by the Moon.

an epicyclic curve that is tangent to the mean lunar limb at the point of contact and departs from the limb by h through

$$h = S\,(m-1)\,(1-\cos[C]), \tag{2-3}$$

where:

h is the departure of Sun's limb from mean lunar limb,
S is the Sun's semi-diameter,
m is the eclipse magnitude, and
C is the angle from the point of contact.

Herald (1983) takes advantage of this geometry in developing a graphic procedure for estimating correction times over a range of position angles. Briefly, a displacement curve of the Sun's limb is constructed on a transparent overlay by way of equation (2-3). For a given position angle, the solar limb overlay is moved radially from the mean lunar limb contact point until it is tangent to the lowest lunar profile feature in the vicinity. The solar limb's distance d (in arc seconds) from the mean lunar limb is then converted to a time correction δ by

$$\delta = d/v \cos\,[X - C], \tag{2-4}$$

where:

δ is the correction to contact time (in seconds),
d is the distance of solar limb from Moon's mean limb (in arc seconds),
v is the angular velocity of the Moon with respect to the Sun (arc seconds per second),
X is the central line position angle of the contact, and
C is the angle from the point of contact.

This operation may be used for predicting the formation and location of Baily's beads. When calculations are performed over a large range of position angles, a contact time correction curve can then be constructed.

Because the limb profile data are available in digital form, an analytical solution to the problem is possible that is quite straightforward and robust. Curves of corrections to the times of second and third contact for most position angles have been computer generated and are plotted in Figure 1–5. The circular protractor scale at the center represents the nominal contact time using a mean lunar limb. The departure of the contact correction curves from this scale graphically illustrates the time correction to the mean predictions for any position angle as a result of the Moon's true limb profile. Time corrections external to the circular scale are added to the mean contact time; time corrections internal to the protractor are subtracted from the mean contact time. The magnitude of the time correction at a given position angle is measured using any of the four radial scales plotted at each cardinal point.

For example, Table 2–13 gives the following data for Grand Island, Nebraska:

Second Contact = 12:58:34.2 CDT P_2 = 105°, and
Third Contact = 13:01:08.5 CDT P_3 = 298°.

Using Figure 1–5, the measured time corrections and the resulting contact times are

C_2= –1.2 seconds;
Second Contact = 12:58:34.2 –1.2 = 12:58:33.0 CDT, and

C_3= –2.0 seconds;
Third Contact = 13:01:08.5 – 2.0 = 13:01:06.5 CDT.

The above corrected contact times are within 0.2 seconds of a rigorous calculation using the Kaguya/Herald limb profile.

Table 2–1: Local Circumstances for Alabama, Alaska, Arizona & Arkansas

Location Name	Latitude	Longitude	First Contact Time h m s	P °	V °	Alt °	Second Contact Time h m s	P °	V °	Third Contact Time h m s	P °	V °	Fourth Contact Time h m s	P °	V °	Alt °	Maximum Eclipse Time h m s	P °	V °	Alt °	Azm °	Eclip. Mag.	Eclip. Obs.	Umbral Depth	Umbral Durat.
Alabama																									
Anniston	33°39'N	085°50'W	12:02:44	297	320	66	—			—			14:59:45	111	62	53	13:33:52	25	0	66	209	0.951	0.945		
Auburn	32°36'N	085°29'W	12:04:37	299	322	67	—			—			15:02:01	110	59	52	13:36:10	25	357	66	212	0.928	0.916		
Bessemer	33°24'N	086°58'W	12:00:29	298	325	66	—			—			14:58:15	110	62	54	13:31:49	25	3	67	206	0.930	0.920		
Birmingham	33°31'N	086°48'W	12:00:44	298	324	66	—			—			14:58:22	111	62	54	13:32:00	25	2	66	206	0.935	0.926		
Decatur	34°36'N	086°59'W	11:59:19	296	322	64	—			—			14:56:19	112	65	54	13:30:06	25	5	66	204	0.962	0.959		
Dothan	31°13'N	085°24'W	12:06:27	301	324	69	—			—			15:04:20	109	55	52	13:38:26	25	354	67	216	0.893	0.872		
Florence	34°48'N	087°41'W	11:57:38	296	323	64	—			—			14:54:49	112	66	54	13:28:20	24	7	66	201	0.959	0.955		
Gadsden	34°01'N	086°01'W	12:01:58	297	320	66	—			—			14:58:51	112	63	53	13:32:58	25	1	66	208	0.958	0.954		
Hoover	33°24'N	086°48'W	12:00:51	298	324	66	—			—			14:58:33	110	62	54	13:32:10	25	2	66	206	0.932	0.922		
Huntsville	34°44'N	086°35'W	12:00:04	296	320	65	—			—			14:56:45	112	66	53	13:30:45	25	4	65	205	0.970	0.969		
Mobile	30°41'N	088°03'W	12:01:12	304	335	67	—			—			15:00:21	106	54	55	13:33:20	25	1	69	207	0.845	0.811		
Montgomery	32°23'N	086°19'W	12:03:00	300	325	67	—			—			15:00:58	109	59	53	13:34:41	25	359	67	210	0.912	0.896		
Prichard	30°45'N	088°06'W	12:01:00	303	335	67	—			—			15:00:10	106	54	56	13:33:07	25	2	69	207	0.846	0.812		
Tuscaloosa	33°12'N	087°34'W	11:59:22	299	327	65	—			—			14:57:32	110	61	54	13:30:47	25	4	67	204	0.918	0.904		
Alaska																									
Anchorage	61°13'N	149°54'W	08:21:37	256	286	13	—			—			10:13:50	130	157	26	09:16:16	193	222	19	100	0.556	0.456		
Fairbanks	64°51'N	147°43'W	08:29:39	251	277	15	—			—			10:16:34	136	158	25	09:21:59	194	219	20	105	0.478	0.369		
Arizona																									
Chandler	33°18'N	111°50'W	09:14:18	305	2	41	—			—			12:01:02	93	109	68	10:34:17	19	66	56	122	0.696	0.625		
Flagstaff	35°12'N	111°39'W	09:13:50	302	356	40	—			—			12:01:31	96	110	66	10:34:23	19	62	55	125	0.757	0.700		
Gilbert	33°21'N	111°47'W	09:14:20	305	1	41	—			—			12:01:11	93	109	68	10:34:23	19	65	56	122	0.698	0.627		
Glendale	33°32'N	112°11'W	09:13:41	305	1	40	—			—			12:00:09	93	110	67	10:33:29	19	66	55	122	0.702	0.632		
Lake Havasu City	34°25'N	114°08'W	09:10:43	303	359	38	—			—			11:55:15	93	116	65	10:29:19	18	66	53	119	0.720	0.654		
Mesa	33°25'N	111°50'W	09:14:14	305	1	40	—			—			12:01:03	93	109	67	10:34:16	19	65	56	122	0.700	0.630		
Nogales	31°20'N	110°56'W	09:16:51	309	7	42	—			—			12:03:13	90	105	70	10:36:46	19	68	58	121	0.640	0.556		
Peoria	33°35'N	112°14'W	09:13:36	305	1	40	—			—			12:00:01	93	110	67	10:33:22	19	65	55	122	0.703	0.633		
Phoenix	33°27'N	112°04'W	09:13:53	305	1	40	—			—			12:00:26	93	110	67	10:33:45	19	66	56	122	0.700	0.629		
Prescott	34°33'N	112°28'W	09:12:55	303	358	39	—			—			11:59:29	94	112	66	10:32:46	19	64	55	122	0.732	0.669		
Scottsdale	33°29'N	111°56'W	09:14:04	305	1	40	—			—			12:00:48	93	109	67	10:34:02	19	65	56	122	0.702	0.631		
Sierra Vista	31°33'N	110°18'W	09:17:38	309	6	43	—			—			12:05:01	91	103	70	10:38:10	20	67	59	123	0.650	0.569		
Tempe	33°25'N	111°56'W	09:14:06	305	1	40	—			—			12:00:47	93	109	67	10:34:03	19	65	56	122	0.700	0.629		
Tucson	32°13'N	110°58'W	09:16:09	307	5	42	—			—			12:03:15	91	105	69	10:36:26	19	66	58	122	0.667	0.590		
Yuma	32°43'N	114°37'W	09:10:44	306	4	38	—			—			11:53:32	90	116	66	10:28:22	18	69	53	117	0.665	0.587		
Arkansas																									
Conway	35°05'N	092°26'W	11:47:23	298	334	60	—			—			14:45:46	109	70	59	13:17:36	24	21	67	183	0.912	0.896		
Fayetteville	36°04'N	094°10'W	11:43:17	297	335	57	—			—			14:41:02	109	75	60	13:12:41	23	27	66	176	0.921	0.908		
Florence	33°46'N	091°39'W	11:50:03	300	337	62	—			—			14:49:07	107	65	58	13:20:56	24	18	68	187	0.885	0.862		
Fort Smith	35°25'N	094°25'W	11:43:10	298	337	58	—			—			14:41:24	108	73	60	13:12:48	23	27	66	175	0.900	0.881		
Hot Springs	34°30'N	093°03'W	11:46:33	299	337	60	—			—			14:45:21	107	69	59	13:16:53	24	23	67	181	0.889	0.867		
Jonesboro	35°50'N	090°42'W	11:50:27	296	328	61	—			—			14:47:55	111	71	57	13:20:32	24	16	66	189	0.952	0.946		
Little Rock	34°45'N	092°17'W	11:47:56	298	335	60	—			—			14:46:31	108	69	59	13:18:19	24	20	67	184	0.905	0.887		
North Little Rock	34°45'N	092°16'W	11:47:58	298	335	60	—			—			14:46:33	108	69	59	13:18:22	24	20	67	184	0.905	0.887		
Pine Bluff	34°13'N	092°01'W	11:48:54	299	336	61	—			—			14:47:47	108	67	59	13:19:33	24	19	68	185	0.893	0.872		
Rogers	36°20'N	094°07'W	11:43:14	296	334	57	—			—			14:40:46	109	76	59	13:12:31	23	27	65	176	0.929	0.918		
Springdale	36°11'N	094°08'W	11:43:17	297	335	57	—			—			14:40:56	109	75	60	13:12:38	23	27	66	176	0.924	0.912		

Table 2–2: Local Circumstances for California

Location Name	Latitude	Longitude	First Contact Time	P	V	Alt	Second Contact Time	P	V	Third Contact Time	P	V	Fourth Contact Time	P	V	Alt	Maximum Eclipse Time	P	V	Alt	Azm	Eclip. Mag.	Eclip. Obs.	Umbral Depth	Umbral Durat.
			h m s	°	°	°	h m s	°	°	h m s	°	°	h m s	°	°	°	h m s	°	°	°	°				
California																									
Alameda	37°46'N	122°15'W	09:01:40	296	350	29	—			—			11:37:29	96	131	57	10:15:31	16	65	43	111	0.801	0.755		
Anaheim	33°50'N	117°55'W	09:06:10	304	1	34	—			—			11:45:32	90	123	63	10:21:48	17	69	49	113	0.688	0.615		
Bakersfield	35°23'N	119°01'W	09:04:40	301	357	32	—			—			11:43:36	93	126	61	10:20:05	17	67	47	113	0.734	0.671		
Berkeley	37°52'N	122°16'W	09:01:41	296	350	29	—			—			11:37:29	96	131	57	10:15:32	16	65	43	112	0.804	0.759		
Calexico	32°40'N	115°30'W	09:09:36	306	5	37	—			—			11:51:11	89	118	66	10:26:31	18	69	52	115	0.660	0.580		
Chula Vista	32°39'N	117°05'W	09:07:36	306	5	35	—			—			11:47:02	89	122	65	10:23:18	17	70	50	113	0.653	0.573		
Concord	37°59'N	122°02'W	09:01:56	296	349	29	—			—			11:38:02	96	131	57	10:15:57	16	65	43	112	0.808	0.764		
Corona	33°53'N	117°34'W	09:06:34	304	1	34	—			—			11:46:26	91	123	63	10:22:29	17	69	49	114	0.691	0.618		
Costa Mesa	33°38'N	117°55'W	09:06:14	304	2	34	—			—			11:45:26	90	123	63	10:21:46	17	69	49	113	0.682	0.607		
Cucamonga	34°06'N	117°36'W	09:06:29	304	1	34	—			—			11:46:27	91	123	63	10:22:27	17	68	49	114	0.697	0.626		
East Los Angeles	34°01'N	118°02'W	09:05:59	304	1	34	—			—			11:45:20	91	124	63	10:21:37	17	69	49	113	0.694	0.621		
El Monte	34°04'N	118°02'W	09:05:59	304	1	34	—			—			11:45:22	91	124	63	10:21:37	17	68	49	113	0.695	0.623		
Escondido	33°07'N	117°05'W	09:07:24	306	3	35	—			—			11:47:17	89	122	64	10:23:20	17	69	50	113	0.668	0.590		
Fairfield	38°15'N	122°03'W	09:02:00	296	349	29	—			—			11:38:07	97	131	56	10:16:03	16	64	43	112	0.817	0.774		
Fontana	34°06'N	117°26'W	09:06:40	304	0	34	—			—			11:46:52	91	123	63	10:22:46	17	68	49	114	0.698	0.627		
Fremont	37°32'N	121°57'W	09:01:53	297	351	29	—			—			11:38:00	95	131	57	10:15:53	16	65	43	112	0.795	0.747		
Fresno	36°42'N	119°47'W	09:03:53	299	353	31	—			—			11:42:23	95	127	59	10:19:07	17	66	46	114	0.773	0.720		
Fullerton	33°53'N	117°56'W	09:06:08	304	1	34	—			—			11:45:31	90	124	63	10:21:47	17	69	49	113	0.690	0.617		
Garden Grove	33°47'N	117°55'W	09:06:11	304	2	34	—			—			11:45:31	90	123	63	10:21:48	17	69	49	113	0.687	0.613		
Glendale	34°09'N	118°15'W	09:05:42	303	0	33	—			—			11:44:52	91	124	62	10:21:14	17	68	48	113	0.697	0.626		
Hayward	37°40'N	122°05'W	09:01:48	297	350	29	—			—			11:37:47	96	131	57	10:15:44	16	65	43	112	0.798	0.752		
Huntington Beach	33°40'N	118°05'W	09:06:01	304	2	34	—			—			11:45:02	90	124	63	10:21:28	17	69	49	113	0.682	0.608		
Inglewood	33°58'N	118°21'W	09:05:38	304	1	33	—			—			11:44:32	90	124	62	10:21:00	17	69	48	113	0.691	0.618		
Irvine	33°41'N	117°46'W	09:06:23	304	2	34	—			—			11:45:50	90	123	63	10:22:04	17	69	49	113	0.684	0.609		
Lancaster	34°42'N	118°08'W	09:05:44	302	359	33	—			—			11:45:25	92	124	62	10:21:33	17	68	48	114	0.715	0.647		
Long Beach	33°47'N	118°11'W	09:05:52	304	2	34	—			—			11:44:51	90	124	63	10:21:17	17	69	49	113	0.686	0.612		
Los Angeles	34°04'N	118°15'W	09:05:43	304	1	33	—			—			11:44:50	91	124	63	10:21:13	17	69	48	113	0.694	0.622		
Modesto	37°39'N	121°00'W	09:02:50	297	350	30	—			—			11:40:05	96	129	58	10:17:27	16	65	44	113	0.800	0.753		
Monterey	36°37'N	121°55'W	09:01:45	299	353	29	—			—			11:37:38	94	130	58	10:15:34	16	66	44	111	0.766	0.711		
Oakland	37°49'N	122°16'W	09:01:40	296	350	29	—			—			11:37:28	96	131	57	10:15:31	16	65	43	111	0.803	0.757		
Oceanside	33°12'N	117°23'W	09:07:00	305	3	34	—			—			11:46:33	89	122	64	10:22:45	17	69	50	113	0.670	0.592		
Ontario	34°04'N	117°39'W	09:06:26	304	1	34	—			—			11:46:19	91	123	63	10:22:21	17	68	49	114	0.696	0.625		
Orange	33°47'N	117°51'W	09:06:16	304	2	34	—			—			11:45:41	90	123	63	10:21:55	17	69	49	113	0.687	0.613		
Oxnard	34°15'N	119°07'W	09:04:42	303	0	32	—			—			11:42:48	91	126	62	10:19:38	17	69	47	112	0.698	0.627		
Pasadena	34°09'N	118°09'W	09:05:49	303	0	33	—			—			11:45:07	91	124	63	10:21:25	17	68	49	113	0.697	0.626		
Pomona	34°04'N	117°45'W	09:06:19	304	1	34	—			—			11:46:04	91	123	63	10:22:09	17	68	49	113	0.696	0.624		
Riverside	33°59'N	117°22'W	09:06:47	304	1	34	—			—			11:46:59	91	123	63	10:22:53	17	68	49	114	0.695	0.623		
Sacramento	38°35'N	121°29'W	09:02:38	295	348	30	—			—			11:39:25	97	130	57	10:17:04	16	64	43	113	0.828	0.789		
Salinas	36°40'N	121°39'W	09:02:00	299	353	30	—			—			11:38:14	94	130	58	10:16:01	16	66	44	111	0.768	0.713		
San Bernardino	34°10'N	117°16'W	09:06:51	304	0	34	—			—			11:47:18	91	122	63	10:23:06	17	68	49	114	0.701	0.630		
San Diego	32°43'N	117°09'W	09:07:29	306	5	35	—			—			11:46:53	89	122	64	10:23:10	17	70	50	113	0.655	0.575		
San Francisco	37°47'N	122°25'W	09:01:31	296	350	29	—			—			11:37:08	96	131	56	10:15:16	16	65	43	111	0.802	0.755		
San Jose	37°20'N	121°53'W	09:01:54	297	351	29	—			—			11:38:03	95	130	57	10:15:54	16	65	43	111	0.788	0.739		
Santa Ana	33°46'N	117°52'W	09:06:15	304	2	34	—			—			11:45:38	90	123	63	10:21:53	17	69	49	113	0.686	0.612		
Santa Barbara	34°26'N	119°44'W	09:03:59	303	360	32	—			—			11:41:25	91	127	61	10:18:34	17	69	47	111	0.702	0.632		
Santa Cruz	36°58'N	122°01'W	09:01:42	298	352	29	—			—			11:37:36	94	130	57	10:15:33	16	66	43	111	0.777	0.724		
Santa Rosa	33°58'N	120°06'W	09:03:40	304	1	31	—			—			11:40:16	90	127	61	10:17:47	17	69	46	110	0.686	0.613		
Simi Valley	34°17'N	118°53'W	09:04:57	303	0	33	—			—			11:43:23	91	125	62	10:20:05	17	69	48	112	0.700	0.629		
Stockton	37°58'N	121°17'W	09:02:38	296	349	30	—			—			11:39:36	96	130	57	10:17:07	16	64	44	113	0.809	0.765		
Sunnyvale	37°23'N	122°02'W	09:01:46	297	351	29	—			—			11:37:46	95	131	57	10:15:41	16	65	43	111	0.790	0.741		
Thousand Oaks	34°10'N	118°50'W	09:05:02	303	0	33	—			—			11:43:27	91	125	62	10:20:09	17	69	48	112	0.696	0.624		
Torrance	33°50'N	118°19'W	09:05:42	304	1	33	—			—			11:44:32	90	124	63	10:21:03	17	69	48	112	0.687	0.613		
Vallejo	38°14'N	122°02'W	09:02:00	296	349	29	—			—			11:38:08	97	131	56	10:16:04	16	64	43	112	0.816	0.774		

Table 2–3: Local Circumstances for Colorado, Connecticut, Delaware, & DC

Location Name	Latitude	Longitude	First Contact Time	P	V	Alt	Second Contact Time	P	V	Third Contact Time	P	V	Fourth Contact Time	P	V	Alt	Maximum Eclipse Time	P	V	Alt	Azm	Eclip. Mag.	Eclip. Obs.	Umbral Depth	Umbral Durat.
			h m s	°	°	°	h m s	°	°	h m s	°	°	h m s	°	°	°	h m s	°	°	°	°				
Colorado																									
Aurora	39°44'N	104°52'W	10:23:34	293	337	45	—			—			13:14:58	107	102	62	11:47:22	20	48	58	144	0.934	0.924		
Boulder	40°01'N	105°17'W	10:22:58	293	337	45	—			—			13:13:52	107	103	62	11:46:27	20	48	57	143	0.940	0.931		
Broomfield	39°55'N	105°05'W	10:23:15	293	337	45	—			—			13:14:22	107	103	62	11:46:52	20	48	57	144	0.938	0.929		
Colorado Springs	38°50'N	104°49'W	10:23:33	295	340	46	—			—			13:15:48	106	100	63	11:47:45	20	49	58	143	0.908	0.891		
Denver	39°44'N	104°59'W	10:23:22	293	338	45	—			—			13:14:43	107	102	62	11:47:07	20	48	57	144	0.933	0.923		
Englewood	39°39'N	104°59'W	10:23:22	293	338	45	—			—			13:14:48	107	102	62	11:47:09	20	48	58	144	0.931	0.920		
Fort Collins	40°35'N	105°05'W	10:23:22	291	335	45	—			—			13:13:49	108	104	61	11:46:41	20	47	57	144	0.957	0.953		
Grand Junction	39°04'N	108°33'W	10:18:01	294	342	42	—			—			13:07:32	104	108	63	11:40:11	19	54	55	135	0.891	0.870		
Greeley	40°25'N	104°42'W	10:23:55	292	335	45	—			—			13:14:44	108	103	61	11:47:29	20	47	57	145	0.955	0.951		
Lakewood	39°44'N	105°05'W	10:23:14	293	338	45	—			—			13:14:31	107	102	62	11:46:56	20	48	57	144	0.933	0.922		
Littleton	39°37'N	105°00'W	10:23:21	293	338	45	—			—			13:14:47	107	102	62	11:47:08	20	48	58	144	0.930	0.919		
Longmont	40°10'N	105°06'W	10:23:16	292	336	45	—			—			13:14:08	108	103	62	11:46:46	20	48	57	144	0.945	0.938		
Loveland	40°24'N	105°05'W	10:23:20	292	336	45	—			—			13:13:58	108	104	61	11:46:44	20	47	57	144	0.952	0.947		
Northglenn	39°53'N	104°58'W	10:23:26	293	337	45	—			—			13:14:38	107	102	62	11:47:07	20	48	57	144	0.938	0.929		
Pueblo	38°14'N	104°36'W	10:23:55	296	342	46	—			—			13:16:45	105	98	63	11:48:26	20	49	59	143	0.892	0.871		
Thornton	39°52'N	104°58'W	10:23:25	293	337	45	—			—			13:14:39	107	102	62	11:47:07	20	48	57	144	0.937	0.928		
Westminster	39°50'N	105°02'W	10:23:19	293	337	45	—			—			13:14:32	107	103	62	11:47:00	20	48	57	144	0.936	0.926		
Wheat Ridge	39°46'N	105°05'W	10:23:14	293	338	45	—			—			13:14:29	107	102	62	11:46:55	20	48	57	144	0.934	0.923		
Connecticut																									
Bridgeport	41°11'N	073°12'W	13:24:40	278	267	60	—			—			16:00:25	129	83	40	14:45:30	204	169	52	227	0.750	0.691		
Danbury	41°24'N	073°28'W	13:23:59	278	268	60	—			—			15:59:43	129	84	40	14:44:45	204	170	52	226	0.748	0.689		
Hartford	41°46'N	072°41'W	13:25:25	277	266	59	—			—			15:59:34	129	85	40	14:45:23	204	169	51	227	0.732	0.669		
New Britain	41°41'N	072°47'W	13:25:15	277	266	59	—			—			15:59:40	129	84	40	14:45:21	204	169	52	227	0.735	0.672		
New Haven	41°18'N	072°55'W	13:25:12	278	267	60	—			—			16:00:24	129	83	40	14:45:45	204	169	52	228	0.745	0.684		
New London	41°22'N	072°06'W	13:26:53	278	264	60	—			—			16:00:55	129	84	39	14:46:53	204	168	51	229	0.734	0.671		
Norwalk	41°07'N	073°22'W	13:24:22	279	268	60	—			—			16:00:25	128	83	40	14:45:21	204	169	52	227	0.754	0.695		
Stamford	41°03'N	073°32'W	13:24:03	279	269	60	—			—			16:00:26	128	83	40	14:45:12	204	169	52	227	0.757	0.700		
Waterbury	41°33'N	073°03'W	13:24:46	278	267	60	—			—			15:59:44	129	84	40	14:45:10	204	169	52	227	0.741	0.679		
West Hartford	41°46'N	072°45'W	13:25:17	277	266	59	—			—			15:59:30	129	85	40	14:45:17	204	169	51	227	0.733	0.670		
Delaware																									
Dover	39°10'N	075°32'W	13:20:58	283	276	62	—			—			16:02:37	125	78	42	14:44:56	204	169	55	226	0.821	0.781		
Wilmington	39°45'N	075°33'W	13:20:32	282	276	62	—			—			16:01:23	126	79	42	14:44:00	204	170	55	225	0.808	0.764		
District of Columbia																									
Washington	38°55'N	076°59'W	13:17:59	284	281	63	—			—			16:01:41	124	77	44	14:42:53	204	171	56	223	0.844	0.809		

Table 2–4: Local Circumstances for Florida & Georgia

Location Name	Latitude	Longitude	First Contact Time h m s	P °	V °	Alt °	Second Contact Time h m s	P °	V °	Third Contact Time h m s	P °	V °	Fourth Contact Time h m s	P °	V °	Alt °	Maximum Eclipse Time h m s	P °	V °	Alt °	Azm °	Eclip. Mag.	Eclip. Obs.	Umbral Depth	Umbral Durat.
Florida																									
Altamonte Springs	28°40'N	081°23'W	13:19:14	303	310	73	—			—			16:14:45	108	48	48	14:51:03	26	340	64	233	0.880	0.856		
Cape Coral	26°35'N	081°57'W	13:21:26	307	315	75	—			—			16:16:54	105	42	48	14:53:23	26	336	66	237	0.819	0.778		
Clearwater	27°58'N	082°48'W	13:17:06	305	319	73	—			—			16:13:32	106	45	49	14:49:15	26	341	66	231	0.843	0.808		
Coral Springs	26°18'N	080°11'W	13:25:59	306	304	76	—			—			16:20:05	106	41	46	14:57:32	26	333	64	241	0.836	0.800		
Daytona Beach	29°13'N	081°01'W	13:19:14	302	308	73	—			—			16:14:27	109	49	47	14:50:53	26	340	64	233	0.899	0.880		
Fort Lauderdale	26°07'N	080°08'W	13:26:26	306	304	76	—			—			16:20:25	106	41	46	14:57:57	26	332	64	242	0.832	0.795		
Fort Myers	26°38'N	081°51'W	13:21:34	307	315	75	—			—			16:16:59	105	42	48	14:53:30	26	336	66	237	0.821	0.781		
Fort Walton Beach	30°25'N	086°36'W	12:04:47	303	331	69	—			—			15:03:25	107	53	54	13:37:00	25	356	68	213	0.856	0.825		
Gainesville	29°40'N	082°20'W	13:15:32	302	314	72	—			—			16:11:44	109	50	49	14:47:25	26	344	65	228	0.893	0.872		
Hialeah	25°50'N	080°17'W	13:26:37	307	305	76	—			—			16:20:35	105	40	46	14:58:08	26	332	64	242	0.823	0.783		
Hollywood	26°01'N	080°09'W	13:26:35	307	304	76	—			—			16:20:32	106	41	46	14:58:05	26	332	64	242	0.830	0.791		
Jacksonville	30°20'N	081°39'W	13:16:09	300	310	71	—			—			16:11:43	110	53	48	14:47:45	26	344	64	229	0.919	0.905		
Lakeland	28°03'N	081°57'W	13:18:55	304	314	74	—			—			16:14:47	107	46	48	14:50:54	26	340	65	233	0.857	0.826		
Melbourne	28°05'N	080°37'W	13:21:56	303	306	74	—			—			16:16:48	108	46	47	14:53:36	26	337	64	236	0.876	0.850		
Miami	25°47'N	080°11'W	13:26:56	307	304	76	—			—			16:20:48	105	40	46	14:58:25	26	331	64	242	0.823	0.783		
Miami Beach	25°47'N	080°08'W	13:27:03	307	304	76	—			—			16:20:53	105	40	46	14:58:31	26	331	64	243	0.824	0.784		
Orlando	28°33'N	081°23'W	13:19:25	303	310	73	—			—			16:14:55	108	47	48	14:51:15	26	340	64	233	0.877	0.852		
Panama City	30°10'N	085°40'W	12:07:14	303	328	70	—			—			15:05:27	107	52	53	13:39:28	26	353	68	217	0.862	0.832		
Pembroke Pines	26°00'N	080°14'W	13:26:25	307	304	76	—			—			16:20:26	105	41	46	14:57:57	26	332	64	242	0.828	0.789		
Pensacola	30°25'N	087°13'W	12:03:24	304	333	68	—			—			15:02:17	106	53	55	13:35:37	25	358	69	211	0.848	0.815		
Sarasota	27°20'N	082°32'W	13:18:47	306	318	74	—			—			16:14:53	106	44	49	14:50:52	26	339	66	233	0.830	0.792		
St. Petersburg	27°47'N	082°40'W	13:17:43	305	318	74	—			—			16:14:01	106	45	49	14:49:50	26	340	66	232	0.840	0.805		
Tallahassee	30°27'N	084°17'W	13:09:59	302	322	70	—			—			16:07:23	108	53	51	14:42:02	26	350	67	221	0.887	0.865		
Tampa	27°57'N	082°27'W	13:17:56	305	317	74	—			—			16:14:08	106	45	49	14:50:01	26	340	66	232	0.847	0.814		
West Palm Beach	26°43'N	080°03'W	13:25:33	305	303	75	—			—			16:19:41	107	43	46	14:57:05	26	333	63	241	0.849	0.816		
Georgia																									
Albany	31°35'N	084°10'W	13:08:46	300	319	69	—			—			16:05:48	110	56	51	14:40:30	25	352	66	219	0.918	0.904		
Athens	33°57'N	083°23'W	13:07:53	296	311	67	—			—			16:03:05	114	63	50	14:38:31	25	355	64	216	0.989	0.991		
Atlanta	33°45'N	084°23'W	13:05:51	296	315	67	—			—			16:01:54	113	62	51	14:36:46	25	357	65	214	0.971	0.971		
Augusta	33°28'N	081°58'W	13:11:35	296	306	68	—			—			16:05:59	114	62	49	14:42:06	25	351	63	221	0.994	0.996		
Baldwin	34°30'N	083°32'W	13:07:01	295	311	66	14:36:58	186	157	14:37:48	223	194	16:01:54	114	65	50	14:37:23	25	356	64	215	1.030	1.000	0.053	00m51s
Clayton	34°53'N	083°23'W	13:07:00	294	309	66	14:35:49	126	98	14:38:25	284	255	16:01:27	115	66	50	14:37:07	25	356	63	215	1.030	1.000	0.810	02m35s
Cleveland	34°36'N	083°46'W	13:06:24	295	311	66	14:36:26	190	162	14:37:05	219	191	16:01:23	114	65	50	14:36:46	25	357	64	214	1.030	1.000	0.031	00m39s
Columbus	32°28'N	084°59'W	13:05:53	299	320	68	—			—			16:03:04	111	59	52	14:37:25	25	356	66	214	0.931	0.920		
Cornelia	34°31'N	083°32'W	13:07:00	295	310	66	14:36:50	182	153	14:37:53	228	199	16:01:53	114	65	50	14:37:22	25	356	64	215	1.030	1.000	0.080	01m02s
East Point	33°41'N	084°27'W	13:05:46	297	315	67	—			—			16:01:55	112	62	51	14:36:43	25	357	65	214	0.969	0.968		
Hartwell	34°21'N	082°56'W	13:08:30	295	309	67	14:37:53	160	130	14:39:44	250	219	16:03:03	114	64	50	14:38:49	25	354	63	217	1.030	1.000	0.290	01m51s
Lavonia	34°26'N	083°06'W	13:08:03	295	309	67	14:37:25	160	130	14:39:17	250	219	16:02:40	114	64	50	14:38:21	25	355	63	216	1.030	1.000	0.292	01m52s
Macon	32°51'N	083°38'W	13:08:29	297	314	68	—			—			16:04:34	112	60	50	14:39:40	25	353	65	218	0.958	0.954		
Marietta	33°57'N	084°33'W	13:05:17	296	315	66	—			—			16:01:18	113	63	51	14:36:07	25	357	65	213	0.974	0.974		
Peachtree City	33°24'N	084°34'W	13:05:48	297	317	67	—			—			16:02:12	112	61	51	14:36:54	25	356	65	214	0.960	0.957		
Rome	34°15'N	085°10'W	13:03:37	296	317	66	—			—			15:59:50	113	64	52	14:34:24	25	359	65	210	0.974	0.975		
Roswell	34°02'N	084°22'W	13:05:37	296	314	66	—			—			16:01:27	113	63	51	14:36:23	25	357	64	213	0.979	0.980		
Royston	34°17'N	083°07'W	13:08:09	295	309	67	14:38:14	191	161	14:38:52	219	188	16:02:54	114	64	50	14:38:33	25	355	63	216	1.030	1.000	0.030	00m38s
Savannah	32°05'N	081°06'W	13:15:08	297	305	70	—			—			16:09:37	113	58	48	14:46:01	25	346	63	226	0.971	0.970		
Smyrna	33°53'N	084°31'W	13:05:25	296	315	67	—			—			16:01:28	113	63	51	14:36:17	25	357	65	213	0.973	0.973		
Toccoa	34°35'N	083°19'W	13:07:25	294	310	66	14:36:40	154	125	14:38:43	255	226	16:02:05	114	65	50	14:37:42	25	356	63	215	1.030	1.000	0.366	02m03s
Valdosta	30°50'N	083°17'W	13:11:45	301	316	70	—			—			16:08:24	110	54	50	14:43:33	26	348	65	223	0.910	0.894		
Warner Robins	32°37'N	083°36'W	13:08:50	298	314	68	—			—			16:05:01	112	59	50	14:40:06	25	353	65	218	0.952	0.947		

Table 2–5: Local Circumstances for Hawaii & Idaho

Location Name	Latitude	Longitude	First Contact Time h m s	P °	V °	Alt °	Second Contact Time h m s	P °	V °	Third Contact Time h m s	P °	V °	Fourth Contact Time h m s	P °	V °	Alt °	Maximum Eclipse Time h m s	P °	V °	Alt °	Azm °	Eclip. Mag.	Eclip. Obs.	Umbral Depth	Umbral Durat.
Hawaii																									
Hilo	19°44'N	155°05'W	—				—			—			07:22:43	57	130	18	06:35:50	10	81	7	80	0.322	0.209		
Honolulu	21°19'N	157°52'W	—				—			—			07:25:23	62	133	16	06:35:59	10	79	5	79	0.388	0.274		
Kailua	21°24'N	157°45'W	—				—			—			07:25:34	62	133	16	06:36:01	10	79	5	79	0.390	0.276		
Kaneohe	21°25'N	157°48'W	—				—			—			07:25:36	62	133	16	06:36:02	10	79	5	79	0.391	0.276		
Idaho																									
Ammon	43°28'N	111°58'W	10:15:15	287	331	37	11:33:10	149	183	11:34:54	247	280	12:58:13	109	120	58	11:34:02	18	51	50	133	1.029	1.000	0.340	01m44s
Arco	43°38'N	113°18'W	10:13:48	287	331	36	11:31:04	152	187	11:32:41	244	278	12:55:36	109	123	57	11:31:52	18	52	48	130	1.028	1.000	0.303	01m38s
Ashton	44°04'N	111°27'W	10:16:16	286	329	38	11:34:12	62	94	11:35:47	335	7	12:58:52	110	120	57	11:34:59	198	230	50	134	1.029	1.000	0.274	01m35s
Boise	43°37'N	116°13'W	10:10:34	287	332	34	—			—			12:50:07	108	127	56	11:27:14	17	54	46	126	0.994	0.996		
Cascade	44°31'N	116°02'W	10:11:27	285	329	34	11:26:54	78	114	11:28:49	316	352	12:50:13	109	127	55	11:27:51	197	233	46	127	1.028	1.000	0.515	01m55s
Challis	44°30'N	114°14'W	10:13:22	285	329	35	11:30:03	47	82	11:31:10	348	22	12:53:30	110	125	56	11:30:37	198	232	47	130	1.028	1.000	0.133	01m07s
Coeur d'Alene	47°41'N	116°46'W	10:13:41	280	321	33	—			—			12:47:53	114	131	52	11:28:07	197	230	44	128	0.917	0.902		
Council	44°44'N	116°26'W	10:11:13	285	329	33	11:26:47	48	84	11:27:53	346	22	12:49:25	109	128	55	11:27:20	197	233	45	126	1.028	1.000	0.140	01m07s
Driggs	43°44'N	111°06'W	10:16:28	287	330	38	11:34:22	100	132	11:36:40	297	329	12:59:42	110	119	58	11:35:31	198	230	50	134	1.029	1.000	0.851	02m18s
Emmett	43°52'N	116°30'W	10:10:28	286	331	33	11:26:49	192	229	11:27:00	202	239	12:49:32	108	127	56	11:26:54	17	54	46	125	1.028	1.000	0.004	00m12s
Fruitland	44°00'N	116°55'W	10:10:09	286	331	33	11:25:40	160	198	11:26:59	234	271	12:48:44	108	128	55	11:26:19	17	54	45	125	1.028	1.000	0.198	01m18s
Horseshoe Bend	43°55'N	116°12'W	10:10:49	286	331	34	11:26:46	163	200	11:28:00	232	269	12:50:04	108	127	56	11:27:23	17	54	46	126	1.028	1.000	0.177	01m15s
Idaho Falls	43°30'N	112°02'W	10:15:12	287	331	37	11:33:01	146	179	11:34:51	250	284	12:58:04	109	121	58	11:33:56	18	51	50	132	1.029	1.000	0.389	01m49s
Iona	43°32'N	111°56'W	10:15:20	287	331	38	11:33:07	139	172	11:35:06	257	290	12:58:14	109	120	58	11:34:06	18	51	50	133	1.029	1.000	0.487	01m59s
Ketchum	43°41'N	114°22'W	10:12:37	287	331	35	11:29:38	169	204	11:30:43	226	262	12:53:34	108	124	57	11:30:10	18	53	48	129	1.028	1.000	0.124	01m05s
Lewiston	46°25'N	117°01'W	10:12:09	282	324	33	—			—			12:47:54	112	130	53	11:27:12	197	231	44	127	0.954	0.949		
Lewisville	43°42'N	112°01'W	10:15:21	287	330	37	11:32:52	119	153	11:35:07	277	310	12:58:00	110	121	57	11:33:59	18	51	49	133	1.029	1.000	0.805	02m15s
Mackay	43°55'N	113°37'W	10:13:38	286	330	36	11:30:19	119	154	11:32:32	277	311	12:54:53	109	123	57	11:31:26	18	52	48	130	1.028	1.000	0.807	02m13s
Menan	43°43'N	111°59'W	10:15:24	287	330	37	11:32:55	117	150	11:35:11	279	312	12:58:03	110	121	57	11:34:03	18	51	49	133	1.029	1.000	0.849	02m17s
Nampa	43°34'N	116°34'W	10:10:11	287	332	33	—			—			12:49:28	107	127	56	11:26:42	17	55	46	125	0.991	0.993		
New Plymouth	43°58'N	116°49'W	10:10:13	286	331	33	11:25:54	165	203	11:27:02	229	266	12:48:55	108	128	55	11:26:28	17	54	45	125	1.028	1.000	0.148	01m09s
Payette	44°05'N	116°56'W	10:10:12	286	331	33	11:25:29	146	183	11:27:11	248	285	12:48:41	108	128	55	11:26:20	17	54	45	125	1.028	1.000	0.373	01m42s
Pocatello	42°52'N	112°27'W	10:14:17	288	333	37	—			—			12:57:34	108	120	58	11:33:07	18	53	50	131	0.985	0.987		
Rexburg	43°49'N	111°47'W	10:15:42	286	330	38	11:33:15	102	135	11:35:32	295	327	12:58:22	110	121	57	11:34:24	198	231	50	133	1.029	1.000	0.886	02m17s
Rigby	43°40'N	111°55'W	10:15:26	287	330	37	11:33:02	122	155	11:35:16	275	307	12:58:12	110	121	57	11:34:09	18	51	50	133	1.029	1.000	0.766	02m14s
Ririe	43°38'N	111°47'W	10:15:35	287	330	38	11:33:15	123	157	11:35:29	273	306	12:58:28	110	120	58	11:34:22	18	51	50	133	1.029	1.000	0.737	02m14s
Roberts	43°43'N	112°08'W	10:15:13	287	330	37	11:32:40	119	153	11:34:56	277	310	12:57:46	109	121	57	11:33:48	18	51	49	132	1.029	1.000	0.806	02m15s
Shelley	43°23'N	112°07'W	10:15:01	287	331	37	11:33:13	169	203	11:34:20	227	260	12:57:58	109	120	58	11:33:46	18	52	50	132	1.029	1.000	0.123	01m07s
St. Anthony	43°58'N	111°41'W	10:15:55	286	329	38	11:33:33	81	114	11:35:37	315	347	12:58:29	110	121	57	11:34:35	198	231	49	134	1.029	1.000	0.548	02m03s
Sugar City	43°52'N	111°45'W	10:15:46	286	330	38	11:33:20	95	128	11:35:35	301	334	12:58:25	110	121	57	11:34:27	198	231	50	133	1.029	1.000	0.773	02m15s
Sun Valley	43°42'N	114°21'W	10:12:39	287	331	35	11:29:35	164	200	11:30:49	231	266	12:53:36	109	124	57	11:30:12	18	53	48	129	1.028	1.000	0.163	01m14s
Teton	43°53'N	111°40'W	10:15:53	286	330	38	11:33:30	92	124	11:35:42	305	337	12:58:33	110	120	57	11:34:36	198	231	50	133	1.029	1.000	0.714	02m12s
Twin Falls	42°34'N	114°28'W	10:11:47	289	335	35	—			—			12:53:45	107	123	58	11:29:43	18	55	48	127	0.968	0.967		
Ucon	43°36'N	111°58'W	10:15:20	287	331	37	11:33:00	131	164	11:35:07	265	298	12:58:09	109	121	57	11:34:03	18	51	50	133	1.029	1.000	0.614	02m08s
Victor	43°36'N	111°07'W	10:16:22	287	330	38	11:34:19	116	148	11:36:38	281	313	12:59:45	110	119	58	11:35:28	18	51	50	134	1.029	1.000	0.872	02m18s
Weiser	44°45'N	116°58'W	10:10:42	285	329	33	11:25:53	55	92	11:27:13	339	15	12:48:27	109	128	55	11:26:33	197	233	45	125	1.028	1.000	0.213	01m20s

Table 2–6: Local Circumstances for Illinois

Location Name	Latitude	Longitude	First Contact Time	P	V	Alt	Second Contact Time	P	V	Third Contact Time	P	V	Fourth Contact Time	P	V	Alt	Maximum Eclipse Time	P	V	Alt	Azm	Eclip. Mag.	Eclip. Obs.	Umbral Depth	Umbral Durat.
			h m s	°	°	°	h m s	°	°	h m s	°	°	h m s	°	°	°	h m s	°	°	°	°				
Illinois																									
Addison	41°56'N	087°59'W	11:53:37	285	306	57	—			—			14:42:04	120	87	52	13:19:06	203	194	59	192	0.891	0.869		
Anna	37°28'N	089°15'W	11:52:30	293	321	60	13:20:23	130	121	13:22:56	277	266	14:47:51	114	75	55	13:21:39	24	14	64	192	1.031	1.000	0.712	02m34s
Arlington Heights	42°05'N	087°59'W	11:53:35	284	306	57	—			—			14:41:48	121	88	52	13:18:56	203	194	59	192	0.887	0.864		
Aurora	41°45'N	088°19'W	11:52:59	285	307	57	—			—			14:41:55	120	87	52	13:18:40	203	194	60	191	0.899	0.879		
Benton	38°00'N	088°55'W	11:52:55	292	318	60	13:21:03	54	44	13:22:22	354	343	14:47:31	115	76	54	13:21:43	204	193	63	193	1.031	1.000	0.134	01m20s
Bloomington	40°29'N	089°00'W	11:51:52	287	312	58	—			—			14:43:09	118	84	53	13:18:44	203	195	61	190	0.938	0.929		
Bolingbrook	41°44'N	088°03'W	11:53:30	285	307	57	—			—			14:42:20	120	87	52	13:19:11	203	194	60	192	0.897	0.877		
Cahokia	38°34'N	090°11'W	11:50:05	291	320	59	13:18:06	40	34	13:18:51	7	0	14:44:30	115	79	55	13:18:29	203	197	63	188	1.031	1.000	0.042	00m45s
Carbondale	37°44'N	089°13'W	11:52:26	292	320	60	13:20:06	102	93	13:22:43	305	295	14:47:28	114	76	55	13:21:25	204	194	64	192	1.031	1.000	0.806	02m37s
Carterville	37°47'N	089°05'W	11:52:41	292	319	60	13:20:23	92	82	13:22:52	316	305	14:47:36	115	76	55	13:21:38	204	194	64	192	1.031	1.000	0.625	02m28s
Centreville	38°33'N	090°07'W	11:50:14	291	320	59	13:18:20	36	30	13:18:55	11	4	14:44:38	115	79	55	13:18:38	203	197	63	188	1.031	1.000	0.025	00m35s
Champaign	40°07'N	088°15'W	11:53:28	288	311	59	—			—			14:44:56	118	82	53	13:20:35	203	193	61	193	0.940	0.932		
Chester	37°55'N	089°49'W	11:51:06	292	321	60	13:18:38	108	101	13:21:17	298	290	14:46:10	114	77	55	13:19:58	203	196	64	190	1.031	1.000	0.912	02m39s
Chicago	41°53'N	087°38'W	11:54:18	285	305	57	—			—			14:42:40	121	87	52	13:19:48	203	193	59	193	0.889	0.866		
Christopher	37°59'N	089°03'W	11:52:39	292	319	60	13:20:34	66	56	13:22:21	342	331	14:47:19	115	77	54	13:21:27	204	194	63	192	1.031	1.000	0.257	01m47s
Cicero	41°51'N	087°45'W	11:54:05	285	306	57	—			—			14:42:33	120	87	52	13:19:37	203	193	60	192	0.891	0.869		
Columbia	38°27'N	090°12'W	11:50:06	291	320	59	13:17:41	65	59	13:19:27	341	335	14:44:40	115	79	55	13:18:34	203	197	63	188	1.031	1.000	0.255	01m46s
Danville	40°08'N	087°37'W	11:54:45	287	310	59	—			—			14:45:51	118	82	52	13:21:48	203	192	61	194	0.933	0.923		
Decatur	39°51'N	088°57'W	11:52:08	289	314	58	—			—			14:44:19	117	82	54	13:19:31	203	195	62	191	0.954	0.949		
Des Plaines	42°03'N	087°52'W	11:53:49	284	306	57	—			—			14:42:01	121	88	52	13:19:11	203	194	59	192	0.887	0.864		
Du Quoin	38°01'N	089°14'W	11:52:15	292	319	60	13:20:03	71	62	13:22:02	336	326	14:46:58	115	77	55	13:21:03	204	194	63	192	1.031	1.000	0.330	01m58s
Dupo	38°31'N	090°13'W	11:50:02	291	320	59	13:17:46	55	49	13:19:10	351	345	14:44:31	115	79	55	13:18:28	203	197	63	188	1.031	1.000	0.151	01m24s
East St. Louis	38°37'N	090°07'W	11:50:12	291	319	59	—			—			14:44:31	115	79	55	13:18:33	203	197	63	188	0.999	1.000		
Eldorado	37°49'N	088°26'W	11:54:01	292	318	61	13:22:20	50	39	13:23:30	357	345	14:48:36	115	76	54	13:22:55	204	192	63	194	1.031	1.000	0.103	01m10s
Elgin	42°02'N	088°17'W	11:53:00	285	307	57	—			—			14:41:28	120	88	52	13:18:26	203	195	59	191	0.891	0.870		
Evanston	42°03'N	087°41'W	11:54:11	284	305	57	—			—			14:42:17	121	87	52	13:19:32	203	193	59	192	0.885	0.862		
Freeburg	38°26'N	089°55'W	11:50:41	291	319	59	13:18:37	48	41	13:19:43	359	352	14:45:09	115	78	55	13:19:10	203	196	63	189	1.031	1.000	0.089	01m06s
Harrisburg	37°44'N	088°32'W	11:53:51	292	318	61	13:21:50	71	60	13:23:47	337	325	14:48:35	115	75	54	13:22:49	204	192	64	194	1.031	1.000	0.318	01m57s
Herrin	37°48'N	089°02'W	11:52:46	292	319	60	13:20:31	88	78	13:22:54	320	309	14:47:39	115	76	55	13:21:42	204	193	64	192	1.031	1.000	0.561	02m24s
Johnston City	37°49'N	088°56'W	11:52:58	292	319	60	13:20:46	81	71	13:23:01	326	315	14:47:47	115	76	54	13:21:54	204	193	64	193	1.031	1.000	0.461	02m15s
Joliet	41°32'N	088°05'W	11:53:28	285	307	57	—			—			14:42:39	120	86	52	13:19:20	203	194	60	192	0.902	0.883		
Kankakee	41°07'N	087°52'W	11:53:59	286	308	58	—			—			14:43:43	119	85	52	13:20:12	203	193	60	193	0.910	0.894		
Marion	37°44'N	088°56'W	11:53:01	292	319	60	13:20:46	90	80	13:23:13	317	306	14:47:56	115	76	54	13:22:00	204	193	64	193	1.031	1.000	0.604	02m27s
Metropolis	37°09'N	088°44'W	11:53:45	293	320	61	13:21:55	141	130	13:24:18	266	254	14:49:13	114	74	54	13:23:07	24	12	64	194	1.031	1.000	0.539	02m23s
Millstadt	38°28'N	090°06'W	11:50:18	291	320	59	13:18:03	56	50	13:19:28	351	344	14:44:48	115	79	55	13:18:45	203	197	63	188	1.031	1.000	0.158	01m26s
Mount Prospect	42°04'N	087°56'W	11:53:41	284	306	57	—			—			14:41:54	121	88	52	13:19:03	203	194	59	192	0.887	0.864		
Murphysboro	37°46'N	089°20'W	11:52:10	292	320	60	13:19:49	104	95	13:22:27	303	293	14:47:13	114	76	55	13:21:08	204	194	64	192	1.031	1.000	0.831	02m38s
Naperville	41°46'N	088°09'W	11:53:18	285	307	57	—			—			14:42:08	120	87	52	13:18:58	203	194	60	191	0.897	0.877		
Oak Lawn	41°43'N	087°44'W	11:54:08	285	306	57	—			—			14:42:49	120	87	52	13:19:48	203	193	60	193	0.894	0.873		
Oak Park	41°53'N	087°47'W	11:54:01	285	306	57	—			—			14:42:27	120	87	52	13:19:31	203	193	59	192	0.890	0.868		
Peoria	40°42'N	089°36'W	11:50:37	287	313	57	—			—			14:41:51	118	85	54	13:17:20	203	197	61	188	0.939	0.930		
Pinckneyville	38°05'N	089°23'W	11:51:55	292	319	60	13:19:41	71	62	13:21:38	336	327	14:46:37	115	77	55	13:20:40	203	195	63	191	1.031	1.000	0.321	01m57s
Quincy	39°56'N	091°23'W	11:47:17	289	319	57	—			—			14:40:17	116	84	55	13:14:37	203	200	62	183	0.977	0.977		
Red Bud	38°12'N	090°00'W	11:50:36	292	320	59	13:18:04	87	80	13:20:27	320	312	14:45:24	114	78	55	13:19:16	203	196	63	189	1.031	1.000	0.550	02m23s
Rockford	42°16'N	089°06'W	11:51:23	285	308	56	—			—			14:39:52	120	89	53	13:16:40	203	196	59	188	0.893	0.872		
Schaumburg	42°02'N	088°05'W	11:53:24	285	306	57	—			—			14:41:45	120	88	52	13:18:48	203	194	59	191	0.889	0.867		
Skokie	42°03'N	087°45'W	11:54:03	284	305	57	—			—			14:42:11	121	87	52	13:19:24	203	194	59	192	0.886	0.862		
Sparta	38°08'N	089°42'W	11:51:15	292	320	59	13:18:50	81	73	13:21:04	326	318	14:46:00	115	77	55	13:19:57	203	195	63	190	1.031	1.000	0.456	02m14s
Springfield	39°48'N	089°39'W	11:50:44	289	315	58	—			—			14:43:18	117	82	54	13:18:12	203	196	62	189	0.963	0.960		
Waterloo	38°20'N	090°09'W	11:50:15	291	320	59	13:17:43	78	72	13:19:53	329	321	14:44:56	115	78	55	13:18:48	203	197	63	188	1.031	1.000	0.423	02m10s
Waukegan	42°22'N	087°50'W	11:53:51	284	305	57	—			—			14:41:30	121	88	52	13:18:55	203	194	59	192	0.878	0.853		
West Frankfort	37°54'N	088°55'W	11:52:58	292	319	60	13:20:52	70	60	13:22:47	338	327	14:47:41	115	76	54	13:21:50	204	193	63	193	1.031	1.000	0.306	01m55s
Wheaton	41°52'N	088°06'W	11:53:23	285	307	57	—			—			14:42:01	120	87	52	13:18:57	203	194	60	191	0.894	0.873		

Table 2–7: Local Circumstances for Indiana & Iowa

Location Name	Latitude	Longitude	First Contact Time (h m s)	P (°)	V (°)	Alt (°)	Second Contact Time (h m s)	P (°)	V (°)	Third Contact Time (h m s)	P (°)	V (°)	Fourth Contact Time (h m s)	P (°)	V (°)	Alt (°)	Maximum Eclipse Time (h m s)	P (°)	V (°)	Alt (°)	Azm (°)	Eclip. Mag.	Eclip. Obs.	Umbral Depth	Umbral Durat.
Indiana																									
Anderson	40°10'N	085°41'W	12:58:42	287	305	60	—			—			15:48:34	120	81	51	14:25:27	203	188	60	200	0.912	0.896		
Bloomington	39°10'N	086°32'W	12:57:20	289	309	60	—			—			15:49:10	118	79	52	14:25:03	204	189	62	199	0.946	0.939		
Carmel	39°59'N	086°08'W	12:57:50	287	306	60	—			—			15:48:16	119	81	51	14:24:49	203	189	61	199	0.921	0.908		
Columbus	39°13'N	085°55'W	12:58:36	288	308	61	—			—			15:49:59	118	79	51	14:26:11	204	187	61	200	0.938	0.929		
East Chicago	41°38'N	087°29'W	11:54:39	285	306	58	—			—			14:43:20	120	86	52	13:20:21	203	193	60	193	0.893	0.872		
Elkhart	41°41'N	085°58'W	12:57:41	284	302	58	—			—			15:45:20	121	86	51	14:23:06	203	190	59	197	0.877	0.851		
Evansville	37°58'N	087°35'W	11:55:42	291	315	61	—			—			14:49:42	116	76	53	13:24:27	204	190	63	197	0.988	0.990		
Fort Wayne	41°04'N	085°09'W	12:59:31	285	301	59	—			—			15:47:36	121	84	50	14:25:21	203	188	59	200	0.883	0.860		
Gary	41°36'N	087°20'W	11:54:57	285	305	58	—			—			14:43:36	120	86	52	13:20:40	203	192	60	194	0.893	0.872		
Greenwood	39°37'N	086°07'W	12:58:01	288	307	60	—			—			15:48:58	119	80	51	14:25:18	204	188	61	199	0.930	0.919		
Hammond	41°38'N	087°30'W	11:54:37	285	306	58	—			—			14:43:18	120	86	52	13:20:19	203	193	60	193	0.894	0.873		
Indianapolis	39°46'N	086°09'W	12:57:53	287	307	60	—			—			15:48:39	119	80	51	14:25:03	204	188	61	199	0.927	0.915		
Kokomo	40°29'N	086°08'W	12:57:40	286	305	59	—			—			15:47:21	120	82	51	14:24:13	203	189	60	198	0.908	0.892		
Lafayette	40°25'N	086°54'W	12:56:07	287	307	59	—			—			15:46:23	119	82	52	14:22:50	203	191	61	196	0.918	0.904		
Lawrence	39°50'N	086°02'W	12:58:06	287	307	60	—			—			15:48:41	119	80	51	14:25:12	204	188	61	199	0.924	0.911		
Marion	40°32'N	085°40'W	12:58:37	286	304	60	—			—			15:47:54	120	82	51	14:25:02	203	188	60	200	0.902	0.884		
Merrillville	41°29'N	087°20'W	11:54:59	285	306	58	—			—			14:43:49	120	86	52	13:20:48	203	192	60	194	0.896	0.875		
Michigan City	41°43'N	086°54'W	11:55:48	285	304	58	—			—			14:44:00	121	86	51	13:21:21	203	192	59	195	0.885	0.862		
Mishawaka	41°42'N	086°12'W	12:57:13	284	303	58	—			—			15:44:59	121	86	51	14:22:40	203	190	59	197	0.879	0.854		
Muncie	40°12'N	085°23'W	12:59:19	286	304	60	—			—			15:48:55	120	81	51	14:25:59	203	187	60	201	0.907	0.890		
New Albany	38°18'N	085°49'W	12:59:15	290	309	62	—			—			15:51:47	117	76	52	14:27:33	204	186	62	202	0.960	0.957		
Portage	41°34'N	087°11'W	11:55:16	285	305	58	—			—			14:43:52	120	86	52	13:20:59	203	192	60	194	0.892	0.871		
Richmond	39°50'N	084°54'W	13:00:28	287	304	61	—			—			15:50:16	120	80	50	14:27:21	204	186	60	203	0.911	0.895		
South Bend	41°41'N	086°15'W	12:57:07	284	303	58	—			—			15:44:57	121	86	51	14:22:35	203	190	59	197	0.880	0.855		
Terre Haute	39°28'N	087°25'W	12:55:23	288	311	60	—			—			15:47:20	118	80	53	14:22:58	203	191	62	196	0.948	0.942		
Iowa																									
Ames	42°02'N	093°37'W	11:42:51	287	317	53	—			—			14:33:16	117	92	56	13:08:19	202	205	60	176	0.942	0.934		
Bettendorf	41°32'N	090°30'W	11:48:44	286	312	56	—			—			14:39:02	118	88	54	13:14:45	203	199	60	185	0.926	0.914		
Cedar Falls	42°32'N	092°27'W	11:45:01	285	313	54	—			—			14:34:20	118	92	55	13:10:07	202	203	59	179	0.918	0.904		
Cedar Rapids	41°59'N	091°40'W	11:46:29	286	313	55	—			—			14:36:28	118	90	54	13:12:06	202	201	60	181	0.925	0.913		
Clinton	41°51'N	090°12'W	11:49:17	286	311	56	—			—			14:38:57	119	88	54	13:15:01	203	198	60	185	0.915	0.900		
Council Bluffs	41°16'N	095°52'W	11:38:44	288	323	52	—			—			14:30:36	115	92	57	13:04:32	202	210	60	169	0.983	0.984		
Davenport	41°32'N	090°35'W	11:48:34	286	313	56	—			—			14:38:55	118	88	54	13:14:35	203	199	60	185	0.927	0.915		
Des Moines	41°35'N	093°37'W	11:42:50	287	318	54	—			—			14:33:58	116	90	56	13:08:41	202	205	60	176	0.954	0.949		
Dubuque	42°30'N	090°41'W	11:48:20	285	310	55	—			—			14:37:06	119	91	53	13:13:29	202	200	59	184	0.903	0.884		
Iowa City	41°40'N	091°32'W	11:46:45	286	314	55	—			—			14:37:12	118	89	55	13:12:38	202	201	60	182	0.932	0.922		
Mason City	43°09'N	093°12'W	11:43:41	285	313	53	—			—			14:32:09	119	94	55	13:08:12	202	205	59	176	0.909	0.892		
Sioux City	42°30'N	096°24'W	11:37:54	287	320	51	—			—			14:27:55	116	96	57	13:02:38	202	211	59	168	0.953	0.948		
Waterloo	42°30'N	092°21'W	11:45:12	285	313	54	—			—			14:34:33	118	92	55	13:10:20	202	203	59	179	0.918	0.904		
West Des Moines	41°35'N	093°43'W	11:42:39	287	318	54	—			—			14:33:48	116	90	56	13:08:30	202	206	60	175	0.955	0.950		

Table 2–8: Local Circumstances for Kansas & Kentucky

Location Name	Latitude	Longitude	First Contact Time	P	V	Alt	Second Contact Time	P	V	Third Contact Time	P	V	Fourth Contact Time	P	V	Alt	Maximum Eclipse Time	P	V	Alt	Azm	Eclip. Mag.	Eclip. Obs.	Umbral Depth	Umbral Durat.
			h m s	°	°	°	h m s	°	°	h m s	°	°	h m s	°	°	°	h m s	°	°	°	°				
Kansas																									
Atchison	39°34'N	095°07'W	11:40:13	291	326	54	13:06:13	141	148	13:08:33	264	270	14:34:22	113	86	58	13:07:23	22	29	62	172	1.030	1.000	0.523	02m19s
Hiawatha	39°51'N	095°32'W	11:39:24	291	326	54	13:05:01	125	133	13:07:35	280	287	14:33:13	113	88	58	13:06:18	22	30	62	170	1.030	1.000	0.782	02m34s
Horton	39°40'N	095°32'W	11:39:25	291	327	54	13:05:21	146	154	13:07:33	259	266	14:33:28	113	87	58	13:06:27	22	30	62	170	1.030	1.000	0.451	02m12s
Hutchinson	38°05'N	097°56'W	11:35:19	294	335	53	—			—			14:30:57	109	86	61	13:02:40	22	36	63	163	0.941	0.933		
Kansas City	39°07'N	094°38'W	11:41:12	292	327	55	13:08:33	193	199	13:08:59	212	217	14:35:54	113	84	58	13:08:46	22	28	63	173	1.031	1.000	0.014	00m26s
Lansing	39°15'N	094°54'W	11:40:40	292	327	55	13:07:34	178	185	13:08:38	226	232	14:35:13	113	85	58	13:08:06	22	29	62	172	1.031	1.000	0.086	01m05s
Lawrence	38°58'N	095°14'W	11:40:06	292	328	54	—			—			14:35:00	112	85	59	13:07:39	22	29	63	171	0.990	0.993		
Leavenworth	39°19'N	094°55'W	11:40:38	291	327	54	13:07:13	166	172	13:08:48	239	245	14:35:06	113	85	58	13:08:01	22	29	62	172	1.031	1.000	0.195	01m34s
Lenexa	38°58'N	094°45'W	11:41:01	292	327	55	—			—			14:35:54	112	84	58	13:08:40	22	28	63	173	0.995	0.997		
Manhattan	39°11'N	096°35'W	11:37:34	292	330	53	—			—			14:32:11	111	87	59	13:04:40	22	32	62	167	0.984	0.986		
Marysville	39°51'N	096°39'W	11:37:21	291	328	53	13:03:24	173	184	13:04:39	230	241	14:31:10	112	89	59	13:04:01	22	32	61	167	1.030	1.000	0.121	01m15s
Olathe	38°53'N	094°49'W	11:40:55	292	328	55	—			—			14:35:54	112	84	58	13:08:36	22	28	63	173	0.992	0.994		
Overland Park	38°58'N	094°40'W	11:41:10	292	327	55	—			—			14:36:03	112	84	58	13:08:50	22	28	63	173	0.996	0.998		
Sabetha	39°54'N	095°48'W	11:38:54	291	326	53	13:04:27	129	138	13:06:59	275	283	14:32:40	113	88	58	13:05:43	22	30	62	169	1.030	1.000	0.713	02m31s
Salina	38°50'N	097°37'W	11:35:44	293	332	52	—			—			14:30:39	110	87	60	13:02:47	22	35	62	164	0.965	0.963		
Seneca	39°50'N	096°04'W	11:38:25	291	327	53	13:04:08	147	156	13:06:19	258	266	14:32:16	113	88	59	13:05:14	22	31	62	169	1.030	1.000	0.434	02m10s
Shawnee	39°01'N	094°43'W	11:41:04	292	327	55	—			—			14:35:53	112	84	58	13:08:41	22	28	63	173	0.997	0.998		
Topeka	39°03'N	095°40'W	11:39:16	292	329	54	—			—			14:34:05	112	85	59	13:06:41	22	30	62	170	0.989	0.991		
Wichita	37°42'N	097°20'W	11:36:31	295	335	53	—			—			14:32:38	109	84	61	13:04:15	22	35	63	164	0.936	0.926		
Kentucky																									
Benton	36°52'N	088°21'W	11:54:43	293	320	61	13:23:15	156	143	13:25:15	252	239	14:50:19	114	73	54	13:24:15	24	11	64	196	1.031	1.000	0.332	02m00s
Bowling Green	36°59'N	086°27'W	11:58:40	292	314	62	13:27:32	44	27	13:28:27	4	346	14:53:10	115	72	52	13:28:00	204	186	63	202	1.030	1.000	0.061	00m55s
Cadiz	36°52'N	087°50'W	11:55:48	293	318	62	13:24:02	131	117	13:26:36	277	262	14:51:10	114	73	54	13:25:19	24	10	64	198	1.031	1.000	0.711	02m34s
Calvert City	37°02'N	088°21'W	11:54:37	293	319	61	13:22:48	136	124	13:25:17	271	258	14:50:03	114	73	54	13:24:03	24	11	64	196	1.031	1.000	0.617	02m28s
Central City	37°18'N	087°07'W	11:57:04	292	315	62	13:25:57	36	21	13:26:31	12	356	14:51:34	115	74	53	13:26:14	204	188	63	199	1.031	1.000	0.023	00m34s
Covington	39°05'N	084°31'W	13:01:37	288	304	61	—			—			15:52:12	119	78	50	14:29:02	204	184	61	205	0.926	0.914		
Dawson Springs	37°10'N	087°41'W	11:55:56	292	317	62	13:24:00	95	81	13:26:30	313	298	14:50:54	115	73	54	13:25:15	204	189	64	198	1.031	1.000	0.668	02m31s
Earlington	37°16'N	087°30'W	11:56:16	292	317	62	13:24:28	74	60	13:26:31	333	318	14:51:02	115	74	53	13:25:30	204	189	64	198	1.031	1.000	0.363	02m03s
Eddyville	37°03'N	088°04'W	11:55:12	293	319	61	13:23:17	122	109	13:25:56	285	271	14:50:29	114	73	54	13:24:37	24	10	64	197	1.031	1.000	0.852	02m39s
Elkton	36°49'N	087°09'W	11:57:17	293	317	62	13:25:28	106	91	13:28:06	302	285	14:52:21	115	72	53	13:26:47	204	188	64	200	1.031	1.000	0.868	02m38s
Franklin	36°43'N	086°35'W	11:58:34	293	315	63	13:26:51	91	74	13:29:18	317	299	14:53:25	115	72	53	13:28:05	204	186	64	202	1.031	1.000	0.613	02m27s
Greenville	37°12'N	087°11'W	11:56:59	292	316	62	13:25:22	65	49	13:27:06	343	327	14:51:38	115	73	53	13:26:14	204	188	64	199	1.031	1.000	0.242	01m44s
Guthrie	36°39'N	087°10'W	11:57:21	293	317	62	13:25:39	123	108	13:28:17	285	268	14:52:36	114	72	53	13:26:58	24	8	64	200	1.031	1.000	0.837	02m38s
Hopkinsville	36°52'N	087°29'W	11:56:33	293	318	62	13:24:42	116	101	13:27:22	292	276	14:51:44	114	72	53	13:26:02	24	9	64	199	1.031	1.000	0.969	02m40s
Lexington	38°03'N	084°30'W	13:02:12	289	306	62	—			—			15:54:08	118	75	50	14:30:26	204	183	62	206	0.951	0.946		
Louisville	38°15'N	085°46'W	12:59:23	290	309	62	—			—			15:51:57	117	76	52	14:27:42	204	186	62	202	0.961	0.958		
Madisonville	37°20'N	087°30'W	11:56:14	292	316	62	13:24:32	65	51	13:26:17	343	327	14:50:55	115	74	53	13:25:25	204	189	64	198	1.031	1.000	0.249	01m45s
Marion	37°20'N	088°05'W	11:55:00	292	318	61	13:22:58	95	82	13:25:29	313	299	14:49:59	115	74	54	13:24:13	204	191	64	196	1.031	1.000	0.676	02m31s
Oak Grove	36°40'N	087°26'W	11:56:47	293	318	62	13:25:08	133	118	13:27:39	275	258	14:52:09	114	72	53	13:26:24	24	8	64	199	1.031	1.000	0.667	02m31s
Owensboro	37°46'N	087°07'W	11:56:48	291	314	61	—			—			14:50:46	116	75	53	13:25:39	204	189	63	199	0.988	0.990		
Paducah	37°05'N	088°37'W	11:54:01	293	320	61	13:22:16	143	132	13:24:36	264	252	14:49:32	114	74	54	13:23:26	24	12	64	195	1.031	1.000	0.510	02m20s
Princeton	37°07'N	087°53'W	11:55:33	293	318	62	13:23:34	108	95	13:26:14	300	285	14:50:40	114	73	54	13:24:54	204	190	64	197	1.031	1.000	0.900	02m39s
Providence	37°24'N	087°46'W	11:55:38	292	317	61	13:23:48	72	58	13:25:46	336	321	14:50:23	115	74	54	13:24:47	204	190	64	197	1.031	1.000	0.331	01m59s
Russellville	36°51'N	086°53'W	11:57:50	293	316	62	13:26:04	91	75	13:28:31	317	299	14:52:43	115	72	53	13:27:17	204	187	64	201	1.031	1.000	0.613	02m27s
Scottsville	36°45'N	086°11'W	11:59:24	292	314	63	13:27:58	66	48	13:29:44	342	323	14:53:59	115	72	52	13:28:51	204	185	63	203	1.031	1.000	0.255	01m46s
Sturgis	37°33'N	087°59'W	11:55:05	292	317	61	13:23:18	64	51	13:25:01	344	330	14:49:46	115	75	54	13:24:09	204	191	63	196	1.031	1.000	0.234	01m42s

Table 2–9: Local Circumstances for Louisiana, Maine, Maryland & Mass.

Location Name	Latitude	Longitude	First Contact Time	P	V	Alt	Second Contact Time	P	V	Third Contact Time	P	V	Fourth Contact Time	P	V	Alt	Maximum Eclipse Time	P	V	Alt	Azm	Eclip. Mag.	Eclip. Obs.	Umbral Depth	Umbral Durat.
			h m s	°	°	°	h m s	°	°	h m s	°	°	h m s	°	°	°	h m s	°	°	°	°				
Louisiana																									
Alexandria	31°18'N	092°27'W	11:50:52	305	346	63	—			—			14:50:41	103	58	60	13:22:11	24	18	70	187	0.807	0.763		
Baton Rouge	30°27'N	091°11'W	11:54:38	306	345	65	—			—			14:54:25	103	54	59	13:26:20	25	12	71	194	0.799	0.753		
Bossier City	32°31'N	093°44'W	11:46:52	303	345	61	—			—			14:46:28	104	63	61	13:17:37	24	24	69	180	0.826	0.788		
Houma	29°36'N	090°43'W	11:56:47	307	347	66	—			—			14:56:26	102	51	59	13:28:39	25	9	71	198	0.781	0.731		
Kenner	29°59'N	090°15'W	11:57:17	306	344	66	—			—			14:56:56	103	52	58	13:29:13	25	8	71	199	0.798	0.751		
Lafayette	30°14'N	092°01'W	11:53:07	306	348	64	—			—			14:52:54	102	54	60	13:24:38	25	15	71	190	0.783	0.733		
Lake Charles	30°14'N	093°13'W	11:50:33	307	351	63	—			—			14:50:15	101	55	61	13:21:42	24	20	72	185	0.768	0.714		
Metairie	29°58'N	090°10'W	11:57:29	306	344	66	—			—			14:57:08	103	52	58	13:29:26	25	8	71	200	0.798	0.752		
Monroe	32°30'N	092°07'W	11:50:16	302	341	62	—			—			14:49:53	105	61	59	13:21:26	24	18	69	187	0.844	0.811		
New Iberia	30°01'N	091°49'W	11:53:50	307	348	65	—			—			14:53:35	102	53	60	13:25:25	25	14	72	192	0.779	0.728		
New Orleans	29°58'N	090°04'W	11:57:42	306	344	66	—			—			14:57:20	103	52	58	13:29:41	25	7	71	200	0.800	0.754		
Shreveport	32°31'N	093°45'W	11:46:50	303	345	61	—			—			14:46:26	104	63	61	13:17:34	24	24	69	180	0.826	0.787		
Maine																									
Augusta	44°19'N	069°47'W	13:30:02	272	256	56	—			—			15:55:56	134	91	37	14:45:35	203	170	48	229	0.648	0.566		
Bangor	44°48'N	068°46'W	13:31:50	271	254	55	—			—			15:55:28	135	93	37	14:46:12	203	169	47	230	0.628	0.542		
Lewiston	44°06'N	070°13'W	13:29:16	272	258	56	—			—			15:56:08	133	91	38	14:45:19	203	170	49	228	0.657	0.577		
Portland	43°39'N	070°16'W	13:29:23	273	258	57	—			—			15:57:08	133	90	38	14:45:57	203	169	49	229	0.667	0.589		
Maryland																									
Annapolis	38°59'N	076°30'W	13:18:59	284	279	63	—			—			16:02:02	124	77	43	14:43:36	204	170	56	224	0.837	0.800		
Baltimore	39°17'N	076°37'W	13:18:32	283	279	63	—			—			16:01:18	124	78	43	14:42:57	204	171	56	223	0.831	0.793		
Bethesda	38°59'N	077°06'W	13:17:41	284	281	63	—			—			16:01:25	124	77	44	14:42:35	204	171	56	223	0.844	0.809		
Columbia	39°14'N	076°50'W	13:18:05	283	280	63	—			—			16:01:11	124	78	43	14:42:39	204	171	56	223	0.835	0.798		
Cumberland	39°39'N	078°46'W	13:13:38	284	286	62	—			—			15:58:15	123	79	45	14:38:43	204	175	57	218	0.847	0.814		
Dundalk	39°16'N	076°30'W	13:18:48	283	279	63	—			—			16:01:27	124	78	43	14:43:10	204	171	56	224	0.830	0.792		
Hagerstown	39°39'N	077°43'W	13:15:54	283	283	62	—			—			15:59:23	124	79	44	14:40:32	204	173	56	221	0.835	0.798		
Salisbury	38°22'N	075°36'W	13:21:24	284	277	63	—			—			16:04:14	124	75	42	14:46:04	204	168	55	227	0.840	0.805		
Silver Spring	38°59'N	077°02'W	13:17:49	284	281	63	—			—			16:01:29	124	77	44	14:42:42	204	171	56	223	0.843	0.808		
Massachusetts																									
Boston	42°22'N	071°04'W	13:28:27	275	261	58	—			—			15:59:29	131	86	38	14:46:50	203	168	50	229	0.702	0.631		
Brockton	42°05'N	071°01'W	13:28:44	276	261	58	—			—			16:00:09	131	86	38	14:47:21	204	168	50	230	0.707	0.638		
Cambridge	42°22'N	071°06'W	13:28:23	275	261	58	—			—			15:59:27	131	86	38	14:46:47	203	168	50	229	0.702	0.632		
Fall River	41°42'N	071°09'W	13:28:41	276	261	59	—			—			16:00:55	130	85	39	14:47:46	204	168	50	230	0.717	0.650		
Fitchburg	42°35'N	071°48'W	13:26:48	276	263	58	—			—			15:58:26	131	87	39	14:45:25	203	169	50	228	0.705	0.635		
Framingham	42°18'N	071°26'W	13:27:44	276	262	58	—			—			15:59:21	131	86	39	14:46:24	203	169	50	229	0.707	0.638		
Lawrence	42°43'N	071°10'W	13:28:03	275	261	58	—			—			15:58:37	131	87	39	14:46:08	203	169	50	229	0.695	0.624		
Lowell	42°38'N	071°19'W	13:27:47	275	261	58	—			—			15:58:41	131	87	39	14:46:03	203	169	50	228	0.699	0.628		
Lynn	42°28'N	070°57'W	13:28:38	275	260	58	—			—			15:59:21	131	87	38	14:46:50	203	168	50	229	0.698	0.627		
New Bedford	41°38'N	070°56'W	13:29:11	276	261	59	—			—			16:01:14	130	84	38	14:48:11	204	167	50	231	0.716	0.649		
Newton	42°21'N	071°12'W	13:28:11	276	261	58	—			—			15:59:25	131	86	39	14:46:40	203	168	50	229	0.704	0.634		
Pittsfield	42°27'N	073°15'W	13:23:51	277	267	59	—			—			15:57:34	130	86	40	14:43:28	203	171	52	225	0.723	0.658		
Quincy	42°15'N	071°00'W	13:28:40	276	261	58	—			—			15:59:48	131	86	38	14:47:07	203	168	50	230	0.703	0.633		
Somerville	42°23'N	071°06'W	13:28:23	275	261	58	—			—			15:59:25	131	86	38	14:46:45	203	168	50	229	0.702	0.631		
Springfield	42°06'N	072°35'W	13:25:26	277	265	59	—			—			15:58:54	130	85	40	14:45:01	204	170	51	227	0.724	0.658		
Worcester	42°16'N	071°48'W	13:26:59	276	263	59	—			—			15:59:09	131	86	39	14:45:55	204	169	51	228	0.712	0.644		

Table 2–10: Local Circumstances for Michigan, Minnesota & Mississippi

Location Name	Latitude	Longitude	First Contact Time	P	V	Alt	Second Contact Time	P	V	Third Contact Time	P	V	Fourth Contact Time	P	V	Alt	Maximum Eclipse Time	P	V	Alt	Azm	Eclip. Mag.	Eclip. Obs.	Umbral Depth	Umbral Durat.
			h m s	°	°	°	h m s	°	°	h m s	°	°	h m s	°	°	°	h m s	°	°	°	°				
Michigan																									
Ann Arbor	42°17'N	083°45'W	13:02:03	282	295	59	—			—			15:47:04	123	87	49	14:26:24	203	186	58	203	0.839	0.803		
Battle Creek	42°19'N	085°11'W	12:59:08	283	299	58	—			—			15:45:10	123	87	50	14:23:48	203	189	58	199	0.853	0.821		
Dearborn	42°19'N	083°11'W	13:03:12	282	294	59	—			—			15:47:42	124	86	48	14:27:21	203	185	58	204	0.832	0.795		
Dearborn Heights	42°20'N	083°18'W	13:02:58	282	294	59	—			—			15:47:31	124	87	48	14:27:08	203	185	58	204	0.833	0.796		
Detroit	42°20'N	083°03'W	13:03:28	282	293	59	—			—			15:47:50	124	86	48	14:27:34	203	185	57	204	0.831	0.792		
Farmington Hills	42°29'N	083°22'W	13:02:47	282	294	59	—			—			15:47:08	124	87	48	14:26:49	203	186	57	203	0.830	0.792		
Flint	43°01'N	083°41'W	13:02:02	281	294	58	—			—			15:45:42	124	88	48	14:25:37	203	187	57	202	0.821	0.780		
Grand Rapids	42°58'N	085°40'W	12:58:04	282	298	57	—			—			15:43:18	123	89	50	14:22:11	203	190	58	197	0.842	0.807		
Jackson	42°15'N	084°24'W	13:00:44	283	297	58	—			—			15:46:19	123	87	49	14:25:17	203	187	58	201	0.847	0.813		
Kalamazoo	42°17'N	085°35'W	12:58:20	283	300	58	—			—			15:44:43	122	87	50	14:23:07	203	189	58	198	0.858	0.827		
Lansing	42°44'N	084°33'W	13:00:20	282	296	58	—			—			15:45:11	123	88	49	14:24:26	203	188	58	200	0.836	0.800		
Livonia	42°23'N	083°23'W	13:02:47	282	294	59	—			—			15:47:19	124	87	48	14:26:55	203	186	58	203	0.833	0.795		
Muskegon	43°14'N	086°16'W	12:56:51	282	299	57	—			—			15:42:00	123	90	50	14:20:49	203	191	58	195	0.841	0.806		
Pontiac	42°38'N	083°18'W	13:02:53	282	293	58	—			—			15:46:55	124	87	48	14:26:45	203	186	57	203	0.826	0.787		
Royal Oak	42°30'N	083°09'W	13:03:14	282	293	59	—			—			15:47:22	124	87	48	14:27:11	203	185	57	204	0.828	0.789		
Southfield	42°29'N	083°17'W	13:02:57	282	294	59	—			—			15:47:15	124	87	48	14:26:58	203	185	57	204	0.829	0.791		
St. Clair Shores	42°30'N	082°54'W	13:03:44	282	293	59	—			—			15:47:40	124	87	48	14:27:37	203	185	57	205	0.825	0.785		
Sterling Heights	42°35'N	083°02'W	13:03:27	282	293	59	—			—			15:47:21	124	87	48	14:27:17	203	185	57	204	0.824	0.785		
Taylor	42°14'N	083°16'W	13:03:03	282	294	59	—			—			15:47:46	124	86	48	14:27:19	203	185	58	204	0.835	0.798		
Troy	42°37'N	083°09'W	13:03:12	282	293	58	—			—			15:47:08	124	87	48	14:27:02	203	185	57	204	0.825	0.785		
Warren	42°31'N	083°02'W	13:03:28	282	293	59	—			—			15:47:29	124	87	48	14:27:22	203	185	57	204	0.826	0.787		
Westland	42°19'N	083°22'W	13:02:50	282	294	59	—			—			15:47:28	124	86	48	14:27:02	203	185	58	204	0.834	0.797		
Wyoming	42°54'N	085°42'W	12:58:01	282	299	57	—			—			15:43:23	123	89	50	14:22:12	203	190	58	197	0.844	0.809		
Minnesota																									
Bloomington	44°50'N	093°17'W	11:43:52	282	309	52	—			—			14:29:17	121	99	53	13:06:46	202	205	57	176	0.865	0.836		
Brooklyn Park	45°06'N	093°23'W	11:43:46	281	308	51	—			—			14:28:42	121	100	53	13:06:24	202	205	57	175	0.859	0.829		
Coon Rapids	45°09'N	093°19'W	11:43:54	281	308	51	—			—			14:28:43	121	100	53	13:06:29	202	205	57	176	0.857	0.826		
Duluth	46°47'N	092°07'W	11:46:32	278	302	51	—			—			14:27:40	124	103	52	13:07:23	201	203	55	178	0.805	0.761		
Minneapolis	44°59'N	093°16'W	11:43:56	282	308	52	—			—			14:29:04	121	99	53	13:06:41	202	205	57	176	0.861	0.831		
Plymouth	45°02'N	093°27'W	11:43:38	282	308	51	—			—			14:28:43	121	99	53	13:06:20	202	205	57	175	0.861	0.832		
Rochester	44°01'N	092°28'W	11:45:09	283	309	53	—			—			14:31:52	120	96	54	13:08:52	202	203	58	178	0.879	0.855		
St. Cloud	45°34'N	094°09'W	11:42:35	281	308	51	—			—			14:26:48	121	101	53	13:04:43	201	206	56	173	0.853	0.821		
St. Paul	44°57'N	093°07'W	11:44:12	282	308	52	—			—			14:29:20	121	99	53	13:06:59	202	204	57	176	0.861	0.831		
Mississippi																									
Biloxi	30°24'N	088°53'W	11:59:43	304	339	67	—			—			14:59:08	105	53	56	13:31:50	25	4	70	204	0.826	0.788		
Columbus	33°30'N	088°25'W	11:57:12	299	329	64	—			—			14:55:34	109	63	55	13:28:30	25	7	67	200	0.915	0.901		
Greenville	33°24'N	091°04'W	11:51:37	300	336	62	—			—			14:50:46	107	64	58	13:22:43	24	15	68	190	0.881	0.857		
Gulfport	30°22'N	089°06'W	11:59:17	305	340	67	—			—			14:58:46	105	53	57	13:31:22	25	4	70	204	0.823	0.783		
Hattiesburg	31°20'N	089°17'W	11:57:40	303	337	66	—			—			14:57:06	106	56	57	13:29:34	25	7	69	201	0.847	0.813		
Jackson	32°18'N	090°12'W	11:54:34	302	337	64	—			—			14:53:59	106	60	57	13:26:08	25	11	69	195	0.862	0.832		
Laurel	31°41'N	089°08'W	11:57:35	302	336	66	—			—			14:56:54	106	57	56	13:29:25	25	7	69	201	0.858	0.828		
Meridian	32°22'N	088°42'W	11:57:45	301	333	65	—			—			14:56:44	108	59	56	13:29:27	25	6	68	201	0.882	0.858		
Natchez	31°34'N	091°24'W	11:52:48	304	343	64	—			—			14:52:33	105	58	59	13:24:19	25	15	70	191	0.827	0.789		
Pascagoula	30°21'N	088°33'W	12:00:32	304	338	67	—			—			14:59:51	105	53	56	13:32:40	25	2	70	206	0.829	0.791		
Tupelo	34°16'N	088°43'W	11:55:51	298	328	64	—			—			14:53:51	110	65	55	13:26:49	24	9	67	198	0.932	0.922		
Vicksburg	32°21'N	090°53'W	11:53:03	302	339	63	—			—			14:52:34	106	60	58	13:24:29	24	14	69	192	0.855	0.824		

Table 2–11: Local Circumstances for Missouri - 1

Location Name	Latitude	Longitude	First Contact Time h m s	P °	V °	Alt °	Second Contact Time h m s	P °	V °	Third Contact Time h m s	P °	V °	Fourth Contact Time h m s	P °	V °	Alt °	Maximum Eclipse Time h m s	P °	V °	Alt °	Azm °	Eclip. Mag.	Eclip. Obs.	Umbral Depth	Umbral Durat.
Missouri																									
Arnold	38°26'N	090°22'W	11:49:46	291	320	59	13:17:11	76	71	13:19:19	330	324	14:44:24	114	79	55	13:18:15	203	197	63	188	1.031	1.000	0.399	02m08s
Ballwin	38°36'N	090°34'W	11:49:18	291	320	58	13:16:47	64	59	13:18:32	342	336	14:43:48	115	79	55	13:17:39	203	198	63	187	1.031	1.000	0.246	01m45s
Blue Springs	39°01'N	094°17'W	11:41:53	292	327	55	13:09:10	184	189	13:10:00	221	225	14:36:41	113	84	58	13:09:35	23	27	63	174	1.031	1.000	0.052	00m51s
Bonne Terre	37°55'N	090°33'W	11:49:37	292	322	59	13:17:14	138	133	13:19:39	268	262	14:44:56	114	77	56	13:18:27	23	17	64	187	1.031	1.000	0.578	02m25s
Boonville	38°58'N	092°44'W	11:44:53	291	324	56	13:11:30	116	117	13:14:10	290	290	14:39:31	114	82	57	13:12:50	23	23	63	179	1.031	1.000	0.953	02m39s
Brentwood	38°37'N	090°21'W	11:49:44	291	320	59	13:17:36	45	39	13:18:34	2	356	14:44:08	115	79	55	13:18:05	203	197	63	187	1.031	1.000	0.069	00m58s
Buckner	39°08'N	094°12'W	11:42:01	292	326	55	13:08:41	156	161	13:10:37	249	253	14:36:39	113	84	58	13:09:39	23	27	63	175	1.031	1.000	0.312	01m56s
California	38°38'N	092°34'W	11:45:19	292	324	57	13:12:22	145	145	13:14:38	261	261	14:40:19	113	81	57	13:13:30	23	23	63	180	1.031	1.000	0.473	02m16s
Cameron	39°44'N	094°14'W	11:41:50	291	324	55	13:07:49	91	96	13:10:17	314	318	14:35:43	114	86	57	13:09:03	202	207	62	174	1.030	1.000	0.635	02m27s
Cape Girardeau	37°19'N	089°32'W	11:51:59	293	322	60	13:20:21	162	153	13:22:08	245	236	14:47:37	114	75	55	13:21:15	24	14	64	191	1.031	1.000	0.252	01m47s
Carrollton	39°22'N	093°30'W	11:43:18	291	324	56	13:09:34	103	106	13:12:11	302	304	14:37:34	114	84	57	13:10:53	203	205	62	177	1.030	1.000	0.830	02m37s
Centralia	39°13'N	092°08'W	11:45:59	291	322	57	13:13:00	61	61	13:14:38	345	343	14:40:10	114	82	56	13:13:49	203	202	63	181	1.030	1.000	0.215	01m38s
Chesterfield	38°40'N	090°35'W	11:49:14	291	320	58	13:16:52	54	49	13:18:14	352	347	14:43:40	115	80	55	13:17:33	203	198	63	187	1.031	1.000	0.142	01m22s
Chillicothe	39°48'N	093°33'W	11:43:07	290	323	55	13:09:50	47	50	13:10:55	358	1	14:36:49	114	85	57	13:10:23	203	205	62	176	1.030	1.000	0.090	01m06s
Columbia	38°57'N	092°20'W	11:45:40	291	323	57	13:12:22	102	102	13:14:58	304	303	14:40:14	114	82	57	13:13:40	203	202	63	181	1.031	1.000	0.814	02m37s
Crestwood	38°34'N	090°23'W	11:49:41	291	320	59	13:17:19	58	53	13:18:50	348	342	14:44:10	115	79	55	13:18:04	203	197	63	187	1.031	1.000	0.179	01m31s
Creve Coeur	38°40'N	090°27'W	11:49:30	291	320	58	13:17:24	42	36	13:18:14	5	359	14:43:53	115	79	55	13:17:49	203	198	63	187	1.031	1.000	0.051	00m50s
Crystal City	38°14'N	090°23'W	11:49:49	292	321	59	13:17:09	99	94	13:19:44	307	301	14:44:42	114	78	55	13:18:26	203	197	63	188	1.031	1.000	0.758	02m35s
Cuba	38°04'N	091°24'W	11:47:50	292	324	58	13:15:37	161	158	13:17:25	245	241	14:43:14	113	78	56	13:16:31	23	20	64	184	1.031	1.000	0.257	01m47s
De Soto	38°08'N	090°34'W	11:49:29	292	322	59	13:16:50	117	111	13:19:30	290	284	14:44:33	114	78	56	13:18:10	23	18	64	187	1.031	1.000	0.942	02m40s
Des Peres	38°36'N	090°27'W	11:49:32	291	320	58	13:17:10	57	51	13:18:38	350	344	14:44:00	115	79	55	13:17:54	203	198	63	187	1.031	1.000	0.167	01m28s
Desloge	37°52'N	090°31'W	11:49:42	292	322	59	13:17:24	142	137	13:19:44	264	258	14:45:04	114	77	56	13:18:34	23	17	64	187	1.031	1.000	0.514	02m20s
Ellisville	38°36'N	090°36'W	11:49:14	291	321	58	13:16:41	66	61	13:18:30	340	334	14:43:45	115	79	55	13:17:35	203	198	63	187	1.031	1.000	0.269	01m49s
Eureka	38°30'N	090°38'W	11:49:12	291	321	58	13:16:30	81	76	13:18:45	326	320	14:43:51	114	79	56	13:17:38	203	198	63	187	1.031	1.000	0.466	02m15s
Excelsior Springs	39°20'N	094°13'W	11:41:57	291	325	55	13:08:11	132	137	13:10:41	273	276	14:36:20	113	85	58	13:09:26	22	27	62	174	1.031	1.000	0.659	02m30s
Farmington	37°47'N	090°25'W	11:49:57	293	322	59	13:17:46	148	142	13:19:59	259	252	14:45:22	114	77	56	13:18:52	23	17	64	188	1.031	1.000	0.438	02m13s
Fayette	39°09'N	092°41'W	11:44:56	291	323	56	13:11:30	95	96	13:14:01	311	311	14:39:19	114	83	57	13:12:45	203	203	63	179	1.031	1.000	0.691	02m31s
Fenton	38°31'N	090°26'W	11:49:36	291	320	59	13:17:04	69	63	13:18:58	338	331	14:44:10	115	79	55	13:18:01	203	197	63	187	1.031	1.000	0.300	01m54s
Festus	38°13'N	090°24'W	11:49:47	292	321	59	13:17:07	102	96	13:19:44	305	298	14:44:42	114	78	55	13:18:25	203	197	63	188	1.031	1.000	0.798	02m37s
Florissant	38°48'N	090°20'W	11:49:42	291	319	58	—			—			14:43:52	115	80	55	13:17:55	203	197	63	187	0.996	0.998		
Fredericktown	37°34'N	090°18'W	11:50:17	293	323	59	13:18:40	172	166	13:20:03	235	227	14:45:55	113	76	56	13:19:22	23	17	64	189	1.031	1.000	0.144	01m23s
Frontenac	38°38'N	090°25'W	11:49:35	291	320	58	13:17:23	47	42	13:18:28	359	353	14:44:00	115	79	55	13:17:56	203	197	63	187	1.031	1.000	0.086	01m05s
Fulton	38°52'N	091°57'W	11:46:27	291	323	57	13:13:16	96	95	13:15:49	310	308	14:41:02	114	81	56	13:14:32	203	201	63	182	1.031	1.000	0.710	02m33s
Gladstone	39°13'N	094°35'W	11:41:17	292	326	55	13:07:57	164	169	13:09:37	241	246	14:35:51	113	85	58	13:08:47	22	28	63	173	1.031	1.000	0.222	01m40s
Glendale	38°36'N	090°23'W	11:49:40	291	320	59	13:17:24	52	46	13:18:40	355	349	14:44:07	115	79	55	13:18:02	203	197	63	187	1.031	1.000	0.121	01m16s
Grain Valley	39°01'N	094°12'W	11:42:03	292	326	55	13:09:10	176	181	13:10:21	229	233	14:36:50	113	84	58	13:09:45	23	27	63	175	1.031	1.000	0.104	01m11s
Hermann	38°42'N	091°27'W	11:47:30	291	322	58	13:14:29	94	91	13:17:00	312	309	14:42:09	114	80	56	13:15:44	203	200	63	184	1.031	1.000	0.669	02m31s
Higginsville	39°04'N	093°43'W	11:42:58	292	325	56	13:09:34	143	146	13:11:52	263	265	14:37:37	113	83	58	13:10:43	23	26	63	176	1.031	1.000	0.499	02m18s
Holts Summit	38°39'N	092°06'W	11:46:14	292	324	57	13:13:09	124	123	13:15:46	282	280	14:41:06	114	81	57	13:14:27	23	22	63	182	1.031	1.000	0.810	02m37s
Hoover	39°22'N	094°40'W	11:41:05	291	326	55	13:07:23	146	153	13:09:35	258	263	14:35:29	113	85	58	13:08:29	22	28	62	173	1.030	1.000	0.439	02m12s
Independence	39°06'N	094°25'W	11:41:37	292	327	55	13:08:37	174	180	13:09:51	230	235	14:36:19	113	84	58	13:09:14	23	27	63	174	1.031	1.000	0.117	01m15s
Jackson	37°23'N	089°40'W	11:51:41	293	322	60	13:19:58	160	152	13:21:49	247	238	14:47:17	114	75	55	13:20:54	24	15	64	191	1.031	1.000	0.272	01m50s
Jefferson City	38°34'N	092°10'W	11:46:07	292	324	57	13:13:10	135	135	13:15:38	271	269	14:41:07	113	80	57	13:14:24	23	22	63	181	1.031	1.000	0.620	02m28s
Joplin	37°06'N	094°31'W	11:42:05	295	333	56	—			—			14:38:57	110	78	59	13:10:53	23	28	65	174	0.946	0.939		
Kansas City	39°06'N	094°35'W	11:41:18	292	327	55	13:08:40	193	198	13:09:07	212	218	14:36:01	113	84	58	13:08:53	22	28	63	173	1.031	1.000	0.015	00m27s
Kearney	39°22'N	094°22'W	11:41:39	291	326	55	13:07:52	134	140	13:10:20	270	275	14:36:01	113	85	58	13:09:06	22	27	62	174	1.030	1.000	0.626	02m27s
Kirkwood	38°35'N	090°24'W	11:49:38	291	320	59	13:17:18	56	51	13:18:44	350	344	14:44:06	115	79	55	13:18:01	203	197	63	187	1.031	1.000	0.161	01m27s
Ladue	38°38'N	090°23'W	11:49:39	291	320	58	13:17:32	44	38	13:18:28	3	357	14:44:03	115	79	55	13:18:00	203	197	63	187	1.031	1.000	0.063	00m56s
Lake St. Louis	38°47'N	090°49'W	11:48:44	291	321	58	13:16:23	49	44	13:17:31	358	353	14:43:05	115	80	56	13:16:57	203	199	63	186	1.031	1.000	0.097	01m08s

Table 2–12: Local Circumstances for Missouri - 2

Location Name	Latitude	Longitude	First Contact Time	P	V	Alt	Second Contact Time	P	V	Third Contact Time	P	V	Fourth Contact Time	P	V	Alt	Maximum Eclipse Time	P	V	Alt	Azm	Eclip. Mag.	Eclip. Obs.	Umbral Depth	Umbral Durat.
			h m s	°	°	°	h m s	°	°	h m s	°	°	h m s	°	°	°	h m s	°	°	°	°				
Missouri																									
Lee's Summit	38°50'N	094°23'W	11:41:45	292	327	55	—			—			14:36:46	112	83	58	13:09:33	23	27	63	174	0.995	0.997		
Lexington	39°11'N	093°52'W	11:42:39	291	325	55	13:09:04	135	139	13:11:31	270	273	14:37:11	113	84	58	13:10:18	23	26	63	176	1.031	1.000	0.612	02m27s
Liberty	39°15'N	094°25'W	11:41:35	291	326	55	13:08:03	150	156	13:10:09	255	259	14:36:06	113	85	58	13:09:06	22	27	63	174	1.031	1.000	0.386	02m06s
Manchester	38°36'N	090°31'W	11:49:24	291	320	58	13:16:56	61	56	13:18:35	345	339	14:43:53	115	79	55	13:17:46	203	198	63	187	1.031	1.000	0.212	01m38s
Maplewood	38°37'N	090°19'W	11:49:48	291	320	59	13:17:45	41	35	13:18:33	6	360	14:44:12	115	79	55	13:18:09	203	197	63	188	1.031	1.000	0.046	00m48s
Marshall	39°07'N	093°12'W	11:43:57	291	324	56	13:10:24	117	120	13:13:03	288	289	14:38:28	114	83	57	13:11:44	23	24	63	178	1.031	1.000	0.917	02m39s
Mexico	39°10'N	091°53'W	11:46:30	291	322	57	13:13:42	53	52	13:15:03	352	350	14:40:40	114	82	56	13:14:22	203	201	63	182	1.031	1.000	0.139	01m21s
Moberly	39°25'N	092°26'W	11:45:21	291	322	56	13:12:34	42	42	13:13:27	3	3	14:39:20	114	83	56	13:13:00	203	203	62	180	1.030	1.000	0.058	00m53s
North Kansas City	39°09'N	094°34'W	11:41:19	292	327	55	13:08:16	175	180	13:09:29	230	235	14:35:58	113	84	58	13:08:53	22	28	63	173	1.031	1.000	0.113	01m14s
Odessa	39°00'N	093°57'W	11:42:32	292	326	55	13:09:26	163	167	13:11:08	242	246	14:37:18	113	83	58	13:10:17	23	26	63	175	1.031	1.000	0.232	01m42s
Pacific	38°29'N	090°45'W	11:48:58	291	321	58	13:16:12	88	84	13:18:37	318	313	14:43:41	114	79	56	13:17:25	203	198	63	186	1.031	1.000	0.574	02m24s
Parkville	39°12'N	094°41'W	11:41:05	292	327	55	13:07:56	173	179	13:09:15	232	237	14:35:41	113	85	58	13:08:36	22	28	63	173	1.031	1.000	0.130	01m19s
Perryville	37°43'N	089°52'W	11:51:06	292	321	60	13:18:49	131	123	13:21:22	276	268	14:46:24	114	76	55	13:20:05	24	16	64	190	1.031	1.000	0.707	02m33s
Pevely	38°17'N	090°24'W	11:49:46	292	321	59	13:17:05	95	89	13:19:37	312	305	14:44:36	114	78	55	13:18:21	203	197	63	188	1.031	1.000	0.682	02m32s
Platte City	39°22'N	094°47'W	11:40:52	291	326	55	13:07:13	152	158	13:09:16	253	259	14:35:16	113	85	58	13:08:15	22	28	62	173	1.030	1.000	0.367	02m03s
Pleasant Valley	39°13'N	094°29'W	11:41:28	292	326	55	13:08:04	158	164	13:09:55	247	251	14:36:01	113	85	58	13:08:59	22	28	63	174	1.031	1.000	0.284	01m51s
Potosi	37°56'N	090°47'W	11:49:08	292	323	59	13:16:50	147	142	13:19:03	260	254	14:44:30	114	77	56	13:17:57	23	18	64	187	1.031	1.000	0.446	02m14s
Richmond	39°17'N	093°58'W	11:42:26	291	325	55	13:08:43	128	133	13:11:16	277	280	14:36:51	113	84	58	13:10:00	23	26	63	175	1.031	1.000	0.727	02m33s
Richmond Heights	38°38'N	090°19'W	11:49:47	291	320	59	13:17:53	34	28	13:18:23	13	6	14:44:10	115	79	55	13:18:08	203	197	63	187	1.031	1.000	0.018	00m30s
Riverside	39°10'N	094°36'W	11:41:15	292	327	55	13:08:09	174	179	13:09:26	231	236	14:35:53	113	85	58	13:08:48	22	28	63	173	1.031	1.000	0.122	01m16s
Rock Hill	38°37'N	090°22'W	11:49:42	291	320	59	13:17:31	46	41	13:18:34	0	354	14:44:07	115	79	55	13:18:03	203	197	63	187	1.031	1.000	0.081	01m03s
Savannah	39°56'N	094°50'W	11:40:41	290	325	54	13:06:26	91	98	13:08:53	313	319	14:34:21	114	87	58	13:07:39	202	208	62	172	1.030	1.000	0.645	02m28s
Scott City	37°13'N	089°30'W	11:52:07	293	322	60	13:20:50	177	168	13:22:02	230	220	14:47:50	113	74	55	13:21:26	24	14	64	192	1.031	1.000	0.103	01m11s
Sedalia	38°42'N	093°14'W	11:44:00	292	326	56	13:11:20	170	172	13:12:46	235	237	14:39:02	113	82	57	13:12:03	23	24	63	178	1.031	1.000	0.157	01m26s
Shrewsbury	38°35'N	090°20'W	11:49:46	291	320	59	13:17:32	51	45	13:18:46	355	349	14:44:13	115	79	55	13:18:09	203	197	63	187	1.031	1.000	0.116	01m14s
Smithville	39°23'N	094°35'W	11:41:15	291	326	55	13:07:29	141	147	13:09:48	264	269	14:35:36	113	85	58	13:08:38	22	28	62	173	1.030	1.000	0.521	02m20s
Springfield	37°13'N	093°17'W	11:44:27	295	330	57	—			—			14:41:07	111	77	58	13:13:26	23	24	65	178	0.962	0.959		
St. Charles	38°48'N	090°32'W	11:49:18	291	320	58	—			—			14:43:32	115	80	55	13:17:30	203	198	63	187	0.998	0.999		
St. Clair	38°21'N	090°58'W	11:48:35	292	322	58	13:15:47	111	107	13:18:26	296	291	14:43:32	114	79	56	13:17:06	203	199	63	186	1.031	1.000	0.954	02m40s
St. Joseph	39°42'N	094°52'W	11:40:39	291	326	54	13:06:28	117	124	13:09:06	288	293	14:34:38	113	86	58	13:07:47	22	28	62	172	1.030	1.000	0.917	02m38s
St. Louis	38°38'N	090°15'W	11:49:55	291	320	59	—			—			14:44:17	115	79	55	13:18:16	203	197	63	188	1.000	1.000		
St. Peters	38°46'N	090°35'W	11:49:12	291	320	58	—			—			14:43:30	115	80	55	13:17:27	203	198	63	187	1.000	1.000		
Ste. Genevieve	37°59'N	090°03'W	11:50:36	292	321	59	13:18:04	111	104	13:20:44	296	288	14:45:40	114	77	55	13:19:24	203	196	64	189	1.031	1.000	0.960	02m40s
Sugar Creek	39°07'N	094°27'W	11:41:33	292	327	55	13:08:30	173	179	13:09:48	232	236	14:36:14	113	84	58	13:09:09	22	27	63	174	1.031	1.000	0.126	01m18s
Sullivan	38°13'N	091°10'W	11:48:14	292	323	58	13:15:34	132	129	13:18:06	274	270	14:43:24	114	79	56	13:16:50	23	19	64	185	1.031	1.000	0.677	02m32s
Sunset Hills	38°33'N	090°25'W	11:49:37	291	320	59	13:17:10	63	58	13:18:52	344	337	14:44:08	115	79	55	13:18:01	203	197	63	187	1.031	1.000	0.231	01m42s
Town and Country	38°38'N	090°27'W	11:49:31	291	320	58	13:17:15	50	45	13:18:28	356	350	14:43:57	115	79	55	13:17:52	203	198	63	187	1.031	1.000	0.109	01m12s
Union	38°27'N	091°00'W	11:48:29	292	322	58	13:15:37	102	98	13:18:14	305	300	14:43:19	114	79	56	13:16:56	203	199	63	185	1.031	1.000	0.802	02m37s
University City	38°40'N	090°20'W	11:49:45	291	320	59	—			—			14:44:05	115	79	55	13:18:04	203	197	63	187	1.000	1.000		
Valley Park	38°33'N	090°29'W	11:49:29	291	320	58	13:16:58	67	62	13:18:48	340	334	14:44:01	115	79	55	13:17:53	203	198	63	187	1.031	1.000	0.276	01m50s
Warrenton	38°49'N	091°09'W	11:48:03	291	321	58	13:15:20	66	62	13:17:08	341	336	14:42:28	114	80	56	13:16:14	203	199	63	185	1.031	1.000	0.264	01m48s
Washington	38°33'N	091°01'W	11:48:25	291	322	58	13:15:33	92	88	13:18:02	314	310	14:43:07	114	79	56	13:16:47	203	199	63	185	1.031	1.000	0.639	02m29s
Webster Groves	38°35'N	090°22'W	11:49:42	291	320	59	13:17:25	54	48	13:18:45	353	347	14:44:10	115	79	55	13:18:05	203	197	63	187	1.031	1.000	0.139	01m21s
Wentzville	38°49'N	090°51'W	11:48:39	291	320	58	13:16:23	43	39	13:17:18	3	358	14:42:59	115	80	56	13:16:51	203	199	63	186	1.031	1.000	0.062	00m55s

Table 2–13: Local Circumstances for Montana & Nebraska

Location Name	Latitude	Longitude	First Contact Time	P	V	Alt	Second Contact Time	P	V	Third Contact Time	P	V	Fourth Contact Time	P	V	Alt	Maximum Eclipse Time	P	V	Alt	Azm	Eclip. Mag.	Eclip. Obs.	Umbral Depth	Umbral Durat.
			h m s	°	°	°	h m s	°	°	h m s	°	°	h m s	°	°	°	h m s	°	°	°	°				
Montana																									
Billings	45°47'N	108°30'W	10:21:08	283	322	40	—			—			13:03:00	114	119	56	11:40:00	199	226	50	141	0.941	0.932		
Bozeman	45°41'N	111°02'W	10:17:57	283	324	38	—			—			12:58:38	113	122	56	11:35:55	198	228	49	136	0.955	0.950		
Butte	46°00'N	112°32'W	10:16:27	283	324	36	—			—			12:55:48	113	124	55	11:33:37	198	229	48	134	0.952	0.946		
Great Falls	47°30'N	111°17'W	10:19:14	280	319	37	—			—			12:56:58	116	125	54	11:35:53	198	226	47	137	0.904	0.886		
Missoula	46°52'N	114°01'W	10:15:37	281	322	35	—			—			12:52:47	114	127	54	11:31:40	197	229	46	132	0.932	0.921		
Nebraska																									
Alliance	42°06'N	102°52'W	10:27:08	289	329	46	11:49:14	106	128	11:51:44	295	316	13:16:44	112	104	59	11:50:29	200	222	57	151	1.030	1.000	0.922	02m30s
Auburn	40°23'N	095°51'W	11:38:46	290	325	53	13:04:09	78	87	13:06:19	326	334	14:31:53	114	89	58	13:05:14	202	210	61	169	1.030	1.000	0.441	02m10s
Aurora	40°52'N	098°00'W	11:34:56	290	327	51	12:59:19	99	113	13:01:51	304	317	14:27:19	113	93	59	13:00:35	202	215	60	163	1.030	1.000	0.781	02m32s
Beatrice	40°16'N	096°45'W	11:37:09	290	327	52	13:02:13	122	133	13:04:48	282	292	14:30:25	113	90	59	13:03:30	22	32	61	167	1.030	1.000	0.823	02m35s
Bellevue	41°09'N	095°54'W	11:38:40	289	323	52	—			—			14:30:43	115	92	58	13:04:33	202	210	60	169	0.986	0.988		
Bridgeport	41°40'N	103°06'W	10:26:39	289	330	46	11:49:33	170	193	11:50:49	231	253	13:16:44	111	103	60	11:50:11	20	43	57	150	1.030	1.000	0.138	01m17s
Broken Bow	41°24'N	099°38'W	11:32:12	289	327	49	12:55:50	91	108	12:58:15	311	327	14:23:37	113	97	59	12:57:02	201	217	59	159	1.030	1.000	0.664	02m25s
Central City	41°07'N	098°00'W	11:34:57	289	326	51	12:59:29	69	82	13:01:22	334	347	14:26:59	113	94	59	13:00:25	201	215	60	163	1.030	1.000	0.319	01m54s
Cozad	40°52'N	099°59'W	11:31:33	290	329	49	12:55:57	169	186	12:57:20	234	251	14:23:36	112	96	60	12:56:38	21	39	59	157	1.030	1.000	0.156	01m23s
Crete	40°38'N	096°58'W	11:36:45	290	326	52	13:01:36	90	102	13:04:01	313	324	14:29:31	113	91	58	13:02:48	202	213	61	166	1.030	1.000	0.631	02m25s
Fairbury	40°08'N	097°11'W	11:36:23	291	328	52	13:01:45	154	166	13:03:42	250	261	14:29:48	112	90	59	13:02:44	22	33	61	165	1.030	1.000	0.330	01m57s
Falls City	40°03'N	095°36'W	11:39:15	290	326	53	13:04:42	106	115	13:07:19	298	305	14:32:49	113	88	58	13:06:00	202	210	61	170	1.030	1.000	0.896	02m37s
Geneva	40°32'N	097°36'W	11:35:38	290	327	51	13:00:19	122	134	13:02:53	282	294	14:28:29	113	92	59	13:01:36	22	34	60	164	1.030	1.000	0.828	02m34s
Gering	41°50'N	103°40'W	10:25:49	289	330	45	11:48:17	163	187	11:49:48	237	260	13:15:28	111	104	60	11:49:03	20	44	57	148	1.030	1.000	0.200	01m30s
Gothenburg	40°56'N	100°10'W	11:31:15	290	329	49	12:55:29	166	184	12:56:59	237	254	14:23:10	112	97	60	12:56:14	21	39	59	157	1.030	1.000	0.185	01m30s
Grand Island	40°55'N	098°21'W	11:34:20	290	327	51	12:58:34	105	119	13:01:09	298	312	14:26:36	113	94	59	12:59:51	201	215	60	162	1.030	1.000	0.880	02m34s
Hastings	40°35'N	098°23'W	11:34:16	290	328	51	12:58:53	142	157	13:01:07	260	274	14:26:58	112	93	59	13:00:00	21	36	60	162	1.030	1.000	0.485	02m14s
Hebron	40°10'N	097°35'W	11:35:40	291	328	52	13:01:10	169	181	13:02:36	235	247	14:29:00	112	91	59	13:01:53	22	34	61	164	1.030	1.000	0.163	01m26s
Kearney	40°42'N	099°05'W	11:33:03	290	329	50	12:57:34	154	170	12:59:29	249	264	14:25:30	112	94	59	12:58:31	21	37	60	160	1.030	1.000	0.321	01m54s
Lexington	40°47'N	099°45'W	11:31:56	290	329	49	12:56:32	173	190	12:57:46	230	247	14:24:08	112	96	60	12:57:09	21	38	59	158	1.030	1.000	0.122	01m14s
Lincoln	40°49'N	096°41'W	11:37:15	289	325	52	13:02:36	52	62	13:03:53	352	2	14:29:47	114	92	58	13:03:15	202	212	60	167	1.030	1.000	0.132	01m17s
Milford	40°47'N	097°03'W	11:36:36	290	326	52	13:01:29	75	87	13:03:35	328	339	14:29:10	113	92	58	13:02:32	202	213	60	166	1.030	1.000	0.404	02m05s
Minden	40°30'N	098°57'W	11:33:17	291	329	50	12:58:39	190	205	12:59:11	213	229	14:26:00	112	94	60	12:58:55	21	37	60	160	1.030	1.000	0.021	00m32s
Mitchell	41°57'N	103°49'W	10:25:38	289	330	45	11:47:46	150	174	11:49:41	250	274	13:15:04	111	105	60	11:48:43	20	44	56	148	1.030	1.000	0.358	01m55s
North Platte	41°08'N	100°46'W	11:30:17	290	329	48	12:54:06	159	178	12:55:48	243	261	14:21:47	112	98	60	12:54:57	21	40	59	155	1.030	1.000	0.254	01m43s
Omaha	41°17'N	096°01'W	11:38:27	288	323	52	—			—			14:30:19	115	92	57	13:04:14	202	211	60	169	0.983	0.985		
Scottsbluff	41°52'N	103°40'W	10:25:50	289	330	45	11:48:11	158	181	11:49:53	243	266	13:15:26	111	104	60	11:49:02	20	44	56	148	1.030	1.000	0.264	01m42s
Seward	40°55'N	097°06'W	11:36:31	289	325	52	13:01:35	57	69	13:03:06	346	357	14:28:54	113	92	58	13:02:21	202	213	60	166	1.030	1.000	0.188	01m31s
St. Paul	41°13'N	098°28'W	11:34:09	289	326	50	12:58:25	74	88	13:00:28	329	343	14:26:00	113	95	59	12:59:26	201	215	60	162	1.030	1.000	0.388	02m02s
Syracuse	40°39'N	096°11'W	11:38:09	290	325	53	13:03:41	54	63	13:05:03	350	359	14:30:55	114	91	58	13:04:22	202	211	61	168	1.030	1.000	0.151	01m23s
Tecumseh	40°22'N	096°11'W	11:38:10	290	326	53	13:03:21	92	102	13:05:49	311	320	14:31:19	113	90	58	13:04:35	202	211	61	168	1.030	1.000	0.666	02m28s
Wilber	40°29'N	096°58'W	11:36:45	290	326	52	13:01:37	106	118	13:04:13	297	308	14:29:43	113	91	59	13:02:55	202	213	61	166	1.030	1.000	0.906	02m36s
Wymore	40°07'N	096°40'W	11:37:18	291	327	52	13:02:35	136	147	13:04:59	268	278	14:30:46	113	90	59	13:03:47	22	32	61	167	1.030	1.000	0.597	02m24s
York	40°52'N	097°36'W	11:35:38	290	326	51	13:00:13	85	98	13:02:33	318	330	14:28:03	113	93	59	13:01:23	202	214	60	164	1.030	1.000	0.559	02m20s

Table 2–14: Local Circumstances for Nevada, N. Hampshire, N. Jersey & N. Mexico

Location Name	Latitude	Longitude	First Contact Time	P	V	Alt	Second Contact Time	P	V	Third Contact Time	P	V	Fourth Contact Time	P	V	Alt	Maximum Eclipse Time	P	V	Alt	Azm	Eclip. Mag.	Eclip. Obs.	Umbral Depth	Umbral Durat.
			h m s	°	°	°	h m s	°	°	h m s	°	°	h m s	°	°	°	h m s	°	°	°	°				
Nevada																									
Carson City	39°10'N	119°46'W	09:04:29	294	346	31	—			—			11:43:09	99	128	58	10:20:00	17	62	45	116	0.850	0.816		
Henderson	36°02'N	114°59'W	09:09:21	300	354	36	—			—			11:53:23	96	119	63	10:27:42	18	64	51	120	0.767	0.713		
Las Vegas	36°10'N	115°09'W	09:09:08	300	354	36	—			—			11:53:00	96	119	63	10:27:23	18	64	51	120	0.771	0.717		
North Las Vegas	36°12'N	115°07'W	09:09:11	300	354	36	—			—			11:53:05	96	119	63	10:27:27	18	64	51	120	0.772	0.719		
Reno	39°31'N	119°48'W	09:04:35	294	345	31	—			—			11:43:09	100	129	57	10:20:05	17	62	45	116	0.860	0.830		
Sparks	39°32'N	119°45'W	09:04:40	294	345	31	—			—			11:43:17	100	129	57	10:20:11	17	62	45	116	0.861	0.830		
New Hampshire																									
Concord	43°12'N	071°32'W	13:27:02	275	261	58	—			—			15:57:15	132	88	39	14:44:51	203	170	50	227	0.689	0.616		
Manchester	43°00'N	071°28'W	13:27:16	275	261	58	—			—			15:57:45	132	88	39	14:45:15	203	170	50	228	0.693	0.620		
Nashua	42°45'N	071°28'W	13:27:24	275	262	58	—			—			15:58:19	131	87	39	14:45:39	203	169	50	228	0.698	0.627		
Portsmouth	43°05'N	070°45'W	13:28:42	274	260	57	—			—			15:58:05	132	88	38	14:46:09	203	169	49	229	0.683	0.609		
New Jersey																									
Ashland	39°52'N	075°01'W	13:21:37	281	274	62	—			—			16:01:38	126	79	42	14:44:41	204	170	54	226	0.800	0.753		
Atlantic City	39°21'N	074°27'W	13:23:12	282	273	62	—			—			16:03:16	126	78	41	14:46:24	204	168	54	228	0.805	0.760		
Bayonne	40°40'N	074°07'W	13:23:02	280	271	61	—			—			16:00:45	127	82	41	14:44:53	204	170	53	226	0.772	0.718		
Camden	39°56'N	075°07'W	13:21:21	281	274	62	—			—			16:01:24	126	80	42	14:44:25	204	170	54	225	0.799	0.753		
Cherry Hill	39°56'N	075°02'W	13:21:32	281	274	62	—			—			16:01:29	126	80	42	14:44:33	204	170	54	226	0.798	0.752		
Clifton	40°53'N	074°09'W	13:22:50	279	270	61	—			—			16:00:15	128	82	41	14:44:29	204	170	53	226	0.767	0.713		
East Orange	40°46'N	074°13'W	13:22:46	280	271	61	—			—			16:00:27	127	82	41	14:44:34	204	170	53	226	0.771	0.717		
Edison	40°31'N	074°25'W	13:22:29	280	272	61	—			—			16:00:48	127	81	41	14:44:38	204	170	53	226	0.778	0.727		
Elizabeth	40°40'N	074°13'W	13:22:49	280	271	61	—			—			16:00:40	127	82	41	14:44:43	204	170	53	226	0.773	0.720		
Irvington	40°44'N	074°14'W	13:22:45	280	271	61	—			—			16:00:30	127	82	41	14:44:36	204	170	53	226	0.772	0.718		
Jersey City	40°44'N	074°04'W	13:23:06	280	270	61	—			—			16:00:39	128	82	41	14:44:51	204	170	53	226	0.770	0.716		
Middletown	40°24'N	074°08'W	13:23:10	280	271	61	—			—			16:01:19	127	81	41	14:45:16	204	169	53	227	0.778	0.726		
Newark	40°44'N	074°10'W	13:22:53	280	271	61	—			—			16:00:34	127	82	41	14:44:42	204	170	53	226	0.771	0.717		
Passaic	40°51'N	074°07'W	13:22:55	279	270	61	—			—			16:00:21	128	82	41	14:44:36	204	170	53	226	0.768	0.713		
Paterson	40°55'N	074°11'W	13:22:44	279	270	61	—			—			16:00:09	128	82	41	14:44:23	204	170	53	226	0.767	0.712		
Trenton	40°14'N	074°46'W	13:21:55	281	273	61	—			—			16:01:05	127	80	42	14:44:31	204	170	54	226	0.789	0.739		
Vineland	39°29'N	075°02'W	13:21:50	282	274	62	—			—			16:02:26	126	78	42	14:45:15	204	169	54	226	0.808	0.764		
Woodbridge	40°34'N	074°17'W	13:22:44	280	271	61	—			—			16:00:49	127	81	41	14:44:46	204	170	53	226	0.776	0.723		
New Mexico																									
Albuquerque	35°05'N	106°39'W	10:21:19	301	353	45	—			—			13:13:58	99	97	67	11:45:14	20	57	60	135	0.783	0.733		
Clovis	34°24'N	103°12'W	10:27:23	302	353	49	—			—			13:22:48	100	86	67	11:53:27	21	51	63	144	0.788	0.740		
Farmington	36°44'N	108°12'W	10:18:33	299	349	43	—			—			13:09:29	101	104	65	11:41:22	20	56	57	133	0.823	0.783		
Las Cruces	32°19'N	106°47'W	10:22:39	307	2	47	—			—			13:14:37	94	92	70	11:46:11	21	61	62	132	0.698	0.627		
Rio Rancho	35°15'N	106°40'W	10:21:16	301	353	45	—			—			13:13:51	99	97	67	11:45:09	20	56	60	135	0.788	0.739		
Roswell	33°24'N	104°32'W	10:25:41	304	357	49	—			—			13:20:05	98	87	68	11:50:56	21	55	63	139	0.748	0.689		
Santa Fe	35°41'N	105°57'W	10:22:13	300	351	46	—			—			13:15:22	100	96	66	11:46:33	20	55	60	138	0.806	0.762		
South Valley	35°02'N	106°41'W	10:21:19	302	353	45	—			—			13:13:54	99	97	67	11:45:12	20	57	60	135	0.781	0.731		

Table 2–15: Local Circumstances for New York, North Dakota & Ohio

Location Name	Latitude	Longitude	First Contact Time h m s	P °	V °	Alt °	Second Contact Time h m s	P °	V °	Third Contact Time h m s	P °	V °	Fourth Contact Time h m s	P °	V °	Alt °	Maximum Eclipse Time h m s	P °	V °	Alt °	Azm °	Eclip. Mag.	Eclip. Obs.	Umbral Depth	Umbral Durat.
New York																									
Albany	42°39'N	073°45'W	13:22:42	277	268	59	—			—			15:56:43	130	87	41	14:42:24	203	172	52	224	0.724	0.659		
Amherst	42°59'N	078°48'W	13:12:03	279	281	59	—			—			15:51:15	127	87	45	14:33:56	203	179	55	213	0.771	0.717		
Binghamton	42°06'N	075°55'W	13:18:25	279	274	60	—			—			15:55:59	128	85	42	14:39:52	204	174	54	221	0.760	0.703		
Buffalo	42°53'N	078°53'W	13:11:54	279	281	59	—			—			15:51:23	127	87	45	14:33:56	203	179	55	213	0.774	0.721		
Cheektowaga	42°54'N	078°45'W	13:12:10	279	281	59	—			—			15:51:29	127	87	45	14:34:08	203	179	55	214	0.772	0.719		
Elmira	42°06'N	076°48'W	13:16:34	279	277	60	—			—			15:55:09	127	85	43	14:38:27	204	175	54	219	0.769	0.715		
Glens Falls	43°19'N	073°39'W	13:22:35	276	267	58	—			—			15:55:18	131	88	41	14:41:33	203	172	51	223	0.709	0.640		
Greece	43°13'N	077°41'W	13:14:16	278	278	59	—			—			15:51:53	128	88	44	14:35:26	203	177	54	216	0.754	0.696		
Ithaca	42°27'N	076°30'W	13:17:02	278	276	59	—			—			15:54:41	128	86	43	14:38:25	203	175	54	219	0.758	0.701		
Jamestown	42°06'N	079°14'W	13:11:27	280	284	60	—			—			15:52:40	126	85	45	14:34:26	203	178	56	214	0.796	0.748		
Kingston	41°56'N	073°59'W	13:22:35	278	269	60	—			—			15:58:06	129	85	41	14:43:08	204	171	52	225	0.743	0.682		
Middletown	41°27'N	074°25'W	13:21:56	279	271	60	—			—			15:58:47	128	84	41	14:43:12	204	171	53	225	0.758	0.701		
Mount Vernon	40°55'N	073°50'W	13:23:29	279	269	60	—			—			16:00:27	128	82	41	14:44:56	204	170	53	226	0.763	0.707		
New Rochelle	40°55'N	073°47'W	13:23:36	279	269	60	—			—			16:00:30	128	82	41	14:45:01	204	169	53	227	0.763	0.707		
New York	40°43'N	074°00'W	13:23:15	280	270	61	—			—			16:00:45	128	82	41	14:44:59	204	169	53	226	0.769	0.715		
Newburgh	41°30'N	074°01'W	13:22:45	278	270	60	—			—			15:59:01	128	84	41	14:43:45	204	171	53	225	0.752	0.694		
Niagara Falls	43°06'N	079°03'W	13:11:29	279	282	59	—			—			15:50:45	127	88	45	14:33:22	203	179	55	213	0.771	0.717		
Poughkeepsie	41°42'N	073°56'W	13:22:49	278	269	60	—			—			15:58:39	129	84	41	14:43:34	204	171	52	225	0.747	0.687		
Rochester	43°10'N	077°37'W	13:14:26	278	278	59	—			—			15:52:03	128	88	44	14:35:36	203	177	54	216	0.754	0.696		
Schenectady	42°49'N	073°57'W	13:22:12	276	268	59	—			—			15:56:10	130	87	41	14:41:51	203	172	52	223	0.723	0.658		
Syracuse	43°03'N	076°09'W	13:17:31	277	274	59	—			—			15:53:42	129	88	43	14:38:06	203	175	53	219	0.741	0.680		
Tonawanda	43°01'N	078°53'W	13:11:52	279	281	59	—			—			15:51:06	127	88	45	14:33:45	203	179	55	213	0.771	0.717		
Utica	43°06'N	075°14'W	13:19:24	277	271	59	—			—			15:54:26	129	88	42	14:39:28	203	174	53	221	0.730	0.667		
Yonkers	40°56'N	073°54'W	13:23:20	279	270	60	—			—			16:00:22	128	82	41	14:44:48	204	170	53	226	0.763	0.708		
North Dakota																									
Bismarck	46°48'N	100°47'W	11:32:32	280	313	45	—			—			14:14:49	119	111	55	12:52:42	200	215	53	158	0.869	0.841		
Fargo	46°53'N	096°48'W	11:38:46	279	308	48	—			—			14:20:49	122	107	53	12:59:24	201	210	54	167	0.839	0.803		
Grand Forks	47°55'N	097°03'W	11:38:58	278	306	47	—			—			14:18:54	123	110	53	12:58:30	200	210	53	167	0.814	0.771		
Minot	48°14'N	101°18'W	11:32:47	278	309	44	—			—			14:12:10	121	114	53	12:51:29	200	215	52	157	0.833	0.796		
Ohio																									
Akron	41°05'N	081°31'W	13:07:04	283	292	60	—			—			15:52:10	123	83	47	14:31:53	204	181	58	210	0.844	0.810		
Canton	40°48'N	081°23'W	13:07:27	284	292	61	—			—			15:52:54	123	82	47	14:32:30	204	180	58	210	0.849	0.816		
Cincinnati	39°06'N	084°31'W	13:01:36	288	304	61	—			—			15:52:10	119	78	50	14:29:01	204	184	61	205	0.925	0.913		
Cleveland	41°30'N	081°42'W	13:06:31	283	291	60	—			—			15:51:07	124	84	47	14:31:01	203	182	57	209	0.836	0.800		
Columbus	39°58'N	083°00'W	13:04:24	286	298	61	—			—			15:52:33	121	80	49	14:30:43	204	182	59	207	0.887	0.865		
Dayton	39°45'N	084°12'W	13:01:58	287	302	61	—			—			15:51:23	120	80	50	14:28:46	204	184	60	205	0.906	0.888		
Hamilton	39°24'N	084°34'W	13:01:21	287	304	61	—			—			15:51:33	119	79	50	14:28:32	204	185	61	204	0.918	0.904		
Kettering	39°41'N	084°10'W	13:02:04	287	302	61	—			—			15:51:33	120	79	50	14:28:55	204	184	60	205	0.907	0.890		
Lakewood	41°29'N	081°48'W	13:06:19	283	292	60	—			—			15:51:02	124	84	47	14:30:51	203	182	58	208	0.838	0.801		
Lima	40°44'N	084°06'W	13:01:47	285	299	60	—			—			15:49:38	121	82	49	14:27:42	203	185	59	204	0.880	0.856		
Lorain	41°28'N	082°11'W	13:05:31	283	293	60	—			—			15:50:36	123	84	48	14:30:12	203	183	58	208	0.842	0.807		
Mansfield	40°45'N	082°31'W	13:05:06	284	295	60	—			—			15:51:37	122	82	48	14:30:33	204	182	58	208	0.863	0.834		
Middletown	39°31'N	084°24'W	13:01:39	287	303	61	—			—			15:51:33	120	79	50	14:28:42	204	185	60	204	0.914	0.898		
Newark	40°03'N	082°24'W	13:05:38	285	296	61	—			—			15:53:09	122	80	48	14:31:42	204	181	59	209	0.879	0.854		
Parma	41°23'N	081°43'W	13:06:32	283	292	60	—			—			15:51:20	124	84	47	14:31:08	204	182	58	209	0.839	0.803		
Portsmouth	38°44'N	083°00'W	13:05:02	288	300	62	—			—			15:54:55	120	77	49	14:32:23	204	181	60	209	0.917	0.903		
Sandusky	41°27'N	082°42'W	13:04:27	283	294	60	—			—			15:50:01	123	84	48	14:29:19	203	183	58	206	0.848	0.815		
Springfield	39°55'N	083°49'W	13:02:42	286	300	61	—			—			15:51:35	121	80	49	14:29:16	204	184	60	205	0.897	0.878		
Steubenville	40°22'N	080°37'W	13:09:17	284	290	61	—			—			15:54:41	123	81	47	14:34:27	204	179	58	213	0.851	0.819		
Toledo	41°39'N	083°33'W	13:02:38	283	296	59	—			—			15:48:34	123	85	49	14:27:32	203	185	58	204	0.852	0.820		
Youngstown	41°06'N	080°39'W	13:08:53	283	289	61	—			—			15:53:09	124	83	46	14:33:23	204	180	57	212	0.834	0.797		

Table 2–16: Local Circumstances for North Carolina & Oklahoma

Location Name	Latitude	Longitude	First Contact Time	P	V	Alt	Second Contact Time	P	V	Third Contact Time	P	V	Fourth Contact Time	P	V	Alt	Maximum Eclipse Time	P	V	Alt	Azm	Eclip. Mag.	Eclip. Obs.	Umbral Depth	Umbral Durat.
			h m s	°	°	°	h m s	°	°	h m s	°	°	h m s	°	°	°	h m s	°	°	°	°				
North Carolina																									
Asheville	35°36'N	082°33'W	13:08:13	292	305	66	—			—			16:01:22	116	68	49	14:37:43	205	176	62	216	0.990	0.992		
Brevard	35°14'N	082°44'W	13:08:08	293	306	66	14:37:22	49	19	14:38:25	1	331	16:01:46	116	67	49	14:37:53	205	175	63	216	1.030	1.000	0.085	01m03s
Burlington	36°06'N	079°26'W	13:14:44	290	293	66	—			—			16:04:31	119	69	46	14:42:52	205	170	60	223	0.940	0.931		
Cary	35°47'N	078°46'W	13:16:31	290	291	66	—			—			16:05:56	119	68	45	14:44:36	205	169	59	225	0.939	0.930		
Charlotte	35°13'N	080°51'W	13:12:21	292	299	66	—			—			16:04:22	117	67	47	14:41:35	205	172	61	221	0.978	0.979		
Durham	36°00'N	078°54'W	13:16:01	290	291	66	—			—			16:05:21	119	69	45	14:44:01	205	169	59	224	0.935	0.926		
Fayetteville	35°03'N	078°53'W	13:16:56	291	292	67	—			—			16:07:11	118	66	45	14:45:32	205	168	60	226	0.958	0.954		
Franklin	35°11'N	083°23'W	13:06:44	294	309	66	14:35:25	97	69	14:37:55	313	284	16:00:55	115	66	50	14:36:41	205	177	63	214	1.030	1.000	0.691	02m30s
Gastonia	35°16'N	081°11'W	13:11:34	292	301	66	—			—			16:03:50	117	67	48	14:40:52	205	172	61	220	0.981	0.983		
Goldsboro	35°23'N	077°59'W	13:18:38	290	288	66	—			—			16:07:37	119	67	44	14:46:39	205	167	59	227	0.939	0.930		
Greensboro	36°04'N	079°48'W	13:13:57	290	294	66	—			—			16:04:07	119	69	46	14:42:15	205	171	60	222	0.945	0.938		
Greenville	35°37'N	077°23'W	13:19:46	289	286	66	—			—			16:07:52	120	68	44	14:47:21	205	166	58	228	0.926	0.914		
Hickory	35°44'N	081°21'W	13:10:46	292	300	66	—			—			16:02:45	117	68	48	14:39:50	205	173	61	219	0.972	0.971		
High Point	35°57'N	080°00'W	13:13:36	290	295	66	—			—			16:04:06	118	69	46	14:42:03	205	171	60	221	0.950	0.944		
Jacksonville	34°45'N	077°26'W	13:20:30	291	287	67	—			—			16:09:29	118	66	44	14:48:40	205	165	58	229	0.947	0.941		
Kannapolis	35°30'N	080°37'W	13:12:37	291	298	66	—			—			16:04:09	117	67	47	14:41:35	205	172	61	221	0.969	0.967		
Murphy	35°05'N	084°02'W	13:05:23	294	311	66	14:34:17	137	111	14:36:44	272	245	16:00:09	114	66	51	14:35:31	25	358	63	212	1.030	1.000	0.623	02m27s
Raleigh	35°46'N	078°38'W	13:16:49	290	290	66	—			—			16:06:07	119	68	45	14:44:52	205	169	59	225	0.938	0.929		
Rocky Mount	35°57'N	077°48'W	13:18:32	289	287	66	—			—			16:06:44	120	69	44	14:46:05	205	168	58	226	0.923	0.910		
Salisbury	35°40'N	080°29'W	13:12:46	291	297	66	—			—			16:04:01	118	68	47	14:41:35	205	172	61	221	0.963	0.960		
Sylva	35°23'N	083°13'W	13:06:56	293	308	66	14:35:53	64	36	14:37:33	345	317	16:00:49	115	67	50	14:36:43	205	177	63	214	1.030	1.000	0.226	01m40s
Wilmington	34°14'N	077°55'W	13:19:57	292	289	68	—			—			16:09:54	118	64	44	14:48:37	205	165	59	229	0.966	0.964		
Winston-Salem	36°06'N	080°15'W	13:12:55	290	296	66	—			—			16:03:30	118	69	47	14:41:22	205	172	60	221	0.949	0.944		
Oklahoma																									
Broken Arrow	36°03'N	095°48'W	11:40:05	297	338	56	—			—			14:37:47	108	77	61	13:09:04	23	31	66	170	0.904	0.886		
Edmond	35°39'N	097°29'W	11:37:04	299	342	55	—			—			14:34:45	106	78	62	13:05:39	23	36	65	164	0.876	0.850		
Enid	36°24'N	097°53'W	11:35:58	297	340	54	—			—			14:33:03	107	81	62	13:04:07	22	37	65	162	0.894	0.873		
Lawton	34°37'N	098°25'W	11:35:54	301	346	54	—			—			14:33:50	104	76	64	13:04:25	23	39	66	160	0.837	0.801		
Midwest City	35°27'N	097°24'W	11:37:20	299	342	55	—			—			14:35:09	106	77	62	13:06:01	23	36	66	164	0.871	0.844		
Moore	35°21'N	097°30'W	11:37:12	299	343	55	—			—			14:35:03	105	77	63	13:05:53	23	36	66	164	0.867	0.839		
Muskogee	35°45'N	095°22'W	11:41:05	298	338	57	—			—			14:39:03	108	75	61	13:10:20	23	30	66	171	0.900	0.881		
Norman	35°13'N	097°26'W	11:37:24	299	343	55	—			—			14:35:21	105	76	63	13:06:09	23	36	66	164	0.864	0.835		
Oklahoma City	35°30'N	097°30'W	11:37:07	299	342	55	—			—			14:34:53	106	77	63	13:05:45	23	36	66	164	0.872	0.845		
Stillwater	36°07'N	097°04'W	11:37:38	298	340	55	—			—			14:35:05	107	79	62	13:06:11	23	35	65	165	0.893	0.872		
Tulsa	36°10'N	095°55'W	11:39:48	297	338	56	—			—			14:37:24	108	77	61	13:08:42	23	32	65	169	0.906	0.889		

Table 2–17: Local Circumstances for Oregon

Location Name	Latitude	Longitude	First Contact Time h m s	P °	V °	Alt °	Second Contact Time h m s	P °	V °	Third Contact Time h m s	P °	V °	Fourth Contact Time h m s	P °	V °	Alt °	Maximum Eclipse Time h m s	P °	V °	Alt °	Azm °	Eclip. Mag.	Eclip. Obs.	Umbral Depth	Umbral Durat.
Oregon																									
Albany	44°38'N	123°06'W	09:05:06	285	330	28	10:17:06	128	169	10:18:57	263	304	11:37:42	107	135	51	10:18:01	16	56	40	117	1.027	1.000	0.616	01m51s
Aumsville	44°51'N	122°52'W	09:05:29	284	330	28	10:17:28	99	139	10:19:27	293	333	11:38:07	107	135	51	10:18:27	196	236	40	117	1.027	1.000	0.879	02m00s
Beaverton	45°29'N	122°48'W	09:06:08	283	328	28	—			—			11:38:16	108	135	51	10:18:57	196	235	40	118	0.993	0.995		
Corvallis	44°34'N	123°16'W	09:04:54	285	331	28	10:16:56	139	180	10:18:36	252	293	11:37:24	107	135	51	10:17:46	16	57	40	116	1.027	1.000	0.451	01m40s
Dallas	44°55'N	123°19'W	09:05:11	284	330	28	10:16:57	93	133	10:18:54	299	339	11:37:22	107	135	51	10:17:55	196	236	40	117	1.027	1.000	0.774	01m57s
Dayton	45°13'N	123°07'W	09:05:38	284	329	28	10:17:58	40	80	10:18:47	351	31	11:37:43	108	135	51	10:18:22	196	236	40	117	1.027	1.000	0.089	00m49s
Eugene	44°05'N	123°04'W	09:04:38	286	332	28	—			—			11:37:41	106	135	52	10:17:43	16	57	40	116	0.992	0.994		
Gresham	45°30'N	122°26'W	09:06:27	283	328	28	—			—			11:38:53	109	135	51	10:19:25	196	235	40	118	0.992	0.994		
Hillsboro	45°31'N	122°59'W	09:06:01	283	328	28	—			—			11:37:58	108	136	51	10:18:44	196	235	40	118	0.992	0.994		
Hubbard	45°11'N	122°48'W	09:05:51	284	329	28	10:18:15	46	86	10:19:15	346	26	11:38:15	108	135	51	10:18:45	196	236	40	118	1.027	1.000	0.133	01m00s
Independence	44°51'N	123°11'W	09:05:14	284	330	28	10:17:03	101	141	10:19:03	291	331	11:37:35	107	135	51	10:18:03	196	236	40	117	1.027	1.000	0.910	01m59s
Jefferson	44°43'N	123°01'W	09:05:14	285	330	28	10:17:12	117	157	10:19:10	275	315	11:37:51	107	135	51	10:18:11	16	56	40	117	1.027	1.000	0.810	01m58s
John Day	44°25'N	118°57'W	09:08:30	285	330	31	10:22:31	124	162	10:24:32	269	308	11:44:58	108	131	54	10:23:32	17	55	44	122	1.027	1.000	0.705	02m01s
Keizer	45°00'N	123°01'W	09:05:30	284	329	28	10:17:28	80	120	10:19:16	312	352	11:37:52	108	135	51	10:18:21	196	236	40	117	1.027	1.000	0.561	01m48s
Lebanon	44°32'N	122°55'W	09:05:09	285	331	28	10:17:23	142	183	10:19:00	250	290	11:38:00	107	135	51	10:18:12	16	57	40	117	1.027	1.000	0.413	01m38s
Lincoln City	44°57'N	124°01'W	09:04:41	284	330	27	10:16:06	92	133	10:18:02	299	340	11:36:11	107	136	50	10:17:04	196	236	39	116	1.026	1.000	0.761	01m55s
Madras	44°38'N	121°08'W	09:06:43	285	330	29	10:19:36	115	155	10:21:38	277	317	11:41:05	108	133	52	10:20:37	16	56	42	119	1.027	1.000	0.845	02m02s
McMinnville	45°13'N	123°12'W	09:05:34	284	329	28	10:17:50	41	81	10:18:42	350	30	11:37:35	108	136	51	10:18:16	196	236	40	117	1.027	1.000	0.096	00m51s
Medford	42°19'N	122°52'W	09:03:24	289	337	28	—			—			11:37:45	103	134	53	10:16:56	16	59	41	115	0.939	0.930		
Mill City	44°45'N	122°29'W	09:05:42	285	330	28	10:17:53	109	150	10:19:54	282	323	11:38:46	107	135	51	10:18:53	16	56	40	118	1.027	1.000	0.940	02m01s
Molalla	45°09'N	122°35'W	09:05:59	284	329	28	10:18:26	51	91	10:19:35	340	20	11:38:37	108	135	51	10:19:00	196	236	40	118	1.027	1.000	0.185	01m10s
Monmouth	44°51'N	123°14'W	09:05:11	284	330	28	10:17:00	101	141	10:18:59	291	331	11:37:30	107	135	51	10:17:59	196	236	40	117	1.027	1.000	0.915	01m59s
Mount Angel	45°04'N	122°48'W	09:05:44	284	329	28	10:17:53	68	108	10:19:28	323	3	11:38:15	108	135	51	10:18:41	196	236	40	118	1.027	1.000	0.392	01m36s
Newport	44°39'N	124°03'W	09:04:22	285	331	27	10:15:56	131	173	10:17:43	260	301	11:36:06	107	136	50	10:16:49	16	57	39	116	1.026	1.000	0.565	01m47s
Ontario	44°02'N	116°58'W	10:10:07	286	331	33	11:25:32	155	192	11:27:00	239	276	12:48:38	108	128	55	11:26:16	17	54	45	125	1.028	1.000	0.259	01m28s
Philomath	44°32'N	123°22'W	09:04:47	285	331	28	10:16:50	145	186	10:18:23	246	287	11:37:14	107	135	51	10:17:37	16	57	40	116	1.027	1.000	0.367	01m33s
Portland	45°32'N	122°37'W	09:06:20	283	328	28	—			—			11:38:35	109	135	51	10:19:12	196	235	40	118	0.991	0.993		
Prineville	44°18'N	120°51'W	09:06:40	285	331	30	10:20:16	164	204	10:21:22	228	268	11:41:35	107	133	53	10:20:49	16	56	42	119	1.027	1.000	0.153	01m06s
Redmond	44°17'N	121°11'W	09:06:22	285	331	29	10:19:58	175	215	10:20:43	218	258	11:41:00	107	133	53	10:20:21	16	56	42	119	1.027	1.000	0.070	00m45s
Salem	44°56'N	123°02'W	09:05:25	284	330	28	10:17:20	89	129	10:19:15	303	343	11:37:50	107	135	51	10:18:18	196	236	40	117	1.027	1.000	0.710	01m55s
Sheridan	45°06'N	123°24'W	09:05:18	284	329	28	10:17:10	67	107	10:18:43	324	5	11:37:14	108	136	51	10:17:56	196	236	40	117	1.026	1.000	0.374	01m33s
Silverton	45°01'N	122°47'W	09:05:42	284	329	28	10:17:48	76	116	10:19:32	316	356	11:38:16	108	135	51	10:18:40	196	236	40	118	1.027	1.000	0.501	01m44s
Springfield	44°03'N	123°01'W	09:04:38	286	332	28	—			—			11:37:46	106	135	52	10:17:46	16	57	40	116	0.991	0.993		
Stayton	44°48'N	122°48'W	09:05:29	284	330	28	10:17:30	105	145	10:19:31	287	327	11:38:13	107	135	51	10:18:31	196	236	40	117	1.027	1.000	0.983	02m01s
Sublimity	44°50'N	122°47'W	09:05:32	284	330	28	10:17:33	100	141	10:19:33	291	331	11:38:15	107	135	51	10:18:33	196	236	40	117	1.027	1.000	0.907	02m00s
Sweet Home	44°24'N	122°44'W	09:05:11	285	331	28	10:17:50	166	206	10:18:51	226	267	11:38:18	107	135	52	10:18:21	16	57	40	117	1.027	1.000	0.136	01m01s
Tigard	45°26'N	122°46'W	09:06:07	283	328	28	—			—			11:38:19	108	135	51	10:18:57	196	235	40	118	0.994	0.996		
Toledo	44°37'N	123°56'W	09:04:26	285	331	27	10:16:05	136	177	10:17:49	256	297	11:36:17	107	136	51	10:16:57	16	57	39	116	1.026	1.000	0.501	01m43s
Vale	43°59'N	117°15'W	10:09:48	286	331	33	11:25:22	172	210	11:26:17	222	260	12:48:07	108	128	55	11:25:49	17	55	45	124	1.028	1.000	0.093	00m55s
Waldport	44°26'N	124°04'W	09:04:09	285	331	27	10:16:17	173	214	10:17:03	218	260	11:36:02	106	136	51	10:16:40	16	57	39	115	1.026	1.000	0.079	00m46s
Willamina	45°05'N	123°29'W	09:05:13	284	329	27	10:17:01	70	111	10:18:38	321	1	11:37:06	108	136	51	10:17:49	196	236	40	117	1.026	1.000	0.418	01m37s
Woodburn	45°09'N	122°51'W	09:05:46	284	329	28	10:18:03	54	94	10:19:17	338	18	11:38:10	108	135	51	10:18:40	196	236	40	118	1.027	1.000	0.212	01m14s

Table 2–18: Local Circumstances for Penn., Rhode Island, South Dakota & Utah

Location Name	Latitude	Longitude	First Contact Time h m s	P °	V °	Alt °	Second Contact Time h m s	P °	V °	Third Contact Time h m s	P °	V °	Fourth Contact Time h m s	P °	V °	Alt °	Maximum Eclipse Time h m s	P °	V °	Alt °	Azm °	Eclip. Mag.	Eclip. Obs.	Umbral Depth	Umbral Durat.
Pennsylvania																									
Abington	40°07'N	075°07'W	13:21:14	281	274	61	—			—			16:01:01	126	80	42	14:44:08	204	170	54	225	0.795	0.748		
Allentown	40°37'N	075°29'W	13:20:08	281	275	61	—			—			15:59:36	127	81	42	14:42:47	204	171	54	224	0.788	0.739		
Altoona	40°31'N	078°24'W	13:13:56	282	283	61	—			—			15:56:52	125	81	45	14:38:05	204	175	56	218	0.823	0.783		
Bethlehem	40°37'N	075°23'W	13:20:21	281	274	61	—			—			15:59:42	127	81	42	14:42:56	204	171	54	224	0.787	0.737		
Butler	40°52'N	079°54'W	13:10:34	283	287	61	—			—			15:54:29	124	82	46	14:35:01	204	178	57	214	0.831	0.794		
Coatesville	39°59'N	075°49'W	13:19:48	282	276	62	—			—			16:00:37	126	80	42	14:43:12	204	171	55	224	0.806	0.761		
Erie	42°08'N	080°05'W	13:09:40	281	286	60	—			—			15:51:40	125	85	46	14:32:57	203	180	56	212	0.804	0.759		
Harrisburg	40°16'N	076°53'W	13:17:20	282	279	62	—			—			15:58:58	125	80	43	14:41:01	204	173	55	221	0.812	0.769		
Hazleton	40°57'N	075°59'W	13:18:53	280	276	61	—			—			15:58:24	127	82	43	14:41:28	204	172	54	222	0.786	0.736		
Johnstown	40°19'N	078°55'W	13:12:56	283	285	62	—			—			15:56:43	124	80	45	14:37:30	204	176	57	217	0.833	0.796		
Lancaster	40°02'N	076°19'W	13:18:42	282	278	62	—			—			16:00:02	125	80	43	14:42:18	204	172	55	223	0.811	0.767		
Lebanon	40°20'N	076°26'W	13:18:16	282	278	61	—			—			15:59:17	126	81	43	14:41:39	204	172	55	222	0.805	0.760		
New Castle	41°00'N	080°21'W	13:09:33	283	288	61	—			—			15:53:42	124	82	46	14:34:03	204	179	57	213	0.833	0.796		
Philadelphia	39°57'N	075°10'W	13:21:14	281	274	62	—			—			16:01:19	126	80	42	14:44:19	204	170	54	225	0.799	0.753		
Pittsburgh	40°26'N	080°01'W	13:10:32	283	288	61	—			—			15:55:14	124	81	46	14:35:25	204	178	57	214	0.843	0.808		
Pottstown	40°15'N	075°39'W	13:20:00	281	275	61	—			—			16:00:13	126	80	42	14:43:04	204	171	54	224	0.798	0.752		
Reading	40°20'N	075°56'W	13:19:20	281	276	61	—			—			15:59:46	126	81	43	14:42:29	204	171	55	223	0.800	0.753		
Scranton	41°25'N	075°40'W	13:19:18	280	274	60	—			—			15:57:42	127	83	42	14:41:16	204	173	54	222	0.772	0.719		
Sharon	41°14'N	080°31'W	13:09:06	282	289	60	—			—			15:53:02	124	83	46	14:33:26	204	179	57	212	0.830	0.791		
State College	40°48'N	077°52'W	13:14:56	282	282	61	—			—			15:56:50	125	82	44	14:38:35	204	175	56	219	0.811	0.767		
Upper Darby	39°57'N	075°17'W	13:20:59	282	275	62	—			—			16:01:13	126	80	42	14:44:07	204	170	54	225	0.801	0.755		
Washington	40°11'N	080°15'W	13:10:09	284	290	62	—			—			15:55:29	123	80	46	14:35:22	204	178	58	214	0.851	0.819		
Williamsport	41°15'N	077°00'W	13:16:33	281	278	61	—			—			15:56:46	126	83	43	14:39:21	204	174	55	220	0.791	0.742		
York	39°58'N	076°44'W	13:17:50	282	279	62	—			—			15:59:45	125	80	43	14:41:43	204	172	55	222	0.817	0.775		
Rhode Island																									
Cranston	41°47'N	071°26'W	13:28:02	276	262	59	—			—			16:00:31	130	85	39	14:47:13	204	168	50	229	0.718	0.651		
Newport	41°29'N	071°19'W	13:28:28	277	262	59	—			—			16:01:17	130	84	39	14:47:52	204	167	51	230	0.723	0.658		
Pawtucket	41°53'N	071°23'W	13:28:05	276	262	59	—			—			16:00:20	130	85	39	14:47:08	204	168	50	229	0.715	0.648		
Providence	41°49'N	071°24'W	13:28:05	276	262	59	—			—			16:00:28	130	85	39	14:47:13	204	168	50	229	0.717	0.650		
Warwick	41°42'N	071°28'W	13:28:01	277	262	59	—			—			16:00:41	130	85	39	14:47:18	204	168	51	230	0.720	0.654		
South Dakota																									
Rapid City	44°05'N	103°14'W	10:27:20	285	323	45	—			—			13:13:56	114	108	58	11:49:12	200	221	55	151	0.960	0.956		
Sioux Falls	43°33'N	096°44'W	11:37:32	285	317	50	—			—			14:25:50	117	99	56	13:01:20	201	211	58	167	0.928	0.916		
Utah																									
Ogden	41°13'N	111°58'W	10:14:03	291	337	38	—			—			12:59:15	106	118	60	11:33:44	18	54	51	130	0.938	0.929		
Orem	40°19'N	111°42'W	10:14:04	292	340	39	—			—			13:00:10	104	116	61	11:34:10	18	55	52	130	0.912	0.896		
Provo	40°14'N	111°39'W	10:14:05	293	340	39	—			—			13:00:18	104	116	61	11:34:14	18	55	52	130	0.910	0.894		
Salt Lake City	40°45'N	111°53'W	10:13:58	292	339	38	—			—			12:59:36	105	117	60	11:33:51	18	55	51	130	0.925	0.912		
Sandy City	40°36'N	111°36'W	10:14:18	292	339	39	—			—			13:00:16	105	116	61	11:34:22	18	55	52	130	0.921	0.908		
St. George	37°06'N	113°35'W	10:11:06	298	350	38	—			—			12:56:43	98	117	63	11:30:28	18	61	52	123	0.806	0.761		
Taylorsville	40°40'N	111°56'W	10:13:54	292	339	38	—			—			12:59:33	105	117	60	11:33:46	18	55	51	130	0.922	0.908		
West Jordan	40°36'N	111°59'W	10:13:49	292	339	38	—			—			12:59:28	105	117	60	11:33:40	18	55	51	130	0.920	0.906		
West Valley	40°41'N	111°59'W	10:13:50	292	339	38	—			—			12:59:26	105	117	60	11:33:41	18	55	51	130	0.922	0.909		

Table 2–19: Local Circumstances for South Carolina

Location Name	Latitude	Longitude	First Contact Time (h m s)	P (°)	V (°)	Alt (°)	Second Contact Time (h m s)	P (°)	V (°)	Third Contact Time (h m s)	P (°)	V (°)	Fourth Contact Time (h m s)	P (°)	V (°)	Alt (°)	Maximum Eclipse Time (h m s)	P (°)	V (°)	Alt (°)	Azm (°)	Eclip. Mag.	Eclip. Obs.	Umbral Depth	Umbral Durat.
South Carolina																									
Anderson	34°31'N	082°39'W	13:08:59	294	307	67	14:37:51	127	96	14:40:25	283	252	16:03:10	115	65	49	14:39:08	25	354	63	217	1.030	1.000	0.797	02m34s
Andrews	33°27'N	079°34'W	13:17:03	294	297	68	14:45:47	77	39	14:47:49	333	294	16:09:17	115	62	46	14:46:48	205	166	61	227	1.030	1.000	0.386	02m01s
Bamberg	33°18'N	081°02'W	13:13:53	295	303	68	14:43:34	176	140	14:44:50	234	197	16:07:36	114	61	48	14:44:12	25	349	62	224	1.030	1.000	0.126	01m16s
Belton	34°31'N	082°30'W	13:09:19	294	307	67	14:38:07	120	89	14:40:44	290	259	16:03:23	115	65	49	14:39:26	25	354	63	218	1.030	1.000	0.919	02m37s
Cayce	33°59'N	081°04'W	13:13:04	294	302	68	14:41:48	102	67	14:44:20	308	272	16:06:20	115	63	48	14:43:04	205	170	62	222	1.030	1.000	0.777	02m32s
Charleston	32°46'N	079°56'W	13:16:59	295	299	69	14:46:28	170	131	14:47:56	240	200	16:10:02	114	60	46	14:47:12	25	346	62	228	1.030	1.000	0.180	01m29s
Clemson	34°41'N	082°50'W	13:08:25	294	308	66	14:37:12	119	90	14:39:49	290	260	16:02:36	115	65	49	14:38:31	25	355	63	216	1.030	1.000	0.923	02m37s
Clinton	34°29'N	081°53'W	13:10:44	294	304	67	14:39:29	93	61	14:41:54	317	284	16:04:18	115	64	48	14:40:42	205	172	62	219	1.030	1.000	0.630	02m25s
Columbia	34°00'N	081°03'W	13:13:06	294	302	68	14:41:49	100	65	14:44:19	310	275	16:06:19	115	63	48	14:43:04	205	170	62	222	1.030	1.000	0.735	02m30s
Denmark	33°19'N	081°09'W	13:13:36	295	303	68	14:43:31	186	150	14:44:23	224	188	16:07:24	114	61	48	14:43:57	25	349	62	224	1.030	1.000	0.055	00m51s
Easley	34°50'N	082°36'W	13:08:47	294	306	66	14:37:32	94	64	14:39:58	316	285	16:02:40	115	65	49	14:38:45	205	174	63	217	1.030	1.000	0.639	02m27s
Edgefield	33°47'N	081°56'W	13:11:19	295	306	68	14:40:56	171	137	14:42:25	239	205	16:05:29	114	62	49	14:41:40	25	351	63	221	1.030	1.000	0.173	01m28s
Florence	34°12'N	079°46'W	13:15:47	293	296	68	—			—			16:07:40	116	64	46	14:45:14	205	168	61	225	0.990	0.992		
Forest Acres	34°02'N	080°59'W	13:13:13	294	302	68	14:41:57	93	58	14:44:21	317	281	16:06:21	115	63	48	14:43:09	205	170	62	223	1.030	1.000	0.625	02m24s
Fountain Inn	34°42'N	082°12'W	13:09:49	294	305	67	14:38:35	87	56	14:40:54	323	291	16:03:29	115	65	49	14:39:45	205	173	62	218	1.030	1.000	0.533	02m19s
Georgetown	33°23'N	079°17'W	13:17:46	294	295	68	14:46:36	67	28	14:48:18	344	304	16:09:46	116	62	46	14:47:27	205	166	61	228	1.030	1.000	0.253	01m42s
Goose Creek	32°59'N	080°02'W	13:16:30	295	299	69	14:45:35	149	110	14:47:44	262	222	16:09:31	115	60	47	14:46:39	25	346	62	227	1.030	1.000	0.451	02m10s
Greenville	34°51'N	082°24'W	13:09:13	294	306	66	14:38:02	81	51	14:40:13	329	297	16:02:56	115	65	49	14:39:07	205	174	63	217	1.030	1.000	0.445	02m11s
Greenwood	34°12'N	082°10'W	13:10:22	294	306	67	14:39:20	134	102	14:41:48	276	243	16:04:25	115	64	49	14:40:34	25	352	63	219	1.030	1.000	0.669	02m28s
Greer	34°56'N	082°14'W	13:09:31	293	305	66	14:38:35	59	28	14:40:04	351	319	16:03:01	116	66	49	14:39:20	205	174	62	218	1.030	1.000	0.174	01m28s
Hanahan	32°55'N	080°00'W	13:16:40	295	299	69	14:45:49	154	115	14:47:50	256	217	16:09:40	115	60	46	14:46:50	25	346	62	227	1.030	1.000	0.369	02m00s
Honea Path	34°27'N	082°24'W	13:09:36	294	306	67	14:38:25	121	90	14:41:02	289	257	16:03:38	115	64	49	14:39:44	25	353	63	218	1.030	1.000	0.891	02m36s
Irmo	34°05'N	081°11'W	13:12:42	294	302	68	14:41:26	98	64	14:43:55	312	277	16:05:59	115	63	48	14:42:41	205	170	62	222	1.030	1.000	0.709	02m29s
Johnston	33°50'N	081°48'W	13:11:34	295	305	68	14:40:51	154	121	14:42:52	256	221	16:05:35	114	63	48	14:41:52	25	351	63	221	1.030	1.000	0.366	02m01s
Kingstree	33°40'N	079°50'W	13:16:12	294	297	68	14:45:05	67	29	14:46:48	343	305	16:08:33	116	62	46	14:45:57	205	167	61	226	1.030	1.000	0.258	01m43s
Laurens	34°30'N	082°01'W	13:10:25	294	305	67	14:39:09	98	66	14:41:39	312	279	16:04:05	115	64	49	14:40:25	205	173	62	219	1.030	1.000	0.712	02m30s
Lexington	34°00'N	081°14'W	13:12:41	294	303	68	14:41:25	109	74	14:44:00	301	266	16:06:04	115	63	48	14:42:43	205	170	62	222	1.030	1.000	0.888	02m35s
Liberty	34°48'N	082°42'W	13:08:36	294	307	66	14:37:19	102	72	14:39:53	308	277	16:02:35	115	65	49	14:38:36	205	175	63	217	1.030	1.000	0.775	02m33s
Manning	33°42'N	080°13'W	13:15:18	294	299	68	14:44:01	87	50	14:46:18	323	286	16:07:59	115	62	47	14:45:10	205	168	61	225	1.030	1.000	0.528	02m16s
Mauldin	34°47'N	082°19'W	13:09:28	294	306	67	14:38:16	84	54	14:40:31	326	294	16:03:10	115	65	49	14:39:23	205	174	62	218	1.030	1.000	0.490	02m15s
Moncks Corner	33°12'N	080°01'W	13:16:18	295	299	69	14:45:04	126	87	14:47:37	285	246	16:09:09	115	61	47	14:46:21	25	347	61	227	1.030	1.000	0.821	02m32s
Mount Pleasant	32°47'N	079°52'W	13:17:07	295	299	69	14:46:26	163	123	14:48:10	248	208	16:10:05	114	60	46	14:47:18	25	346	61	228	1.030	1.000	0.264	01m45s
Newberry	34°17'N	081°37'W	13:11:32	294	304	67	14:40:16	100	67	14:42:47	310	276	16:05:02	115	64	48	14:41:32	205	171	62	220	1.030	1.000	0.740	02m31s
North Charleston	32°53'N	080°00'W	13:16:42	295	299	69	14:45:56	158	119	14:47:49	252	212	16:09:44	115	60	46	14:46:53	25	346	62	227	1.030	1.000	0.315	01m53s
Orangeburg	33°30'N	080°52'W	13:14:02	295	302	68	14:43:02	139	103	14:45:24	271	234	16:07:28	115	62	47	14:44:13	25	349	62	224	1.030	1.000	0.595	02m23s
Pendleton	34°39'N	082°47'W	13:08:33	294	307	67	14:37:21	120	90	14:39:58	290	259	16:02:44	115	65	49	14:38:40	25	354	63	217	1.030	1.000	0.909	02m37s
Pickens	34°53'N	082°42'W	13:08:31	294	307	66	14:37:15	94	64	14:39:42	316	286	16:02:26	115	66	49	14:38:29	205	175	63	216	1.030	1.000	0.637	02m27s
Rock Hill	34°56'N	081°01'W	13:12:15	293	300	67	—			—			16:04:40	116	66	48	14:41:41	205	171	61	221	0.988	0.990		
Saluda	34°00'N	081°46'W	13:11:28	295	305	67	14:40:26	134	101	14:42:54	276	241	16:05:20	115	63	48	14:41:40	25	351	63	221	1.030	1.000	0.669	02m28s
Seneca	34°41'N	082°57'W	13:08:09	294	308	66	14:36:59	125	95	14:39:35	285	255	16:02:26	115	65	49	14:38:17	25	355	63	216	1.030	1.000	0.829	02m36s
Simpsonville	34°44'N	082°15'W	13:09:40	294	305	67	14:38:27	86	55	14:40:44	324	292	16:03:21	115	65	49	14:39:36	205	174	62	218	1.030	1.000	0.518	02m17s
South Congaree	33°54'N	081°08'W	13:13:00	294	302	68	14:41:46	113	79	14:44:22	297	261	16:06:23	115	63	48	14:43:04	205	170	62	222	1.030	1.000	0.970	02m36s
Spartanburg	34°56'N	081°57'W	13:10:09	293	304	66	—			—			16:03:24	116	66	48	14:39:53	205	173	62	218	0.999	1.000		
Summerville	33°01'N	080°11'W	13:16:08	295	300	69	14:45:18	154	115	14:47:19	257	218	16:09:15	115	60	47	14:46:19	25	347	62	227	1.030	1.000	0.378	02m02s
Sumter	33°55'N	080°21'W	13:14:46	294	299	68	14:43:39	69	33	14:45:28	341	304	16:07:25	116	63	47	14:44:34	205	168	61	224	1.030	1.000	0.285	01m48s
Travelers Rest	34°58'N	082°27'W	13:09:00	293	306	66	14:37:55	70	40	14:39:46	340	309	16:02:39	115	66	49	14:38:51	205	174	62	217	1.030	1.000	0.296	01m51s
Union	34°43'N	081°37'W	13:11:06	293	303	67	14:40:34	38	5	14:41:09	12	339	16:04:15	116	65	48	14:40:52	205	172	62	220	1.030	1.000	0.026	00m35s
Walhalla	34°46'N	083°04'W	13:07:49	294	308	66	14:36:37	122	93	14:39:14	287	257	16:02:07	115	65	50	14:37:55	25	355	63	216	1.030	1.000	0.872	02m37s
West Columbia	34°00'N	081°05'W	13:13:01	294	302	68	14:41:45	101	67	14:44:16	309	273	16:06:17	115	63	48	14:43:01	205	170	62	222	1.030	1.000	0.763	02m31s
Westminster	34°40'N	083°06'W	13:07:50	294	309	66	14:36:45	133	104	14:39:15	276	246	16:02:15	115	65	50	14:38:00	25	355	63	216	1.030	1.000	0.680	02m30s
Williamston	34°37'N	082°29'W	13:09:15	294	306	67	14:38:00	109	79	14:40:37	300	269	16:03:13	115	65	49	14:39:19	205	174	63	218	1.030	1.000	0.902	02m36s
Winnsboro	34°23'N	081°05'W	13:12:38	294	301	67	14:41:46	55	21	14:43:03	355	321	16:05:35	116	64	48	14:42:25	205	171	62	222	1.030	1.000	0.132	01m17s
Woodruff	34°45'N	082°02'W	13:10:08	293	305	67	14:39:03	72	40	14:40:56	338	306	16:03:37	115	65	49	14:40:00	205	173	62	219	1.030	1.000	0.313	01m54s

Table 2–20: Local Circumstances for Tennessee

Location Name	Latitude	Longitude	First Contact Time h m s	P °	V °	Alt °	Second Contact Time h m s	P °	V °	Third Contact Time h m s	P °	V °	Fourth Contact Time h m s	P °	V °	Alt °	Maximum Eclipse Time h m s	P °	V °	Alt °	Azm °	Eclip. Mag.	Eclip. Obs.	Umbral Depth	Umbral Durat.
Tennessee																									
Alcoa	35°48'N	083°59'W	13:04:53	293	309	65	14:33:54	56	30	14:35:16	353	328	15:58:57	115	68	50	14:34:35	205	179	63	211	1.030	1.000	0.144	01m22s
Algood	36°12'N	085°27'W	12:01:22	293	313	64	13:29:51	92	71	13:32:18	317	295	14:56:03	115	70	52	13:31:04	204	183	64	206	1.030	1.000	0.620	02m27s
Ashland City	36°17'N	087°04'W	11:57:49	294	318	63	13:26:43	159	143	13:28:36	249	231	14:53:23	114	71	53	13:27:39	24	7	64	201	1.031	1.000	0.290	01m53s
Athens	35°27'N	084°36'W	13:03:49	294	312	65	14:32:33	127	103	14:35:08	282	257	15:58:39	115	67	51	14:33:51	25	360	64	210	1.030	1.000	0.786	02m35s
Brentwood	36°00'N	086°48'W	11:58:36	294	318	63	13:28:23	196	177	13:28:47	213	194	14:54:17	114	70	53	13:28:35	24	6	64	202	1.031	1.000	0.011	00m24s
Bristol	36°36'N	082°11'W	13:08:13	291	302	65	—			—			16:00:02	118	70	49	14:36:57	204	176	61	215	0.961	0.957		
Carthage	36°15'N	085°57'W	12:00:15	293	315	63	13:28:40	110	90	13:31:19	299	278	14:55:12	115	70	52	13:29:59	204	184	64	204	1.031	1.000	0.920	02m39s
Chattanooga	35°03'N	085°19'W	13:02:34	295	315	65	—			—			15:58:15	114	66	52	14:32:57	25	1	64	208	0.993	0.995		
Clarksville	36°32'N	087°21'W	11:57:03	293	318	62	13:25:35	144	128	13:27:54	264	248	14:52:31	114	71	53	13:26:45	24	8	64	200	1.031	1.000	0.502	02m19s
Cleveland	35°10'N	084°53'W	13:03:26	294	314	65	14:33:07	180	156	14:34:13	229	204	15:58:43	114	67	51	14:33:40	25	0	64	210	1.031	1.000	0.089	01m06s
Cookeville	36°10'N	085°30'W	12:01:17	293	313	64	13:29:45	98	77	13:32:17	311	289	14:56:02	115	70	52	13:31:01	204	183	64	206	1.031	1.000	0.715	02m33s
Crossville	35°57'N	085°02'W	12:02:28	293	313	64	13:30:59	98	76	13:33:32	311	287	14:57:08	115	69	51	13:32:16	204	182	63	208	1.030	1.000	0.717	02m32s
Dayton	35°30'N	085°01'W	13:02:51	294	313	65	14:31:45	142	119	14:34:06	267	243	15:57:56	114	68	51	14:32:56	25	1	64	209	1.031	1.000	0.543	02m22s
Dunlap	35°23'N	085°23'W	12:02:09	294	315	65	13:31:53	184	162	13:32:48	225	201	14:57:34	114	67	52	13:32:21	24	2	64	208	1.031	1.000	0.061	00m55s
Etowah	35°20'N	084°32'W	13:04:03	294	312	65	14:32:54	135	111	14:35:23	274	248	15:58:57	114	67	51	14:34:09	25	359	64	210	1.030	1.000	0.643	02m29s
Gallatin	36°24'N	086°27'W	12:59:04	293	316	63	13:27:26	117	99	13:30:06	291	272	14:54:10	115	71	53	13:28:46	24	6	64	203	1.031	1.000	0.953	02m40s
Goodlettsville	36°19'N	086°43'W	12:58:33	293	317	63	13:27:06	137	120	13:29:34	271	253	14:53:53	114	71	53	13:28:20	24	6	64	202	1.031	1.000	0.610	02m27s
Harriman	35°56'N	084°33'W	13:03:32	293	311	64	14:32:14	75	52	14:34:17	334	309	15:57:53	115	69	51	14:33:16	204	180	63	209	1.030	1.000	0.367	02m03s
Hendersonville	36°18'N	086°37'W	12:58:46	293	317	63	13:27:18	134	116	13:29:49	274	255	14:54:04	114	70	53	13:28:34	24	6	64	202	1.031	1.000	0.657	02m30s
Jackson	35°37'N	088°49'W	12:54:34	295	324	62	—			—			14:51:34	112	69	55	13:24:51	24	11	66	196	0.967	0.965		
Johnson City	36°19'N	082°21'W	13:08:04	291	303	65	—			—			16:00:20	117	70	49	14:37:02	205	176	62	215	0.970	0.969		
Kingsport	36°33'N	082°33'W	13:07:27	291	303	65	—			—			15:59:37	117	70	49	14:36:19	204	177	62	214	0.966	0.964		
Kingston	35°52'N	084°31'W	13:03:39	293	311	65	14:32:19	81	58	14:34:31	328	303	15:58:03	115	69	51	14:33:25	204	180	63	209	1.030	1.000	0.453	02m13s
Knoxville	35°58'N	083°55'W	13:04:54	293	309	65	—			—			15:58:45	116	69	50	14:34:29	204	179	63	211	0.997	0.999		
La Vergne	36°01'N	086°35'W	11:59:03	294	317	63	13:28:12	167	148	13:29:49	241	222	14:54:36	114	70	53	13:29:00	24	5	64	203	1.031	1.000	0.204	01m37s
Lafayette	36°31'N	086°02'W	11:59:53	293	314	63	13:28:17	86	67	13:30:38	322	302	14:54:37	115	71	52	13:29:28	204	185	64	204	1.031	1.000	0.534	02m21s
Lebanon	36°12'N	086°18'W	11:59:32	293	316	63	13:28:03	130	111	13:30:37	278	259	14:54:45	114	70	53	13:29:21	24	5	64	203	1.031	1.000	0.729	02m34s
Lenoir City	35°48'N	084°16'W	13:04:16	293	310	65	14:33:00	75	51	14:35:02	334	308	15:58:32	115	68	51	14:34:01	205	180	63	210	1.030	1.000	0.367	02m03s
Livingston	36°23'N	085°19'W	12:01:31	293	312	64	13:30:17	62	41	13:31:54	347	325	14:55:56	115	70	52	13:31:05	204	183	63	206	1.030	1.000	0.210	01m37s
Loudon	35°45'N	084°20'W	13:04:09	293	311	65	14:32:48	85	60	14:35:06	324	299	15:58:32	115	68	51	14:33:57	205	180	63	210	1.030	1.000	0.503	02m17s
Madisonville	35°31'N	084°22'W	13:04:16	294	311	65	14:32:54	110	85	14:35:32	299	273	15:58:53	115	68	51	14:34:13	205	179	63	211	1.030	1.000	0.919	02m38s
Maryville	35°46'N	083°58'W	13:04:57	293	309	65	14:33:54	60	35	14:35:26	349	323	15:59:02	115	68	50	14:34:40	205	179	63	211	1.030	1.000	0.186	01m32s
McMinnville	35°41'N	085°46'W	12:01:04	294	315	64	13:30:14	161	140	13:32:03	247	225	14:56:28	114	68	52	13:31:08	24	3	64	206	1.031	1.000	0.267	01m49s
Memphis	35°08'N	090°03'W	11:52:18	297	329	62	—			—			14:50:09	110	68	56	13:22:48	24	14	66	192	0.940	0.932		
Monterey	36°09'N	085°16'W	12:01:48	293	313	64	13:30:19	88	67	13:32:42	320	298	14:56:25	115	70	52	13:31:31	204	182	63	207	1.030	1.000	0.563	02m23s
Mount Juliet	36°12'N	086°31'W	11:59:04	293	316	63	13:27:42	140	122	13:30:06	268	249	14:54:24	114	70	53	13:28:54	24	5	64	203	1.031	1.000	0.564	02m24s
Murfreesboro	35°51'N	086°24'W	11:59:34	294	317	64	13:29:09	184	165	13:30:04	225	205	14:55:11	114	69	53	13:29:37	24	5	64	204	1.031	1.000	0.062	00m56s
Nashville	36°10'N	086°47'W	11:58:30	294	317	63	13:27:26	158	140	13:29:21	250	232	14:54:02	114	70	53	13:28:24	24	6	64	202	1.031	1.000	0.307	01m56s
Oak Hill	36°05'N	086°47'W	11:58:34	294	318	63	13:27:46	171	153	13:29:14	238	219	14:54:11	114	70	53	13:28:30	24	6	64	202	1.031	1.000	0.165	01m28s
Oak Ridge	36°01'N	084°16'W	13:04:05	293	310	65	14:33:34	31	6	14:33:51	18	354	15:58:09	116	69	51	14:33:43	204	180	63	210	1.030	1.000	0.006	00m17s
Oliver Springs	36°03'N	084°20'W	13:03:55	293	310	64	14:33:26	29	5	14:33:38	20	356	15:58:00	116	69	51	14:33:32	204	180	63	210	1.030	1.000	0.003	00m12s
Portland	36°35'N	086°31'W	11:58:48	293	316	63	13:27:05	102	84	13:29:41	306	288	14:53:45	115	71	53	13:28:23	204	186	64	202	1.031	1.000	0.788	02m36s
Rockwood	35°52'N	084°41'W	13:03:18	293	312	64	14:31:53	90	66	14:34:17	319	295	15:57:48	115	69	51	14:33:05	204	181	63	209	1.030	1.000	0.583	02m24s
Smithville	35°58'N	085°49'W	12:00:45	294	315	64	13:29:22	131	111	13:31:55	277	256	14:55:54	114	69	52	13:30:39	24	3	64	205	1.031	1.000	0.705	02m33s
Smyrna	35°59'N	086°31'W	11:59:13	294	317	63	13:28:23	168	149	13:29:59	241	222	14:54:46	114	69	53	13:29:11	24	5	64	203	1.031	1.000	0.198	01m36s
Sparta	35°56'N	085°28'W	12:01:32	293	314	64	13:30:05	119	98	13:32:44	290	267	14:56:30	115	69	52	13:31:24	24	3	64	206	1.031	1.000	0.919	02m39s
Springfield	36°31'N	086°53'W	11:58:03	293	317	63	13:26:25	124	108	13:29:02	284	266	14:53:17	114	71	53	13:27:44	24	7	64	201	1.031	1.000	0.823	02m38s
Sweetwater	35°36'N	084°28'W	13:03:59	293	311	65	14:32:35	106	82	14:35:12	303	277	15:58:36	115	68	51	14:33:54	205	180	63	210	1.030	1.000	0.859	02m37s
White House	36°28'N	086°39'W	11:58:35	293	316	63	13:26:56	119	102	13:29:36	289	271	14:53:44	114	71	53	13:28:16	24	6	64	202	1.031	1.000	0.914	02m39s

Table 2–21: Local Circumstances for Texas

Location Name	Latitude	Longitude	First Contact Time h m s	P °	V °	Alt °	Second Contact Time h m s	P °	V °	Third Contact Time h m s	P °	V °	Fourth Contact Time h m s	P °	V °	Alt °	Maximum Eclipse Time h m s	P °	V °	Alt °	Azm °	Eclip. Mag.	Eclip. Obs.	Umbral Depth	Umbral Durat.
Texas																									
Abilene	32°28'N	099°43'W	11:35:03	305	355	54	—			—			14:32:46	100	72	66	13:03:13	23	45	68	154	0.762	0.707		
Amarillo	35°13'N	101°50'W	11:29:20	300	349	51	—			—			14:25:27	102	85	65	12:56:04	22	47	64	149	0.824	0.785		
Arlington	32°44'N	097°07'W	11:39:50	304	350	57	—			—			14:38:42	102	68	64	13:09:17	23	36	69	165	0.796	0.749		
Austin	30°17'N	097°45'W	11:41:09	309	359	58	—			—			14:39:26	98	61	66	13:10:18	23	39	71	162	0.718	0.652		
Baytown	29°43'N	094°59'W	11:47:31	309	356	62	—			—			14:46:42	99	55	64	13:17:58	24	28	72	176	0.733	0.670		
Beaumont	30°05'N	094°06'W	11:48:53	307	353	62	—			—			14:48:25	100	55	62	13:19:43	24	24	72	180	0.754	0.696		
Brownsville	25°54'N	097°30'W	11:48:28	317	14	62	—			—			14:42:59	91	44	68	13:16:10	24	40	75	163	0.592	0.500		
Bryan	30°40'N	096°22'W	11:43:29	307	356	59	—			—			14:42:29	99	60	65	13:13:21	24	34	71	169	0.744	0.685		
Carrollton	32°57'N	096°55'W	11:40:02	303	349	57	—			—			14:38:57	102	68	64	13:09:32	23	35	68	166	0.804	0.760		
Corpus Christi	27°47'N	097°24'W	11:45:22	313	7	60	—			—			14:42:11	94	51	67	13:14:05	24	39	74	164	0.649	0.568		
Dallas	32°47'N	096°49'W	11:40:23	304	350	57	—			—			14:39:21	102	68	64	13:09:57	23	35	69	166	0.800	0.755		
Denton	33°13'N	097°08'W	11:39:23	303	349	57	—			—			14:38:10	103	70	64	13:08:45	23	36	68	165	0.810	0.766		
El Paso	31°45'N	106°29'W	11:23:34	308	4	47	—			—			14:15:33	94	89	70	12:47:11	21	61	63	132	0.682	0.608		
Fort Worth	32°45'N	097°18'W	11:39:28	304	351	57	—			—			14:38:16	102	68	64	13:08:49	23	37	68	164	0.795	0.747		
Freeport	28°57'N	095°21'W	11:47:50	310	360	62	—			—			14:46:32	98	52	64	13:18:00	24	29	73	175	0.706	0.638		
Galveston	29°18'N	094°48'W	11:48:29	309	357	62	—			—			14:47:33	99	53	64	13:18:57	24	27	73	177	0.723	0.658		
Garland	32°55'N	096°38'W	11:40:38	303	349	57	—			—			14:39:37	103	68	64	13:10:15	23	34	68	167	0.806	0.762		
Grand Prairie	32°45'N	097°00'W	11:40:03	304	350	57	—			—			14:38:58	102	68	64	13:09:33	23	36	69	166	0.798	0.751		
Harlingen	26°12'N	097°42'W	11:47:29	316	13	61	—			—			14:42:16	91	45	68	13:15:14	24	41	75	162	0.599	0.508		
Houston	29°46'N	095°22'W	11:46:40	309	357	61	—			—			14:45:45	99	55	64	13:16:56	24	30	72	174	0.730	0.666		
Irving	32°49'N	096°56'W	11:40:08	304	350	57	—			—			14:39:03	102	68	64	13:09:38	23	35	69	166	0.800	0.755		
Killeen	31°07'N	097°44'W	11:40:14	307	357	57	—			—			14:38:46	99	63	66	13:09:27	23	39	70	162	0.743	0.682		
Laredo	27°30'N	099°30'W	11:41:36	314	11	58	—			—			14:36:34	92	54	70	13:08:46	24	48	73	153	0.617	0.530		
Lewisville	33°03'N	097°00'W	11:39:47	303	349	57	—			—			14:38:39	102	69	64	13:09:14	23	36	68	165	0.806	0.762		
Longview	32°30'N	094°44'W	11:44:50	303	347	60	—			—			14:44:20	103	64	62	13:15:15	24	28	69	175	0.815	0.773		
Lubbock	33°35'N	101°51'W	11:30:14	303	354	51	—			—			14:26:41	100	80	67	12:57:10	22	49	65	148	0.776	0.723		
McAllen	26°12'N	098°14'W	11:46:24	317	14	61	—			—			14:40:46	91	46	69	13:13:46	24	44	75	159	0.593	0.501		
Mesquite	32°46'N	096°36'W	11:40:50	304	349	58	—			—			14:39:51	102	67	64	13:10:29	23	34	69	167	0.802	0.757		
Midland	32°00'N	102°05'W	11:31:04	306	359	52	—			—			14:27:09	97	76	68	12:57:49	22	52	66	146	0.726	0.662		
Missouri City	29°37'N	095°32'W	11:46:31	309	358	61	—			—			14:45:29	99	55	64	13:16:42	24	30	72	173	0.723	0.659		
Odessa	31°52'N	102°23'W	11:30:39	307	360	52	—			—			14:26:27	97	77	69	12:57:10	22	53	66	144	0.719	0.654		
Pasadena	29°43'N	095°13'W	11:47:02	309	357	61	—			—			14:46:09	99	55	64	13:17:23	24	29	72	175	0.730	0.667		
Plano	33°01'N	096°42'W	11:40:25	303	349	57	—			—			14:39:22	103	68	64	13:09:59	23	35	68	167	0.808	0.765		
Port Arthur	29°54'N	093°56'W	11:49:29	308	354	63	—			—			14:48:59	100	54	62	13:20:22	24	23	72	181	0.750	0.692		
Richardson	32°57'N	096°44'W	11:40:24	303	349	57	—			—			14:39:22	103	68	64	13:09:59	23	35	68	167	0.806	0.762		
Round Rock	30°31'N	097°41'W	11:41:01	308	358	58	—			—			14:39:24	98	61	66	13:10:13	23	39	71	162	0.726	0.661		
San Angelo	31°28'N	100°26'W	11:34:37	307	359	54	—			—			14:31:44	98	70	68	13:02:21	23	48	68	151	0.726	0.661		
San Antonio	29°25'N	098°30'W	11:40:46	310	3	58	—			—			14:38:10	96	59	67	13:09:19	23	43	71	158	0.685	0.611		
Sherman	33°38'N	096°36'W	11:40:06	302	347	57	—			—			14:38:56	104	70	63	13:09:36	23	34	68	167	0.827	0.789		
Temple	31°06'N	097°21'W	11:41:00	307	356	58	—			—			14:39:43	99	63	65	13:10:25	23	38	70	164	0.746	0.687		
Texarkana	33°26'N	094°03'W	11:45:22	301	343	60	—			—			14:44:41	105	66	61	13:15:48	24	26	68	178	0.849	0.816		
Tyler	32°21'N	095°18'W	11:43:50	304	348	59	—			—			14:43:14	103	64	63	13:14:04	24	30	69	173	0.804	0.760		
Victoria	28°48'N	097°00'W	11:44:39	311	3	60	—			—			14:42:31	96	54	66	13:13:55	24	37	73	166	0.683	0.609		
Waco	31°33'N	097°09'W	11:40:55	306	354	58	—			—			14:39:47	100	64	65	13:10:26	23	37	70	165	0.761	0.706		
Wichita Falls	33°54'N	098°30'W	11:36:13	302	349	55	—			—			14:34:22	103	74	64	13:04:52	23	40	67	160	0.816	0.774		

Table 2–22: Local Circumstances for Vermont, Virginia & Washington

Location Name	Latitude	Longitude	First Contact Time h m s	First Contact P °	First Contact V °	First Contact Alt °	Second Contact Time h m s	Second Contact P °	Second Contact V °	Third Contact Time h m s	Third Contact P °	Third Contact V °	Fourth Contact Time h m s	Fourth Contact P °	Fourth Contact V °	Fourth Contact Alt °	Maximum Eclipse Time h m s	Maximum Eclipse P °	Maximum Eclipse V °	Maximum Eclipse Alt °	Maximum Eclipse Azm °	Eclip. Mag.	Eclip. Obs.	Umbral Depth	Umbral Durat.
Vermont																									
Burlington	44°29'N	073°12'W	13:23:01	274	265	57	—			—			15:53:03	132	91	40	14:40:29	203	173	50	223	0.680	0.604		
Montpelier	44°16'N	072°35'W	13:24:22	274	263	57	—			—			15:54:01	132	91	40	14:41:42	203	172	50	224	0.678	0.602		
Rutland	43°37'N	072°58'W	13:23:52	275	265	58	—			—			15:55:11	131	89	40	14:42:07	203	172	51	224	0.696	0.624		
Virginia																									
Alexandria	38°48'N	077°03'W	13:17:55	284	281	63	—			—			16:01:51	123	76	44	14:42:57	204	171	56	223	0.847	0.814		
Arlington	38°53'N	077°07'W	13:17:42	284	282	63	—			—			16:01:36	124	77	44	14:42:42	204	171	56	223	0.846	0.812		
Blacksburg	37°14'N	080°25'W	13:11:38	289	295	64	—			—			16:01:07	120	72	47	14:39:20	204	174	60	218	0.924	0.911		
Charlottesville	38°02'N	078°30'W	13:15:16	286	287	64	—			—			16:01:49	122	74	45	14:41:36	204	172	58	221	0.882	0.858		
Chesapeake	36°50'N	076°17'W	13:21:08	287	281	65	—			—			16:06:40	122	71	43	14:47:20	205	167	57	228	0.884	0.860		
Danville	36°36'N	079°23'W	13:14:25	289	292	65	—			—			16:03:36	119	70	46	14:42:12	205	171	59	222	0.927	0.915		
Hampton	37°02'N	076°21'W	13:20:49	287	281	65	—			—			16:06:11	122	72	43	14:46:54	205	167	57	228	0.880	0.855		
Harrisonburg	38°27'N	078°52'W	13:14:10	286	288	63	—			—			16:00:34	122	75	45	14:40:19	204	173	58	220	0.877	0.851		
Lynchburg	37°25'N	079°09'W	13:14:18	288	290	64	—			—			16:02:18	121	73	46	14:41:23	204	172	59	221	0.904	0.887		
Manassas	38°45'N	077°29'W	13:17:00	285	283	63	—			—			16:01:29	123	76	44	14:42:17	204	171	57	222	0.853	0.822		
Martinsville	36°41'N	079°52'W	13:13:17	289	294	65	—			—			16:02:51	119	71	46	14:41:11	205	172	60	221	0.931	0.920		
Newport News	36°59'N	076°25'W	13:20:42	287	281	65	—			—			16:06:13	122	72	43	14:46:52	205	167	57	228	0.882	0.858		
Norfolk	36°51'N	076°17'W	13:21:07	287	281	65	—			—			16:06:38	122	71	43	14:47:19	205	167	57	228	0.883	0.860		
North Springfield	38°48'N	077°13'W	13:17:33	284	282	63	—			—			16:01:40	123	76	44	14:42:40	204	171	56	223	0.849	0.816		
Petersburg	37°14'N	077°24'W	13:18:19	287	284	65	—			—			16:04:39	121	72	44	14:44:46	204	169	57	225	0.888	0.865		
Portsmouth	36°50'N	076°18'W	13:21:06	287	281	65	—			—			16:06:39	122	71	43	14:47:19	205	167	57	228	0.884	0.860		
Richmond	37°33'N	077°27'W	13:17:57	286	284	64	—			—			16:03:58	122	73	44	14:44:11	204	170	57	224	0.881	0.857		
Roanoke	37°16'N	079°56'W	13:12:41	288	293	65	—			—			16:01:39	120	72	46	14:40:11	204	173	59	219	0.917	0.903		
Suffolk	36°44'N	076°35'W	13:20:33	287	282	65	—			—			16:06:33	121	71	43	14:46:59	205	167	57	228	0.890	0.868		
Virginia Beach	36°51'N	075°59'W	13:21:47	287	280	65	—			—			16:06:57	122	71	43	14:47:50	205	166	56	229	0.880	0.855		
Washington																									
Auburn	47°18'N	122°14'W	09:08:29	280	323	28	—			—			11:39:11	111	136	50	10:20:50	196	233	40	120	0.939	0.930		
Bellevue	47°37'N	122°12'W	09:08:52	280	322	28	—			—			11:39:13	112	136	49	10:21:05	196	232	40	121	0.930	0.918		
Bellingham	48°46'N	122°29'W	09:10:01	278	319	28	—			—			11:38:41	114	137	48	10:21:31	196	231	39	121	0.897	0.877		
Bremerton	47°34'N	122°38'W	09:08:29	280	322	28	—			—			11:38:32	112	136	49	10:20:31	196	233	39	120	0.932	0.921		
Edmonds	47°49'N	122°23'W	09:08:57	280	322	28	—			—			11:38:55	112	136	49	10:20:59	196	232	39	121	0.924	0.911		
Everett	47°59'N	122°12'W	09:09:18	279	321	28	—			—			11:39:11	113	136	49	10:21:19	196	232	39	121	0.919	0.905		
Federal Way	47°18'N	122°19'W	09:08:25	280	323	28	—			—			11:39:03	111	136	50	10:20:43	196	233	40	120	0.939	0.930		
Kennewick	46°12'N	119°07'W	09:09:58	282	325	31	—			—			11:44:24	111	132	52	10:24:11	197	233	43	124	0.966	0.963		
Kent	47°23'N	122°14'W	09:08:35	280	323	28	—			—			11:39:10	112	136	50	10:20:53	196	233	40	120	0.937	0.927		
Kirkland	47°41'N	122°13'W	09:08:56	280	322	28	—			—			11:39:11	112	136	49	10:21:06	196	232	40	121	0.928	0.916		
Lakes District	47°10'N	122°32'W	09:08:06	281	323	28	—			—			11:38:43	111	136	50	10:20:22	196	233	40	120	0.943	0.935		
Lakewood	48°09'N	122°13'W	09:09:28	279	321	28	—			—			11:39:08	113	137	49	10:21:25	196	232	39	121	0.915	0.899		
Longview	46°08'N	122°57'W	09:06:40	282	326	28	—			—			11:38:03	109	136	50	10:19:11	196	235	40	118	0.974	0.973		
Lynnwood	47°50'N	122°18'W	09:09:02	280	322	28	—			—			11:39:02	112	136	49	10:21:06	196	232	39	121	0.924	0.910		
Olympia	47°03'N	122°53'W	09:07:42	281	324	28	—			—			11:38:10	111	136	49	10:19:52	196	233	39	119	0.947	0.940		
Pasco	46°14'N	119°06'W	09:10:01	282	325	31	—			—			11:44:25	111	132	52	10:24:14	197	233	43	124	0.965	0.962		
Redmond	47°41'N	122°07'W	09:09:01	280	322	28	—			—			11:39:20	112	136	49	10:21:13	196	232	40	121	0.928	0.916		
Renton	47°29'N	122°12'W	09:08:43	280	322	28	—			—			11:39:13	112	136	49	10:20:59	196	233	40	120	0.934	0.923		
Richland	46°17'N	119°18'W	09:09:53	282	325	31	—			—			11:44:05	111	132	52	10:23:59	196	233	42	123	0.964	0.961		
Seattle	47°36'N	122°20'W	09:08:45	280	322	28	—			—			11:39:00	112	136	49	10:20:54	196	233	39	120	0.931	0.919		
Spokane	47°40'N	117°24'W	09:13:03	280	321	32	—			—			11:46:51	114	132	52	10:27:15	197	230	43	127	0.919	0.904		
Tacoma	47°14'N	122°26'W	09:08:15	281	323	28	—			—			11:38:52	111	136	50	10:20:32	196	233	40	120	0.941	0.933		
University Place	47°13'N	122°32'W	09:08:10	281	323	28	—			—			11:38:42	111	136	49	10:20:24	196	233	39	120	0.942	0.933		
Vancouver	45°38'N	122°40'W	09:06:23	283	328	28	—			—			11:38:30	109	135	51	10:19:12	196	235	40	118	0.988	0.990		
Yakima	46°36'N	120°31'W	09:09:08	282	325	30	—			—			11:42:00	111	134	51	10:22:33	196	233	41	122	0.957	0.952		
Yakima	46°36'N	120°31'W	09:09:08	282	325	30	—			—			11:42:00	111	134	51	10:22:33	196	233	41	122	0.957	0.952		

Table 2–23: Local Circumstances for West Virginia, Wisconsin & Wyoming

Location Name	Latitude	Longitude	First Contact Time h m s	P °	V °	Alt °	Second Contact Time h m s	P °	V °	Third Contact Time h m s	P °	V °	Fourth Contact Time h m s	P °	V °	Alt °	Maximum Eclipse Time h m s	P °	V °	Alt °	Azm °	Eclip. Mag.	Eclip. Obs.	Umbral Depth	Umbral Durat.
West Virginia																									
Beckley	37°47'N	081°11'W	13:09:33	288	296	64	—			—			15:59:05	120	74	47	14:37:07	204	176	60	215	0.920	0.906		
Charleston	38°21'N	081°38'W	13:08:12	288	297	63	—			—			15:57:26	120	75	48	14:35:28	204	178	60	213	0.911	0.895		
Clarksburg	39°17'N	080°21'W	13:10:25	285	291	62	—			—			15:57:10	122	78	47	14:36:28	204	177	58	215	0.874	0.848		
Fairmont	39°29'N	080°09'W	13:10:45	285	290	62	—			—			15:57:00	122	78	46	14:36:32	204	177	58	215	0.867	0.839		
Huntington	38°25'N	082°27'W	13:06:24	288	299	63	—			—			15:56:15	120	76	49	14:33:51	204	179	60	211	0.919	0.905		
Morgantown	39°38'N	079°57'W	13:11:05	285	290	62	—			—			15:56:56	123	79	46	14:36:40	204	177	58	216	0.861	0.831		
Parkersburg	39°16'N	081°34'W	13:07:49	286	295	62	—			—			15:55:43	121	78	48	14:34:17	204	179	59	212	0.888	0.866		
Wheeling	40°04'N	080°43'W	13:09:13	284	291	62	—			—			15:55:10	123	80	47	14:34:42	204	178	58	213	0.859	0.829		
Wisconsin																									
Appleton	44°16'N	088°25'W	11:52:41	281	301	55	—			—			14:37:12	123	94	51	13:15:55	202	196	57	189	0.836	0.800		
Beloit	42°31'N	089°02'W	11:51:30	284	307	56	—			—			14:39:31	120	89	52	13:16:32	203	196	59	188	0.886	0.863		
Brookfield	43°04'N	088°09'W	11:53:11	283	304	56	—			—			14:39:46	122	90	52	13:17:36	203	195	58	190	0.864	0.835		
Eau Claire	44°49'N	091°30'W	11:47:03	281	306	53	—			—			14:31:55	122	97	53	13:09:58	202	202	57	180	0.850	0.818		
Fond du Lac	43°47'N	088°27'W	11:52:36	282	303	55	—			—			14:38:03	122	93	51	13:16:20	202	196	58	189	0.849	0.816		
Green Bay	44°31'N	088°00'W	11:53:30	281	300	55	—			—			14:37:18	124	94	51	13:16:25	202	195	57	190	0.826	0.787		
Greenfield	42°58'N	088°00'W	11:53:29	283	304	56	—			—			14:40:10	122	90	51	13:17:58	203	195	59	191	0.865	0.836		
Janesville	42°41'N	089°01'W	11:51:31	284	306	56	—			—			14:39:14	121	90	52	13:16:24	203	196	59	188	0.882	0.858		
Kenosha	42°35'N	087°49'W	11:53:52	284	304	57	—			—			14:41:07	121	89	52	13:18:43	203	194	59	192	0.873	0.846		
La Crosse	43°48'N	091°15'W	11:47:21	283	308	54	—			—			14:34:02	121	94	53	13:11:16	202	201	58	181	0.874	0.848		
Madison	43°04'N	089°24'W	11:50:46	283	306	55	—			—			14:38:00	121	91	52	13:15:19	202	197	59	187	0.876	0.850		
Manitowish	46°08'N	090°01'W	11:50:03	279	300	52	—			—			14:31:39	124	100	51	13:11:27	202	199	56	184	0.804	0.759		
Manitowoc	44°05'N	087°40'W	11:54:07	281	300	55	—			—			14:38:33	123	93	51	13:17:26	202	194	57	191	0.834	0.797		
Menomonee Falls	43°10'N	088°07'W	11:53:15	283	303	56	—			—			14:39:38	122	91	51	13:17:33	203	195	58	190	0.861	0.831		
Milwaukee	43°02'N	087°55'W	11:53:38	283	303	56	—			—			14:40:09	122	90	51	13:18:03	203	194	58	191	0.862	0.833		
New Berlin	42°59'N	088°06'W	11:53:17	283	304	56	—			—			14:40:00	122	90	52	13:17:46	203	195	59	191	0.865	0.837		
Oshkosh	44°01'N	088°33'W	11:52:25	282	302	55	—			—			14:37:29	123	93	51	13:15:56	202	196	58	189	0.844	0.809		
Racine	42°44'N	087°48'W	11:53:53	283	304	56	—			—			14:40:52	121	89	51	13:18:35	203	194	59	192	0.869	0.841		
Sheboygan	43°46'N	087°45'W	11:53:57	282	301	56	—			—			14:39:02	123	92	51	13:17:36	203	194	58	191	0.842	0.808		
Waukesha	43°01'N	088°14'W	11:53:01	283	304	56	—			—			14:39:45	122	90	52	13:17:30	203	195	59	190	0.866	0.837		
Wausau	44°58'N	089°38'W	11:50:30	280	302	54	—			—			14:34:16	123	96	51	13:13:08	202	198	57	185	0.830	0.792		
Wauwatosa	43°03'N	088°00'W	11:53:28	283	303	56	—			—			14:40:01	122	90	51	13:17:53	203	195	58	191	0.863	0.834		
West Allis	43°01'N	088°00'W	11:53:29	283	304	56	—			—			14:40:04	122	90	51	13:17:55	203	195	58	191	0.864	0.835		
Wyoming																									
Casper	42°51'N	106°19'W	10:22:18	288	329	43	11:42:40	106	133	11:45:06	293	319	13:09:26	111	111	59	11:43:53	199	226	54	143	1.029	1.000	0.940	02m26s
Cheyenne	41°08'N	104°49'W	10:23:52	291	333	45	—			—			13:13:52	109	105	61	11:47:01	20	46	56	145	0.975	0.975		
Douglas	42°45'N	105°24'W	10:23:34	288	329	43	11:44:23	96	122	11:46:46	303	328	13:11:16	111	109	59	11:45:34	200	225	55	145	1.029	1.000	0.767	02m23s
Dubois	43°33'N	109°38'W	10:18:12	287	329	40	11:36:51	94	125	11:39:08	303	333	13:02:36	110	117	58	11:37:59	199	229	51	137	1.029	1.000	0.753	02m17s
Evansville	42°52'N	106°16'W	10:22:23	288	329	43	11:42:46	103	130	11:45:11	296	322	13:09:31	111	111	59	11:43:58	199	226	54	143	1.029	1.000	0.888	02m25s
Glenrock	42°52'N	105°52'W	10:22:57	288	329	43	11:43:31	94	120	11:45:52	305	331	13:10:17	111	110	59	11:44:42	200	226	54	144	1.029	1.000	0.728	02m21s
Guernsey	42°16'N	104°45'W	10:24:20	289	330	44	11:45:46	136	161	11:48:00	264	289	13:12:58	111	107	59	11:46:53	20	45	55	146	1.030	1.000	0.563	02m14s
Jackson	43°29'N	110°46'W	10:16:44	287	330	39	11:34:56	123	156	11:37:11	274	305	13:00:29	110	119	58	11:36:03	18	50	50	135	1.029	1.000	0.743	02m15s
Lander	42°50'N	108°44'W	10:18:59	288	331	41	11:39:00	171	202	11:40:06	227	257	13:04:48	110	114	59	11:39:33	19	49	52	138	1.029	1.000	0.114	01m07s
Lusk	42°46'N	104°27'W	10:24:58	288	328	44	11:46:24	68	93	11:48:14	331	355	13:13:03	112	107	59	11:47:19	200	224	55	147	1.030	1.000	0.336	01m51s
Mills	42°50'N	106°22'W	10:22:13	288	329	43	11:42:34	109	136	11:45:01	290	317	13:09:21	111	111	59	11:43:48	199	226	54	143	1.029	1.000	0.993	02m26s
Mountain View	42°51'N	106°23'W	10:22:12	288	329	43	11:42:33	108	135	11:44:59	291	318	13:09:19	111	111	59	11:43:46	199	226	54	143	1.029	1.000	0.967	02m26s
Riverton	43°02'N	108°23'W	10:19:33	288	330	41	11:39:03	130	160	11:41:17	268	297	13:05:20	110	114	59	11:40:10	19	49	53	139	1.029	1.000	0.641	02m14s
Shoshoni	43°14'N	108°07'W	10:20:01	287	329	41	11:39:27	101	130	11:41:49	297	326	13:05:42	111	114	59	11:40:38	199	228	53	140	1.029	1.000	0.862	02m22s
Thermopolis	43°39'N	108°13'W	10:20:06	287	328	41	11:40:01	40	68	11:40:52	358	27	13:05:12	111	115	58	11:40:26	199	228	52	140	1.029	1.000	0.064	00m50s
Torrington	42°04'N	104°11'W	10:25:07	289	330	45	11:46:59	145	170	11:49:01	255	279	13:14:15	111	106	60	11:48:00	20	44	56	147	1.030	1.000	0.421	02m02s
Wheatland	42°03'N	104°58'W	10:23:56	289	331	44	11:46:09	182	208	11:46:54	218	243	13:12:45	111	107	60	11:46:31	20	45	56	146	1.030	1.000	0.047	00m45s

Table 2–24: Local Circumstances for Canada - 1

Location Name	Latitude	Longitude	First Contact Time	P	V	Alt	Second Contact Time	P	V	Third Contact Time	P	V	Fourth Contact Time	P	V	Alt	Maximum Eclipse Time	P	V	Alt	Azm	Eclip. Mag.	Eclip. Obs.	Umbral Depth	Umbral Durat.
			h m s	°	°	°	h m s	°	°	h m s	°	°	h m s	°	°	°	h m s	°	°	°	°				
Alberta																									
Banff	51°10'N	115°34'W	10:18:55	275	311	33	—			—			12:48:06	119	133	49	11:31:22	197	225	42	133	0.816	0.773		
Calgary	51°03'N	114°05'W	10:20:11	275	310	34	—			—			12:50:18	120	132	50	11:33:11	197	224	43	135	0.814	0.772		
Edmonton	53°33'N	113°28'W	10:24:08	271	303	34	—			—			12:49:29	124	134	47	11:35:04	197	221	42	138	0.745	0.684		
Lethbridge	49°42'N	112°50'W	10:19:51	277	313	35	—			—			12:53:01	118	129	51	11:34:19	198	225	45	136	0.848	0.814		
Medicine Hat	50°03'N	110°40'W	10:22:35	276	311	37	—			—			12:56:03	120	127	51	11:37:26	198	223	46	140	0.830	0.792		
Red Deer	52°16'N	113°48'W	10:22:03	273	307	34	—			—			12:49:55	122	133	48	11:34:05	197	223	43	137	0.780	0.729		
British Columbia																									
Abbotsford	49°03'N	122°17'W	09:10:31	278	318	28	—			—			11:38:57	114	137	48	10:21:57	196	231	39	122	0.889	0.866		
Burnaby	49°15'N	122°57'W	09:10:16	277	318	28	—			—			11:37:56	114	138	48	10:21:19	196	231	38	121	0.884	0.860		
Fort Nelson	58°49'N	122°39'W	09:25:30	263	292	27	—			—			11:35:59	129	145	40	10:29:11	196	220	34	129	0.627	0.540		
Kamloops	50°40'N	120°20'W	09:14:07	275	313	30	—			—			11:41:28	117	137	48	10:25:19	196	228	40	126	0.840	0.804		
Kelowna	49°53'N	119°29'W	09:13:48	277	315	30	—			—			11:42:56	116	135	49	10:25:49	196	229	40	126	0.861	0.830		
Nanaimo	49°10'N	123°56'W	09:09:27	277	318	27	—			—			11:36:30	114	139	47	10:20:09	196	231	38	120	0.888	0.864		
Penticton	49°30'N	119°35'W	09:13:14	277	316	30	—			—			11:42:53	116	135	49	10:25:28	196	229	41	126	0.872	0.844		
Prince George	53°55'N	122°45'W	09:17:02	270	305	27	—			—			11:37:23	122	141	44	10:25:03	196	225	36	125	0.755	0.696		
Prince Rupert	54°19'N	130°19'W	09:13:16	269	305	23	—			—			11:28:23	121	146	40	10:18:33	195	227	32	116	0.749	0.689		
Richmond	49°10'N	123°10'W	09:10:00	277	318	28	—			—			11:37:38	114	138	48	10:21:01	196	231	38	121	0.887	0.863		
Vancouver	49°16'N	123°07'W	09:10:10	277	318	28	—			—			11:37:41	114	138	47	10:21:09	196	231	38	121	0.884	0.860		
Victoria	48°25'N	123°22'W	09:08:55	279	320	27	—			—			11:37:22	113	138	48	10:20:15	196	232	38	120	0.908	0.891		
Manitoba																									
Brandon	49°50'N	099°57'W	11:36:01	275	304	44	—			—			14:12:00	124	116	52	12:53:14	200	212	51	161	0.783	0.732		
Churchill	58°46'N	094°10'W	11:53:01	260	275	40	—			—			14:04:46	138	130	42	12:58:47	199	203	43	173	0.534	0.432		
Riverton	50°59'N	096°59'W	11:41:18	273	297	45	—			—			14:14:18	127	116	50	12:57:24	200	208	50	167	0.735	0.672		
Selkirk	50°09'N	096°52'W	11:40:48	274	299	46	—			—			14:15:45	126	114	51	12:57:53	200	208	51	167	0.755	0.697		
The Pas	53°50'N	101°15'W	11:38:13	269	294	41	—			—			14:04:49	129	125	48	12:50:46	199	211	47	159	0.687	0.614		
Winnipeg	49°53'N	097°09'W	11:40:10	275	300	46	—			—			14:15:46	125	114	51	12:57:33	200	209	51	167	0.764	0.708		
New Brunswick																									
Chatham	47°02'N	065°28'W	14:37:10	265	245	52	—			—			16:52:00	139	99	34	15:46:51	202	169	44	232	0.553	0.454		
Edmundston	47°22'N	068°20'W	14:31:33	267	251	53	—			—			16:49:39	138	99	36	15:42:47	202	172	46	227	0.574	0.478		
Moncton	46°06'N	064°47'W	14:38:57	266	244	52	—			—			16:54:36	139	96	34	15:49:11	203	167	44	234	0.564	0.466		
Saint John	45°16'N	066°03'W	14:36:58	268	247	53	—			—			16:55:57	137	94	35	15:48:58	203	167	45	233	0.592	0.499		
Saint Stephen	45°12'N	067°17'W	14:34:34	269	250	54	—			—			16:55:25	136	94	36	15:47:31	203	168	46	232	0.605	0.515		
Newfoundland																									
Gander	48°57'N	054°37'W	15:25:26	256	226	44	—			—			17:20:54	147	106	27	16:25:02	202	164	36	244	0.423	0.311		
Knob Lake	54°48'N	066°49'W	15:02:31	256	243	45	—			—			17:02:18	146	116	34	16:03:42	201	178	41	221	0.427	0.315		
Saint John's	47°34'N	052°43'W	15:29:32	257	223	43	—			—			17:24:43	146	103	25	16:29:05	202	162	34	248	0.429	0.317		
Northwest Territories																									
Aklavik	68°12'N	135°00'W	10:39:56	248	269	20	—			—			12:24:22	141	157	28	11:31:20	195	214	24	121	0.414	0.301		
Fort Simpson	61°52'N	121°23'W	10:32:08	258	283	27	—			—			12:35:55	134	147	37	11:32:50	196	217	33	133	0.550	0.451		
Inuvik	68°25'N	133°30'W	10:40:54	248	269	20	—			—			12:25:13	142	157	28	11:32:16	195	213	25	123	0.410	0.296		
Yellowknife	62°27'N	114°21'W	10:38:08	257	280	30	—			—			12:41:48	136	145	38	11:39:02	196	213	35	143	0.524	0.421		
Nunavut Territory																									
Alert	82°31'N	062°20'W	13:39:43	220	217	19	—			—			14:30:36	171	167	18	14:05:11	196	192	18	209	0.092	0.033		
Iqaluit	63°44'N	068°28'W	13:30:46	244	237	37	—			—			15:10:00	154	135	32	14:20:56	199	185	35	212	0.299	0.189		
Resolute	74°41'N	094°54'W	12:14:44	236	241	27	—			—			13:39:06	157	155	27	12:56:52	197	198	27	173	0.237	0.134		
Grise Fiord	76°25'N	082°54'W	13:25:05	232	232	25	—			—			14:39:35	161	157	25	14:02:22	197	195	25	188	0.185	0.094		

Table 2–25: Local Circumstances for Canada - 2

Location Name	Latitude	Longitude	First Contact Time	P	V	Alt	Second Contact Time	P	V	Third Contact Time	P	V	Fourth Contact Time	P	V	Alt	Maximum Eclipse Time	P	V	Alt	Azm	Eclip. Mag.	Eclip. Obs.	Umbral Depth	Umbral Durat.
			h m s	°	°	°	h m s	°	°	h m s	°	°	h m s	°	°	°	h m s	°	°	°	°				
Ontario																									
Barrie	44°24'N	079°40'W	13:09:53	277	281	57	—			—			15:47:21	128	91	45	14:30:38	203	182	54	210	0.748	0.688		
Brampton	43°41'N	079°46'W	13:09:51	278	283	58	—			—			15:48:46	127	89	45	14:31:25	203	181	55	210	0.765	0.710		
Brantford	43°08'N	080°16'W	13:08:58	279	285	59	—			—			15:49:23	127	88	46	14:31:19	203	181	56	210	0.783	0.732		
Burlington	43°19'N	079°47'W	13:09:55	279	283	58	—			—			15:49:31	127	88	45	14:31:52	203	181	55	211	0.773	0.720		
Cambridge (Galt)	43°22'N	080°19'W	13:08:48	279	284	58	—			—			15:48:51	127	89	46	14:30:55	203	181	55	210	0.778	0.726		
Chatham	42°24'N	082°11'W	13:05:14	281	291	59	—			—			15:48:44	125	86	47	14:29:00	203	184	57	206	0.820	0.779		
Cornwall	45°02'N	074°44'W	13:19:43	274	268	57	—			—			15:50:33	132	93	41	14:37:25	203	176	51	219	0.684	0.609		
East York	43°41'N	079°20'W	13:10:44	278	281	58	—			—			15:49:13	128	89	45	14:32:07	203	180	55	211	0.760	0.704		
Etobicoke	43°39'N	079°34'W	13:10:16	278	282	58	—			—			15:49:03	128	89	45	14:31:47	203	181	55	211	0.764	0.708		
Gloucester	45°22'N	075°35'W	13:17:54	274	270	56	—			—			15:49:05	132	93	42	14:35:41	203	177	51	217	0.685	0.611		
Guelph	43°33'N	080°15'W	13:08:54	279	284	58	—			—			15:48:32	127	89	46	14:30:47	203	182	55	210	0.773	0.720		
Hamilton	43°15'N	079°51'W	13:09:48	279	283	59	—			—			15:49:36	127	88	45	14:31:51	203	181	55	211	0.776	0.723		
Kapuskasing	49°25'N	082°26'W	13:04:39	271	278	52	—			—			15:33:55	132	105	45	14:20:28	202	189	51	199	0.661	0.582		
Kingston	44°14'N	076°30'W	13:16:22	276	273	58	—			—			15:50:47	130	91	43	14:35:53	203	177	53	217	0.719	0.653		
Kitchener	43°27'N	080°29'W	13:08:27	279	285	58	—			—			15:48:30	127	89	46	14:30:32	203	182	55	209	0.778	0.726		
London	42°59'N	081°14'W	13:07:02	280	287	59	—			—			15:48:38	126	88	46	14:29:53	203	183	56	208	0.796	0.749		
Markham	43°52'N	079°16'W	13:10:50	278	281	58	—			—			15:48:54	128	90	45	14:31:59	203	180	54	211	0.756	0.698		
Mississauga	43°35'N	079°37'W	13:10:11	278	282	58	—			—			15:49:08	127	89	45	14:31:47	203	181	55	211	0.766	0.711		
Nepean	45°18'N	075°47'W	13:17:31	274	270	56	—			—			15:49:04	132	93	42	14:35:29	203	177	52	217	0.689	0.615		
Niagara Falls	43°06'N	079°04'W	13:11:27	279	282	59	—			—			15:50:44	127	88	45	14:33:21	203	179	55	213	0.771	0.717		
North Bay	46°19'N	079°28'W	13:10:01	274	277	55	—			—			15:43:26	131	96	44	14:28:31	203	183	53	208	0.703	0.632		
North York	43°46'N	079°25'W	13:10:33	278	281	58	—			—			15:48:57	128	90	45	14:31:52	203	181	55	211	0.759	0.703		
Oakville	43°27'N	079°41'W	13:10:05	279	283	58	—			—			15:49:21	127	89	45	14:31:51	203	181	55	211	0.769	0.715		
Oshawa	43°54'N	078°51'W	13:11:40	277	280	58	—			—			15:49:15	128	90	44	14:32:37	203	180	54	212	0.751	0.692		
Ottawa	45°25'N	075°42'W	13:17:39	274	270	56	—			—			15:48:52	132	94	42	14:35:27	203	177	51	217	0.685	0.611		
Peterborough	44°18'N	078°19'W	13:12:39	277	278	58	—			—			15:48:55	129	91	44	14:32:56	203	180	54	213	0.736	0.674		
Point Pelee N.P.	41°57'N	082°30'W	13:04:43	282	293	59	—			—			15:49:15	124	85	48	14:29:01	203	184	58	206	0.834	0.797		
Saint Catharines	43°10'N	079°15'W	13:11:03	279	282	59	—			—			15:50:24	127	88	45	14:32:57	203	180	55	212	0.771	0.718		
Saint Thomas	42°47'N	081°12'W	13:07:09	280	288	59	—			—			15:49:05	126	87	47	14:30:12	203	182	56	208	0.801	0.755		
Sarnia	42°58'N	082°23'W	13:04:41	281	290	58	—			—			15:47:21	125	88	47	14:27:56	203	184	57	205	0.809	0.765		
Sault Sainte Marie	46°31'N	084°20'W	13:00:38	276	288	54	—			—			15:37:54	128	98	47	14:20:34	202	190	54	197	0.744	0.684		
Scarborough	43°44'N	079°16'W	13:10:52	278	281	58	—			—			15:49:11	128	89	45	14:32:10	203	180	55	211	0.759	0.702		
Sudbury	46°30'N	081°00'W	13:07:01	275	281	55	—			—			15:41:31	130	97	45	14:25:54	202	185	53	205	0.713	0.645		
Thunder Bay	48°23'N	089°15'W	12:52:09	275	293	51	—			—			15:28:30	128	105	49	14:10:54	201	198	53	184	0.742	0.682		
Toronto	43°39'N	079°23'W	13:10:39	278	282	58	—			—			15:49:14	128	89	45	14:32:05	203	180	55	211	0.762	0.706		
Waterloo	43°28'N	080°31'W	13:08:22	279	285	58	—			—			15:48:25	127	89	46	14:30:27	203	182	55	209	0.778	0.726		
Welland	42°59'N	079°15'W	13:11:07	279	282	59	—			—			15:50:47	127	88	45	14:33:12	203	179	55	212	0.776	0.723		
Windsor	42°18'N	083°01'W	13:03:33	282	293	59	—			—			15:47:56	124	86	48	14:27:40	203	185	57	204	0.831	0.793		
Woodstock	43°08'N	080°45'W	13:07:59	280	286	59	—			—			15:48:52	126	88	46	14:30:30	203	182	56	209	0.788	0.738		
York	43°41'N	079°29'W	13:10:26	278	282	58	—			—			15:49:04	128	89	45	14:31:53	203	181	55	211	0.762	0.706		

Table 2–26: Local Circumstances for Canada - 3

Location Name	Latitude	Longitude	First Contact Time	P	V	Alt	Second Contact Time	P	V	Third Contact Time	P	V	Fourth Contact Time	P	V	Alt	Maximum Eclipse Time	P	V	Alt	Azm	Eclip. Mag.	Eclip. Obs.	Umbral Depth	Umbral Durat.
			h m s	°	°	°	h m s	°	°	h m s	°	°	h m s	°	°	°	h m s	°	°	°	°				
Nova Scotia																									
Bridgewater	44°23'N	064°31'W	14:40:29	269	244	53	—			—			16:58:51	137	93	33	15:52:19	203	165	44	237	0.594	0.501		
Dartmouth	44°40'N	063°34'W	14:42:09	268	242	53	—			—			16:58:38	138	93	33	15:52:59	203	165	43	238	0.579	0.484		
Glace Bay	46°12'N	059°57'W	14:47:58	263	235	49	—			—			16:56:22	141	98	30	15:54:30	202	164	40	241	0.516	0.412		
Guysborough	45°23'N	061°30'W	14:45:37	265	238	51	—			—			16:57:47	140	96	31	15:54:11	203	164	42	240	0.546	0.445		
Halifax	44°39'N	063°36'W	14:42:06	268	242	53	—			—			16:58:39	138	93	33	15:52:59	203	165	43	238	0.579	0.485		
Kentville	45°05'N	064°30'W	14:40:05	268	244	53	—			—			16:57:11	138	94	33	15:51:11	203	166	44	236	0.580	0.485		
Liverpool	44°02'N	064°43'W	14:40:19	269	245	54	—			—			16:59:35	136	92	34	15:52:39	203	165	44	237	0.602	0.511		
New Glasgow	45°35'N	062°39'W	14:43:19	266	240	51	—			—			16:56:50	139	96	32	15:52:32	203	165	42	238	0.553	0.454		
Port Hawkesbury	45°37'N	061°21'W	14:45:45	265	238	50	—			—			16:57:17	140	96	31	15:53:57	202	164	41	240	0.540	0.439		
Sable Island	43°55'N	059°50'W	14:49:47	266	235	51	—			—			17:01:58	139	93	30	15:58:31	203	161	41	243	0.556	0.458		
Shelburne	43°46'N	065°19'W	14:39:18	270	246	54	—			—			16:59:54	136	91	34	15:52:20	203	165	45	236	0.613	0.525		
Sydney	46°09'N	060°11'W	14:47:34	263	236	49	—			—			16:56:24	141	98	30	15:54:20	202	164	40	240	0.519	0.416		
Truro	45°22'N	063°16'W	14:42:17	266	242	52	—			—			16:57:05	139	95	33	15:52:11	203	165	43	237	0.563	0.465		
Windsor	44°59'N	064°08'W	14:40:51	267	243	53	—			—			16:57:36	138	94	33	15:51:47	203	165	44	236	0.578	0.483		
Yarmouth	43°50'N	066°07'W	14:37:40	270	248	55	—			—			16:59:19	135	91	35	15:51:13	203	166	46	235	0.620	0.533		
Québec																									
Beauport	46°52'N	071°11'W	13:26:12	269	258	54	—			—			15:49:02	136	97	38	14:39:48	203	174	48	223	0.611	0.522		
Brossard	45°26'N	073°29'W	13:22:06	272	264	56	—			—			15:50:40	133	94	40	14:38:41	203	175	50	221	0.663	0.584		
Cap Chat	49°06'N	066°42'W	13:34:00	263	247	50	—			—			15:46:21	141	103	35	14:42:09	202	172	44	228	0.526	0.423		
Charlesbourg	46°51'N	071°16'W	13:26:02	269	258	54	—			—			15:49:01	136	97	38	14:39:43	203	174	48	223	0.612	0.523		
Chicoutimi	48°26'N	071°04'W	13:25:59	267	256	53	—			—			15:45:25	138	101	38	14:37:41	202	175	47	221	0.579	0.484		
Drummondville	45°53'N	072°29'W	13:23:57	271	262	55	—			—			15:50:23	134	95	39	14:39:27	203	174	49	222	0.643	0.560		
Gatineau	45°29'N	075°38'W	13:17:46	273	270	56	—			—			15:48:47	132	94	42	14:35:27	203	177	51	217	0.683	0.609		
Hull	45°26'N	075°43'W	13:17:37	274	270	56	—			—			15:48:49	132	94	42	14:35:24	203	177	51	217	0.685	0.611		
La Salle	45°26'N	073°38'W	13:21:48	272	265	56	—			—			15:50:32	133	94	40	14:38:28	203	175	50	221	0.664	0.586		
Laval	45°35'N	073°45'W	13:21:31	272	265	56	—			—			15:50:06	133	94	40	14:38:05	203	175	50	220	0.662	0.583		
Longueuil	45°32'N	073°30'W	13:22:02	272	264	56	—			—			15:50:25	133	94	40	14:38:31	203	175	50	221	0.661	0.581		
Montréal	45°31'N	073°34'W	13:21:54	272	265	56	—			—			15:50:24	133	94	40	14:38:26	203	175	50	221	0.662	0.583		
Montréal Nord	45°36'N	073°38'W	13:21:45	272	265	56	—			—			15:50:10	133	94	40	14:38:13	203	175	50	220	0.661	0.581		
Québec	46°49'N	071°14'W	13:26:07	269	258	54	—			—			15:49:07	136	97	38	14:39:48	203	174	48	223	0.612	0.523		
Sainte Foy	46°47'N	071°17'W	13:26:01	269	258	54	—			—			15:49:09	136	97	38	14:39:47	203	174	48	223	0.613	0.525		
Saint Hubert	45°30'N	073°25'W	13:22:13	272	264	56	—			—			15:50:34	133	94	40	14:38:41	203	175	50	221	0.661	0.581		
Saint Laurent	45°30'N	073°40'W	13:21:43	272	265	56	—			—			15:50:22	133	94	40	14:38:19	203	175	50	220	0.663	0.584		
Saint Léonard	45°35'N	073°35'W	13:21:51	272	264	56	—			—			15:50:14	133	94	40	14:38:19	203	175	50	221	0.661	0.581		
Shawinigan	46°33'N	072°45'W	13:23:13	270	262	55	—			—			15:48:39	135	96	40	14:38:06	203	175	49	221	0.632	0.547		
Sherbrooke	45°24'N	071°54'W	13:25:17	272	261	56	—			—			15:51:56	134	94	39	14:40:59	203	173	49	224	0.647	0.565		
Trois Riviäres	46°21'N	072°33'W	13:23:40	271	261	55	—			—			15:49:16	135	96	39	14:38:41	203	175	49	221	0.635	0.550		
Verdun	45°27'N	073°34'W	13:21:56	272	265	56	—			—			15:50:33	133	94	40	14:38:32	203	175	50	221	0.663	0.584		
Saskatchewan																									
Moose Jaw	50°23'N	105°32'W	10:29:01	275	307	40	—			—			13:03:20	122	123	51	11:44:48	199	218	48	150	0.799	0.752		
Regina	50°25'N	104°39'W	10:30:10	275	306	41	—			—			13:04:34	123	122	51	11:46:06	199	217	48	151	0.793	0.745		
Saskatoon	52°07'N	106°38'W	10:29:35	273	303	39	—			—			12:59:55	124	126	50	11:43:24	198	217	46	149	0.758	0.700		
Yukon Territory																									
Dawson	64°04'N	139°25'W	09:29:12	254	280	18	—			—			11:21:27	135	156	29	10:24:07	194	219	24	113	0.504	0.398		
Whitehorse	60°43'N	135°03'W	09:23:16	259	288	20	—			—			11:24:17	130	152	33	10:22:14	194	221	27	116	0.585	0.490		

Table 2–27: Local Circumstances for Mexico & Central America

Location Name	Latitude	Longitude	First Contact UT1 h m s	P °	V °	Alt °	Second Contact UT1 h m s	P °	V °	Third Contact UT1 h m s	P °	V °	Fourth Contact UT1 h m s	P °	V °	Alt °	Maximum Eclipse UT1 h m s	P °	V °	Alt °	Azm °	Eclip. Mag.	Eclip. Obs.	Umbral Depth	Umbral Durat.
Mexico																									
Acapulco	16°51'N	099°55'W	17:09:31	339	54	67	—			—			19:32:47	71	5	77	18:21:33	25	71	83	134	0.293	0.184		
Aguascalientes	21°53'N	102°18'W	16:48:02	328	35	59	—			—			19:27:49	79	40	77	18:07:14	24	70	75	131	0.417	0.305		
Cancún	21°10'N	086°50'W	17:22:08	320	355	79	—			—			20:13:31	93	23	55	18:51:34	27	331	73	240	0.603	0.513		
Chihuahua	28°38'N	106°05'W	16:27:36	314	14	49	—			—			19:17:03	88	81	73	17:50:07	21	66	65	129	0.589	0.496		
Culiacán	24°48'N	107°23'W	16:31:35	322	28	50	—			—			19:11:38	80	81	77	17:49:20	21	75	66	120	0.459	0.349		
Ecatepec	19°36'N	099°03'W	17:01:12	331	40	65	—			—			19:38:21	78	18	74	18:20:15	25	55	81	149	0.387	0.274		
Guadalajara	20°40'N	103°21'W	16:49:29	331	41	58	—			—			19:23:08	76	40	79	18:05:27	23	76	75	124	0.369	0.256		
Guadalupe	25°41'N	100°16'W	16:43:18	318	19	58	—			—			19:34:58	88	49	72	18:08:45	24	54	74	147	0.554	0.456		
Hermosillo	29°04'N	110°58'W	16:18:56	314	15	43	—			—			19:02:24	86	103	72	17:37:26	20	72	59	118	0.568	0.472		
Juárez	31°44'N	106°29'W	16:23:35	308	4	47	—			—			19:15:32	94	89	70	17:47:11	21	61	63	132	0.682	0.608		
León	21°07'N	101°41'W	16:51:21	329	37	60	—			—			19:29:36	78	33	77	18:10:02	24	69	77	133	0.401	0.288		
López Mateos	19°34'N	099°16'W	17:00:53	331	41	65	—			—			19:37:35	78	18	74	18:19:38	25	57	81	147	0.383	0.270		
Mérida	20°58'N	089°37'W	17:16:38	322	11	76	—			—			20:07:00	90	21	59	18:44:55	27	339	76	230	0.557	0.459		
Mexicali	32°39'N	115°28'W	16:09:38	306	5	37	—			—			18:51:15	89	118	66	17:26:35	18	69	52	115	0.659	0.580		
Mexico City	19°26'N	099°08'W	17:01:35	332	41	65	—			—			19:37:56	78	17	74	18:20:13	25	56	81	148	0.380	0.267		
Monterrey	25°40'N	100°19'W	16:43:12	318	19	58	—			—			19:34:48	88	49	72	18:08:35	24	54	74	147	0.553	0.455		
Naucalpan	19°29'N	099°14'W	17:01:14	332	41	65	—			—			19:37:38	78	18	74	18:19:52	25	57	81	147	0.381	0.268		
Nezahualcóyotl	19°25'N	099°02'W	17:01:52	332	41	65	—			—			19:38:20	78	17	74	18:20:35	25	56	81	149	0.381	0.268		
Puebla	19°03'N	098°13'W	17:04:39	332	42	67	—			—			19:41:04	78	14	73	18:23:36	25	48	82	157	0.381	0.268		
Querétaro	20°35'N	100°24'W	16:55:27	330	38	62	—			—			19:33:58	79	26	75	18:14:42	24	63	79	139	0.400	0.287		
Saltillo	25°25'N	101°00'W	16:42:21	319	20	58	—			—			19:32:45	87	50	73	18:06:57	23	57	73	143	0.539	0.438		
San Luis Potosí	22°10'N	100°59'W	16:49:52	326	32	60	—			—			19:32:33	81	36	75	18:10:53	24	63	77	139	0.440	0.329		
San Nicolás	25°45'N	100°18'W	16:43:06	318	18	58	—			—			19:34:50	88	49	72	18:08:33	24	54	74	147	0.556	0.458		
Tijuana	32°32'N	117°01'W	16:07:44	307	5	35	—			—			18:47:08	88	122	65	17:23:25	17	70	50	113	0.650	0.568		
Tlalnepantla	19°32'N	099°11'W	17:01:07	332	41	65	—			—			19:37:49	78	18	74	18:19:54	25	57	81	147	0.383	0.270		
Veracruz	19°11'N	096°08'W	17:08:19	331	37	70	—			—			19:48:20	80	13	69	18:29:41	26	22	83	184	0.413	0.300		
Zapopan	20°43'N	103°24'W	16:49:14	331	41	58	—			—			19:22:59	76	41	79	18:05:14	23	76	75	124	0.370	0.257		
Costa Rica																									
San José	09°56'N	084°05'W	18:04:54	340	233	83	—			—			20:22:52	76	347	50	19:17:28	28	296	66	277	0.338	0.226		
El Salvador																									
San Salvador	13°41'N	089°11'W	17:41:32	337	45	85	—			—			20:09:29	78	354	58	18:58:37	28	306	76	264	0.360	0.247		
Soyapango	13°42'N	089°10'W	17:41:33	337	45	85	—			—			20:09:34	78	355	58	18:58:41	28	306	75	264	0.360	0.247		
Guatemala																									
Guatemala	14°38'N	090°33'W	17:35:26	336	45	82	—			—			20:05:32	78	357	61	18:53:21	28	312	78	258	0.365	0.251		
Mixco	14°38'N	090°36'W	17:35:19	336	46	82	—			—			20:05:22	78	357	61	18:53:12	28	312	78	258	0.364	0.251		
Villa Nueva	14°32'N	090°35'W	17:35:46	337	46	82	—			—			20:05:22	78	356	61	18:53:25	28	311	78	259	0.361	0.248		
Honduras																									
San Pedro Sula	15°28'N	088°02'W	17:36:56	332	23	84	—			—			20:13:05	82	2	56	18:58:30	28	312	74	259	0.428	0.316		
Tegucigalpa	14°05'N	087°13'W	17:43:27	334	18	87	—			—			20:15:19	81	358	55	19:02:57	28	307	73	265	0.402	0.290		
Nicaragua																									
Managua	12°09'N	086°16'W	17:52:34	338	258	89	—			—			20:17:37	78	352	53	19:08:38	28	301	70	271	0.364	0.251		
Panama																									
Panamá	08°58'N	079°32'W	18:15:01	337	235	76	—			—			20:33:40	79	350	43	19:28:23	28	296	59	278	0.386	0.273		
San Miguelito	09°02'N	079°30'W	18:14:50	336	235	76	—			—			20:33:45	79	351	43	19:28:21	28	296	59	278	0.388	0.275		

Table 2–28: Local Circumstances for South America

Location Name	Latitude	Longitude	First Contact UT1				Second Contact UT1			Third Contact UT1			Fourth Contact UT1				Maximum Eclipse UT1					Eclip. Mag.	Eclip. Obs.	Umbral Depth	Umbral Durat.
			h m s	P °	V °	Alt °	h m s	P °	V °	h m s	P °	V °	h m s	P °	V °	Alt °	h m s	P °	V °	Alt °	Azm °				
Brazil																									
Belém	01°26'S	048°30'W	19:13:33	324	226	30	—			—			21:03:28	81	349	3	20:11:25	23	288	16	283	0.496	0.388		
Brasília	15°47'S	047°55'W	19:55:33	2	252	15	—			—			20:39:20	42	295	5	20:17:46	22	273	10	285	0.065	0.020		
Fortaleza	03°47'S	038°35'W	19:20:33	322	224	18	—			—			—				20:14:01	21	286	5	282	0.509	0.403		
Goiânia	16°43'S	049°16'W	20:04:04	10	259	14	—			—			20:31:15	35	286	8	20:17:42	22	272	11	286	0.024	0.005		
Jaboatão	08°07'S	035°01'W	19:28:14	328	227	12	—			—			—				20:15:33	20	282	1	282	0.402	0.288		
João Pessoa	07°07'S	034°52'W	19:26:26	326	226	12	—			—			—				20:15:09	20	283	1	282	0.433	0.320		
Maceió	09°39'S	035°45'W	19:31:19	332	230	11	—			—			—				20:16:10	20	280	1	282	0.349	0.235		
Manaus	03°07'S	060°01'W	19:13:00	339	234	41	—			—			20:56:07	71	334	16	20:07:09	25	285	28	285	0.323	0.211		
Natal	05°48'S	035°13'W	19:24:10	323	225	13	—			—			—				20:14:40	20	284	1	282	0.471	0.361		
Recife	08°05'S	034°55'W	19:28:10	328	227	12	—			—			—				20:15:31	20	282	1	282	0.403	0.289		
Salvador	12°58'S	038°30'W	19:39:35	342	236	11	—			—			—				20:17:23	21	277	2	283	0.227	0.126		
São Luís	02°30'S	044°18'W	19:17:06	323	225	25	—			—			—				20:13:00	22	287	11	283	0.502	0.395		
Teresina	05°06'S	042°48'W	19:22:29	327	227	21	—			—			—				20:14:38	21	284	9	283	0.436	0.324		
Colombia																									
Barranquilla	10°58'N	074°48'W	18:15:19	327	236	72	—			—			20:42:24	87	2	36	19:33:39	27	300	53	275	0.515	0.410		
Bogotá	04°38'N	074°05'W	18:37:39	339	234	65	—			—			20:42:55	76	343	34	19:43:56	27	291	49	283	0.352	0.239		
Bucaramanga	07°08'N	073°08'W	18:29:59	333	233	66	—			—			20:45:22	81	352	33	19:41:56	27	294	49	280	0.436	0.324		
Cali	03°26'N	076°31'W	18:40:13	345	236	66	—			—			20:37:03	71	336	38	19:41:45	28	288	51	285	0.281	0.172		
Cartagena	10°24'N	075°30'W	18:15:59	329	236	73	—			—			20:41:24	85	359	37	19:33:22	27	299	54	276	0.489	0.382		
Cúcuta	07°53'N	072°31'W	18:28:21	331	233	66	—			—			20:46:22	83	355	33	19:41:48	27	296	48	279	0.465	0.356		
Medellín	06°17'N	075°32'W	18:30:00	337	234	68	—			—			20:40:59	78	347	37	19:39:24	28	293	52	281	0.376	0.262		
Ecuador																									
Guayaquil	02°13'S	079°54'W	19:11:24	7	251	60	—			—			20:13:32	49	303	45	19:43:13	28	278	52	293	0.068	0.021		
Quito	00°11'S	078°30'W	18:57:02	357	243	62	—			—			20:25:50	59	318	42	19:43:10	28	282	52	290	0.148	0.067		
Peru																									
Iquitos	03°45'S	073°12'W	19:14:39	358	246	52	—			—			20:32:25	55	313	34	19:54:53	27	280	43	290	0.128	0.054		
Venezuela																									
Barquisimeto	10°03'N	069°18'W	18:26:28	323	232	64	—			—			20:50:10	90	5	29	19:43:11	27	299	45	277	0.572	0.475		
Caracas	10°32'N	066°56'W	18:28:40	320	230	61	—			—			20:52:28	92	8	26	19:45:30	26	300	42	276	0.618	0.530		
Ciudad Guayana	08°22'N	062°37'W	18:40:00	320	227	54	—			—			20:57:03	91	6	20	19:53:08	25	297	36	279	0.617	0.528		
Maracaibo	10°44'N	071°40'W	18:20:57	325	234	68	—			—			20:47:00	89	4	32	19:38:53	27	300	49	276	0.556	0.457		
Valencia	10°14'N	067°59'W	18:27:56	322	231	62	—			—			20:51:33	91	6	27	19:44:38	26	299	43	277	0.595	0.503		

Table 2–29: Local Circumstances for Europe

Location Name	Latitude	Longitude	First Contact UT1 h m s	P °	V °	Alt °	Second Contact UT1 h m s	P °	V °	Third Contact UT1 h m s	P °	V °	Fourth Contact UT1 h m s	P °	V °	Alt °	Maximum Eclipse UT1 h m s	P °	V °	Alt °	Azm °	Eclip. Mag.	Eclip. Obs.	Umbral Depth	Umbral Durat.
Belgium																									
Antwerp	51°13'N	004°25'E	18:40:42	222	184	1	—			—			—				18:49:40	–	–	0	290	0.060	0.017		
Brussels	50°50'N	004°20'E	18:40:49	222	184	1	—			—			—				18:49:08	–	–	0	290	0.058	0.017		
France																									
Nantes	47°14'N	001°34'W	18:42:13	230	187	3	—			—			—				19:05:11	–	–	0	288	0.150	0.068		
Paris	48°52'N	002°20'E	18:41:35	226	186	2	—			—			—				18:52:49	–	–	0	289	0.084	0.029		
Ireland																									
Dublin	53°20'N	006°15'W	18:38:30	223	186	8	—			—			19:26:57	171	135	1	19:02:58	197	161	5	284	0.101	0.038		
Netherlands																									
Amsterdam	52°22'N	004°53'E	18:40:20	220	184	1	—			—			—				18:50:31	–	–	0	290	0.060	0.018		
Rotterdam	51°56'N	004°29'E	18:40:29	221	184	1	—			—			—				18:51:06	–	–	0	290	0.065	0.020		
Utrecht	52°06'N	005°07'E	18:40:25	220	184	1	—			—			—				18:48:59	–	–	0	290	0.054	0.015		
Portugal																									
Lisbon	38°43'N	009°08'W	18:45:58	244	192	6	—			—			—				19:21:16	–	–	0	286	0.301	0.189		
Spain																									
Bilbao	43°15'N	002°56'W	18:44:00	235	189	3	—			—			—				19:03:28	–	–	0	287	0.173	0.084		
Cordoba	37°53'N	004°46'W	18:46:39	243	191	3	—			—			—				19:02:36	–	–	0	286	0.185	0.093		
Gijon	43°32'N	005°40'W	18:43:43	236	189	5	—			—			—				19:14:54	–	–	0	287	0.217	0.118		
Las Palmas	28°06'N	015°26'W	18:51:45	261	198	8	—			—			—				19:33:03	–	–	0	284	0.509	0.402		
Madrid	40°25'N	003°43'W	18:45:22	240	190	3	—			—			—				19:02:04	–	–	0	286	0.175	0.086		
Málaga	36°43'N	004°25'W	18:47:13	245	192	2	—			—			—				18:59:35	–	–	0	285	0.154	0.071		
Sevilla	37°24'N	005°59'W	18:46:54	244	192	3	—			—			—				19:06:46	–	–	0	285	0.225	0.124		
Valladolid	41°39'N	004°44'W	18:44:44	238	190	4	—			—			—				19:08:06	–	–	0	286	0.211	0.113		
Vigo	42°13'N	008°43'W	18:44:02	239	191	7	—			—			—				19:20:32	197	151	1	286	0.248	0.143		
Zaragoza	41°39'N	000°53'W	18:44:41	237	189	1	—			—			—				18:52:42	–	–	0	286	0.088	0.031		
United Kingdom																									
Belfast	54°36'N	005°56'W	18:37:56	221	185	8	—			—			19:23:24	173	138	2	19:00:52	197	162	5	284	0.088	0.031		
Birmingham	52°29'N	001°55'W	18:39:52	222	185	5	—			—			—				19:03:31	197	160	2	288	0.098	0.036		
Bradford	53°48'N	001°45'W	18:39:20	221	184	5	—			—			—				19:01:25	197	161	2	287	0.084	0.029		
Bristol	51°28'N	002°36'W	18:40:13	224	185	5	—			—			—				19:05:15	197	159	2	287	0.110	0.043		
Cardiff	51°29'N	003°11'W	18:40:07	224	186	6	—			—			—				19:05:21	197	159	2	287	0.111	0.044		
Coventry	52°25'N	001°30'W	18:39:58	222	185	5	—			—			—				19:03:31	197	160	2	288	0.098	0.036		
Edinburgh	55°57'N	003°13'W	18:38:02	218	184	7	—			—			19:18:12	175	142	2	18:58:15	197	163	4	285	0.069	0.021		
Glasgow	55°52'N	004°16'W	18:37:47	219	184	7	—			—			19:19:01	175	142	2	18:58:33	197	163	5	285	0.072	0.023		
Kingston upon Hull	53°45'N	000°22'W	18:39:34	220	184	5	—			—			—				19:01:11	197	161	2	288	0.082	0.028		
Leeds	53°49'N	001°33'W	18:39:21	220	184	5	—			—			—				19:01:22	197	161	2	287	0.084	0.029		
Leicester	52°38'N	001°08'W	18:39:55	222	185	5	—			—			—				19:03:05	197	160	2	288	0.094	0.034		
Liverpool	53°25'N	002°59'W	18:39:15	222	185	6	—			—			—				19:02:16	197	161	3	286	0.091	0.033		
London	51°31'N	000°06'W	18:40:30	223	185	4	—			—			—				19:04:33	197	160	1	289	0.104	0.040		
Manchester	53°29'N	002°15'W	18:39:22	221	185	6	—			—			—				19:02:02	197	161	3	287	0.089	0.031		
Sheffield	53°23'N	001°29'W	18:39:33	221	184	5	—			—			—				19:02:00	197	161	2	288	0.088	0.031		
Stoke-on-Trent	53°01'N	002°11'W	18:39:36	222	185	5	—			—			—				19:02:45	197	161	2	287	0.093	0.034		
Wolverhampton	52°35'N	002°09'W	18:39:47	222	185	5	—			—			—				19:03:24	197	160	2	287	0.097	0.036		

Table 2–30: Local Circumstances for Africa & North Atlantic

Location Name	Latitude	Longitude	First Contact UT1 h m s	P °	V °	Alt °	Second Contact UT1 h m s	P °	V °	Third Contact UT1 h m s	P °	V °	Fourth Contact UT1 h m s	P °	V °	Alt °	Maximum Eclipse UT1 h m s	P °	V °	Alt °	Azm °	Eclip. Mag.	Eclip. Obs.	Umbral Depth	Umbral Durat.
Gambia																									
Serre Kunda	13°27'N	016°41'W	19:02:23	282	205	5	—			—			—				19:23:34	–	–	0	282	0.375	0.260		
Guinea																									
Conakry	09°33'N	013°40'W	19:05:10	287	207	1	—			—			—				19:08:04	–	–	0	282	0.053	0.014		
Guinea-Bissau																									
Bissau	11°52'N	015°36'W	19:03:35	284	206	3	—			—			—				19:17:50	–	–	0	282	0.255	0.149		
Mauritania																									
Nouakchott	18°05'N	015°59'W	18:58:49	275	202	6	—			—			—				19:25:02	–	–	0	283	0.444	0.332		
Morocco																									
Agadir	30°25'N	009°37'W	18:50:46	255	195	4	—			—			—				19:12:25	–	–	0	284	0.298	0.187		
Casablanca	33°36'N	007°37'W	18:48:57	250	194	4	—			—			—				19:08:18	–	–	0	285	0.249	0.144		
Fes	34°03'N	005°00'W	18:48:37	249	193	2	—			—			—				18:58:24	–	–	0	285	0.133	0.057		
Kenitra	34°16'N	006°35'W	18:48:34	249	193	3	—			—			—				19:05:00	–	–	0	285	0.212	0.114		
Marrakech	31°38'N	008°00'W	18:50:03	253	195	3	—			—			—				19:07:24	–	–	0	284	0.239	0.135		
Meknes	33°54'N	005°34'W	18:48:44	249	193	2	—			—			—				19:00:27	–	–	0	285	0.158	0.074		
Rabat	34°01'N	006°50'W	18:48:43	249	193	3	—			—			—				19:05:43	–	–	0	285	0.220	0.120		
Safi	32°19'N	009°14'W	18:49:39	253	195	4	—			—			—				19:13:12	–	–	0	284	0.303	0.192		
Tangier	35°47'N	005°49'W	18:47:45	247	193	3	—			—			—				19:03:54	–	–	0	285	0.200	0.104		
Tetouan	35°35'N	005°22'W	18:47:51	247	193	2	—			—			—				19:01:51	–	–	0	285	0.178	0.088		
Senegal																									
Dakar	14°43'N	017°29'W	19:01:23	281	204	6	—			—			—				19:27:55	–	–	0	282	0.463	0.352		
Diourbel	14°40'N	016°14'W	19:01:26	280	204	5	—			—			—				19:22:54	–	–	0	282	0.377	0.263		
Kaolack	14°09'N	016°06'W	19:01:49	281	205	4	—			—			—				19:21:52	–	–	0	282	0.354	0.240		
Mbour	14°26'N	016°58'W	19:01:37	281	205	5	—			—			—				19:25:36	–	–	0	282	0.421	0.308		
Saint-Louis	16°02'N	016°29'W	19:00:22	278	204	5	—			—			—				19:25:09	–	–	0	282	0.429	0.316		
Thiès	14°49'N	016°56'W	19:01:19	280	204	5	—			—			—				19:25:48	–	–	0	282	0.428	0.315		
Ziguinchor	12°35'N	016°17'W	19:03:03	283	205	4	—			—			—				19:21:14	–	–	0	282	0.323	0.211		
Western Sahara																									
al-'Ayun	27°10'N	013°12'W	18:52:38	261	198	6	—			—			—				19:23:06	–	–	0	284	0.429	0.316		
Cape Verde																									
Mindelo	16°53'N	024°59'W	18:58:27	281	205	14	—			—			—				19:56:01	199	126	1	282	0.840	0.800		
Praia	14°56'N	023°32'W	19:00:34	283	206	12	—			—			—				19:52:21	–	–	0	282	0.849	0.812		
Iceland																									
Reykjavík	64°08'N	021°55'W	18:21:38	219	193	18	—			—			19:05:41	176	149	13	18:43:48	198	171	16	265	0.070	0.022		

Section 3: Detailed Maps of the Umbral Path

3.1 Introduction

The path of totality is plotted on a series of detailed maps appearing in Figures 3–1 to 3–13. The maps were carefully chosen with a small (~10%) overlap between each, and cover the entire umbral path across the USA at a fixed scale.

The map scale is ~1:1,400,000, which corresponds to 1 cm ≈ 14 km (1 inch ≈ 22 miles).[1] This is a large enough scale to show major roads and highways, towns and cities, rivers, lakes, parks, national forests, wilderness areas and mountain ranges. A 50 km (30 mi) reference scale appears at the bottom of each map.

The umbral path across the USA ranges in width from 99 km (62 mi) along the Oregon coast to a maximum of 115 km (71 mi) in Kentucky. It is depicted on each map as a lightly shaded region with the northern and southern path limits clearly labeled. The total eclipse can be seen only inside this path (a partial eclipse is visible outside the path). The closer one gets to the central line, the longer the total eclipse lasts. Curves of constant total eclipse duration are marked and labeled in 30 second intervals. This makes it easy to quickly estimate the duration from any location inside the umbral path.

The Universal Time (UT1) of mid-eclipse is marked by a series of yellow lines crossing the umbral eclipse path at 10 minute intervals. The eclipse circumstances on the central line are labeled with the UT1 time of mid-eclipse, the duration of totality (minutes and seconds) and the altitude of the Sun. Addition information and local circumstances at each point can be found in Tables 1–1 through 1–6.

Among the larger cities in the umbral path, there are 3 located on or just inside the northern limit (Bowling Green, SC; Lincoln, NE; St. Louis, MO), and another 4 on or just inside the southern limit (Charleston, SC; Kansas City, KS; Nashville, TN; North Platte, NE). As a consequence of this close proximity to the path limits, the duration of totality in these cities varies significantly depending on one's exact geographic location. In fact, some parts of Lincoln, Kansas City and St. Louis actually lie outside the path, in which case only a partial eclipse will be visible from those areas. On the other hand, Knoxville, TN, is completely outside the path so its citizens will witness a partial eclipse. Fortunately, Knoxville lies within 10 km (6 miles) of the northern limit, so it's just a short trip to get into the path of totality.

Figures 3–14 to 3–21 consist of maps for each of these cities in alphabetic order. The map scale has been increased to ~1:163,000, which corresponds to 1 cm ≈ 1.6 km (1 inch ≈ 2.6 miles). A 15 km (10 mi) reference scale appears at the bottom of each map. The figures include a series of lines of constant duration of totality at 10 second intervals. This makes it possible to easily determine whether a location is in the path of totality and how long the total eclipse will last.

All the maps in Section 3 were produced using Google Maps as the underlying map with overlying eclipse graphics generated using Javascript code. A web page is available to the user for examining any part of the 2017 eclipse path at a range of zoom magnifications. An added benefit of the web page is that it automatically calculates the local circumstances for any point the user chooses, with just the click of a mouse. For more information and to access the interactive 2017 eclipse path plotted on Google Maps, visit:

eclipsewise.com/solar/SEgmap/2001-2100/SE2017Aug21Tgmap.html

[1] Because of the Mercator map projection, the actual scale on a given map can vary by up to 10% of this value.

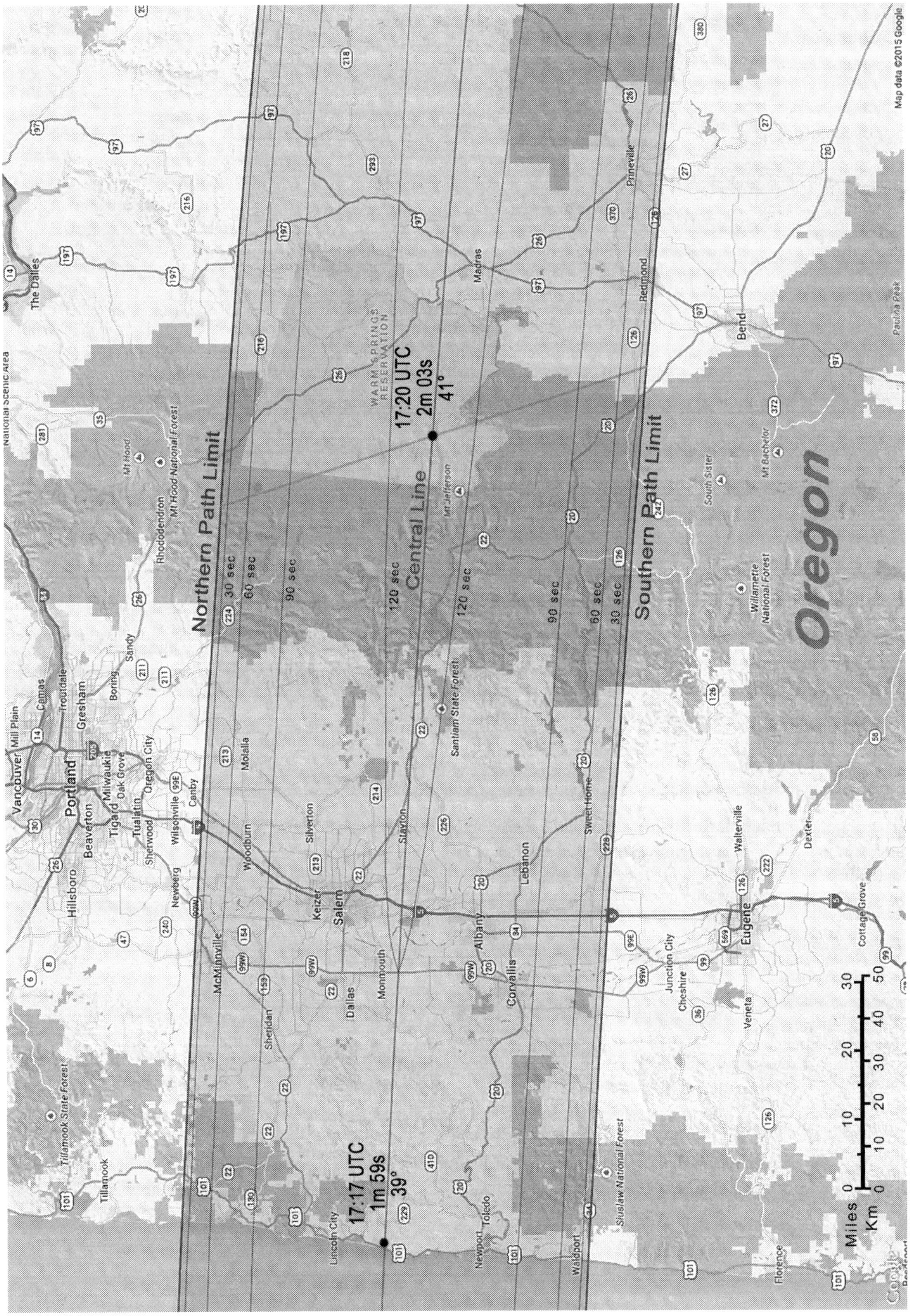

Figure 3–1: Detailed Path Map – Western Oregon

Total Solar Eclipse of 2017 August 21 (Espenak & Anderson)

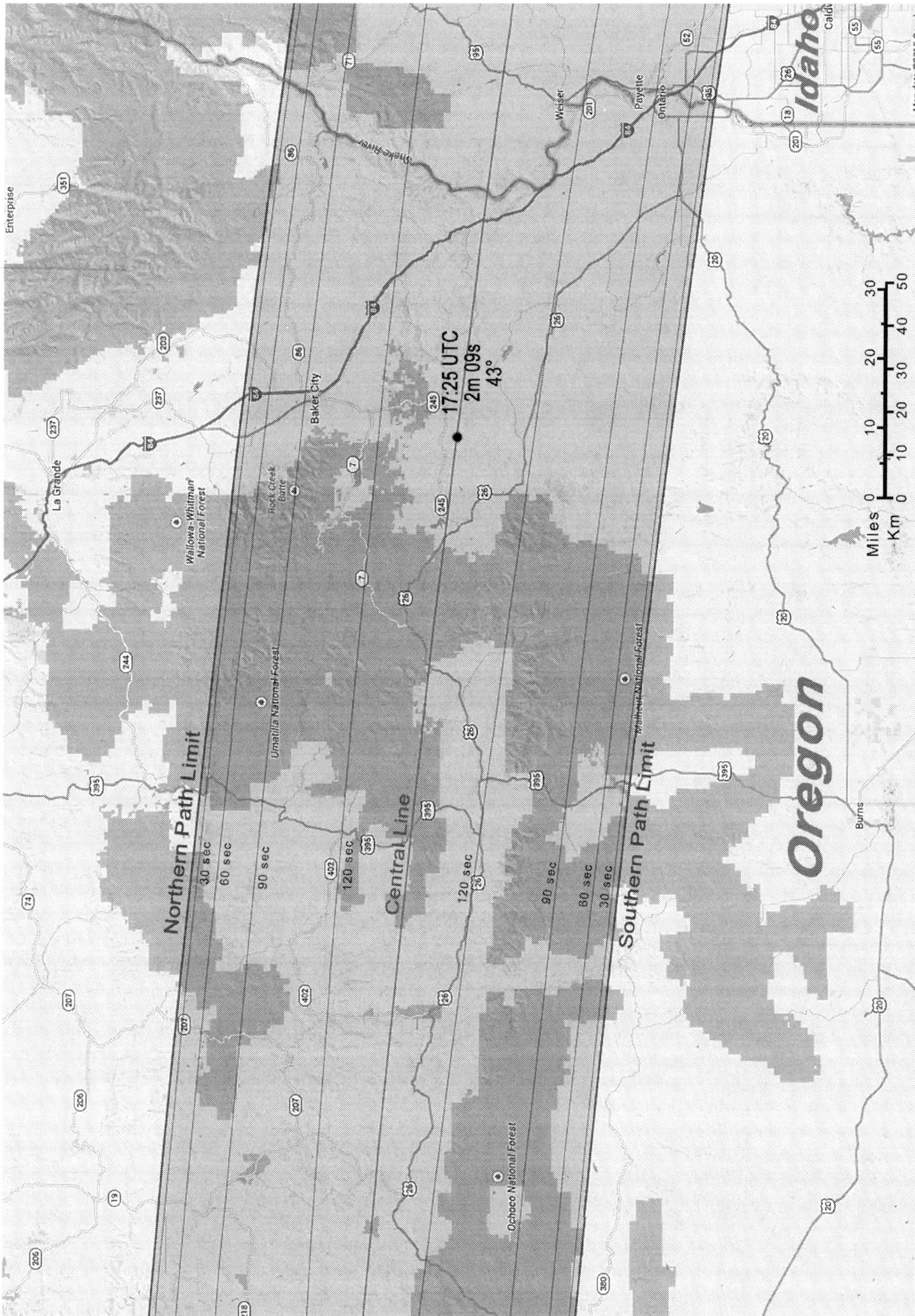

Total Solar Eclipse of 2017 August 21 (Espenak & Anderson)

Figure 3–2: Detailed Path Map – Eastern Oregon

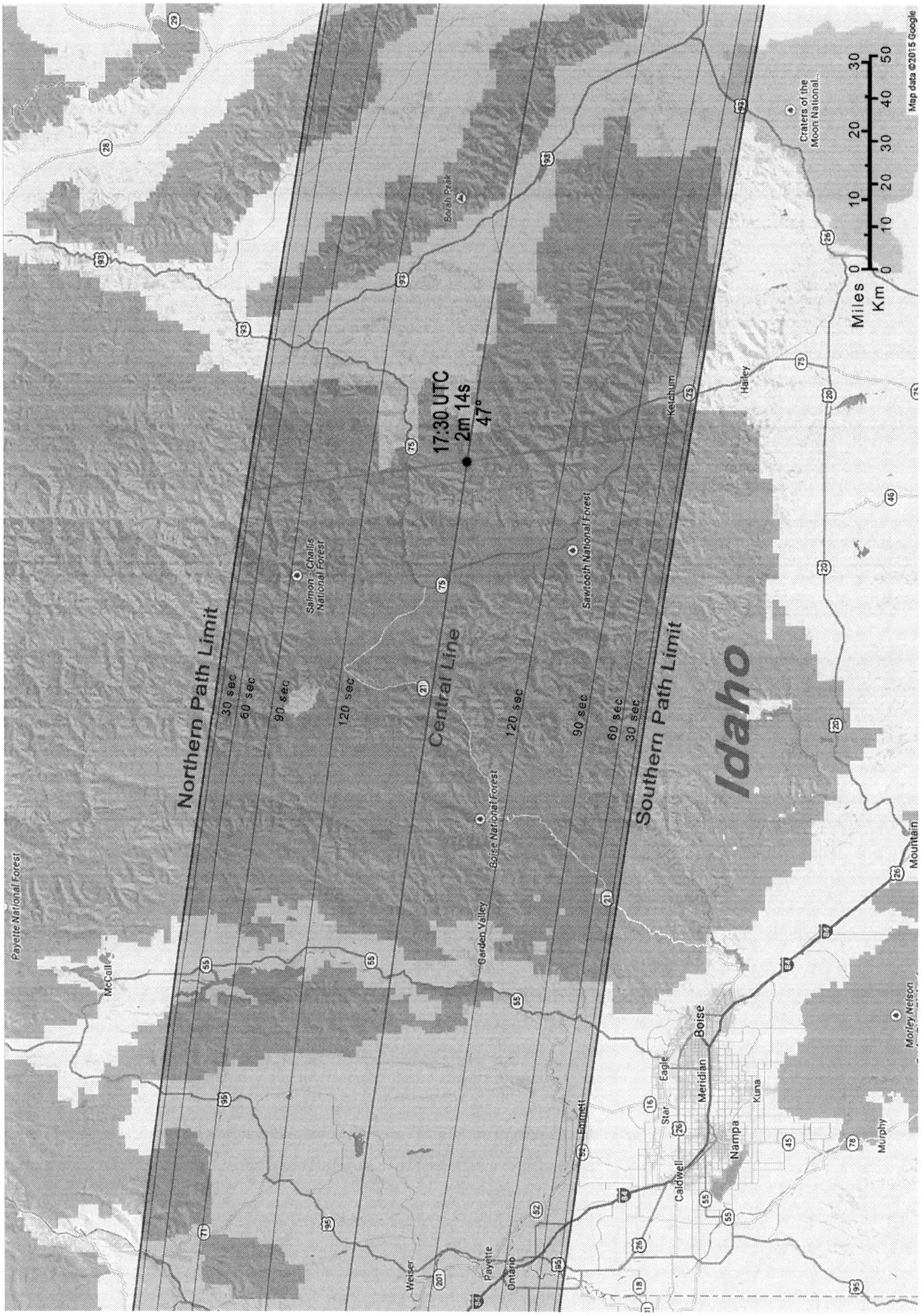

Figure 3–3: Detailed Path Map – Western Idaho

Total Solar Eclipse of 2017 August 21 (Espenak & Anderson)

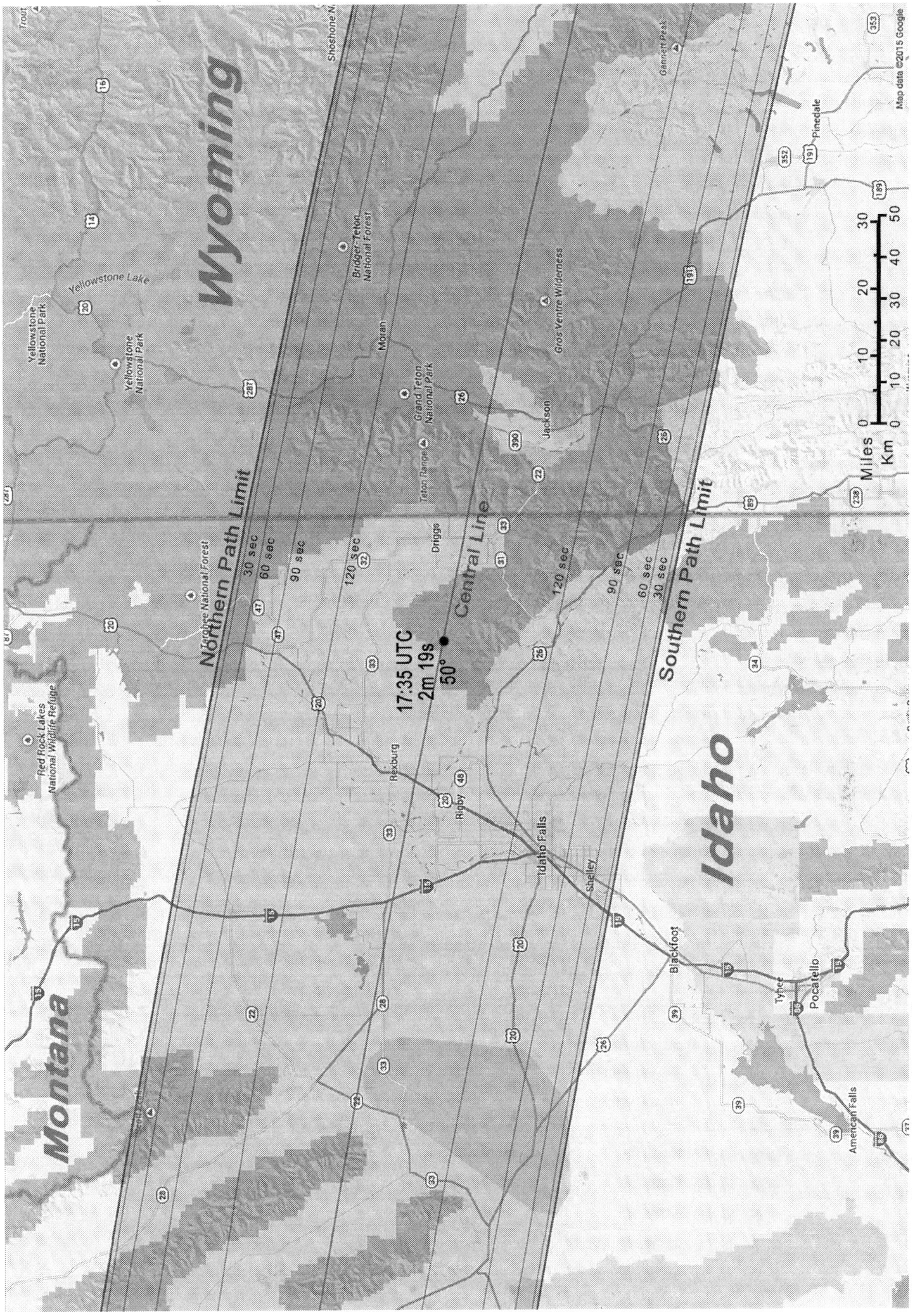

Figure 3–4: Detailed Path Map – Idaho & Wyoming

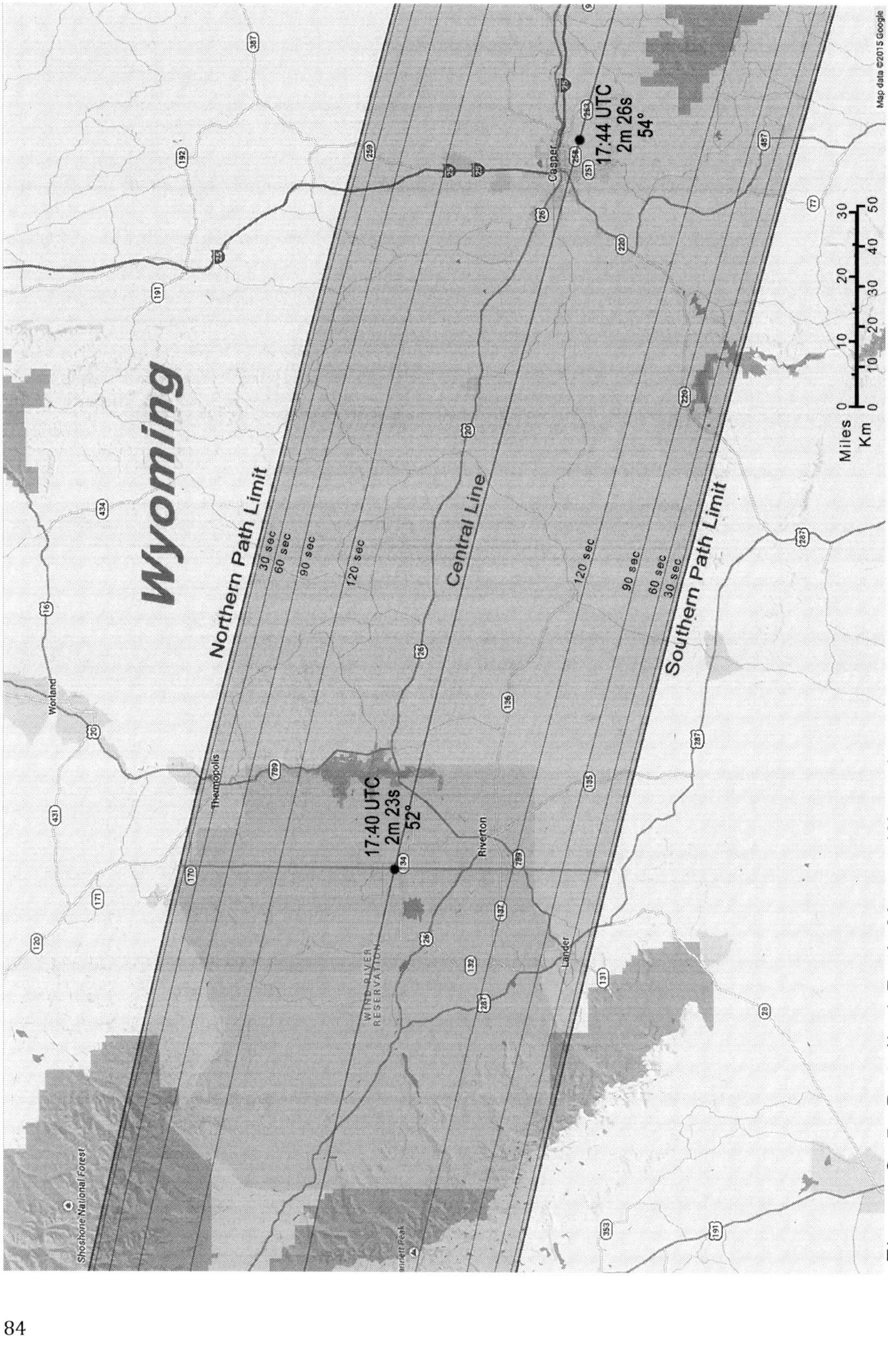

Figure 3–5: Detailed Path Map – Wyoming

Total Solar Eclipse of 2017 August 21 (Espenak & Anderson)

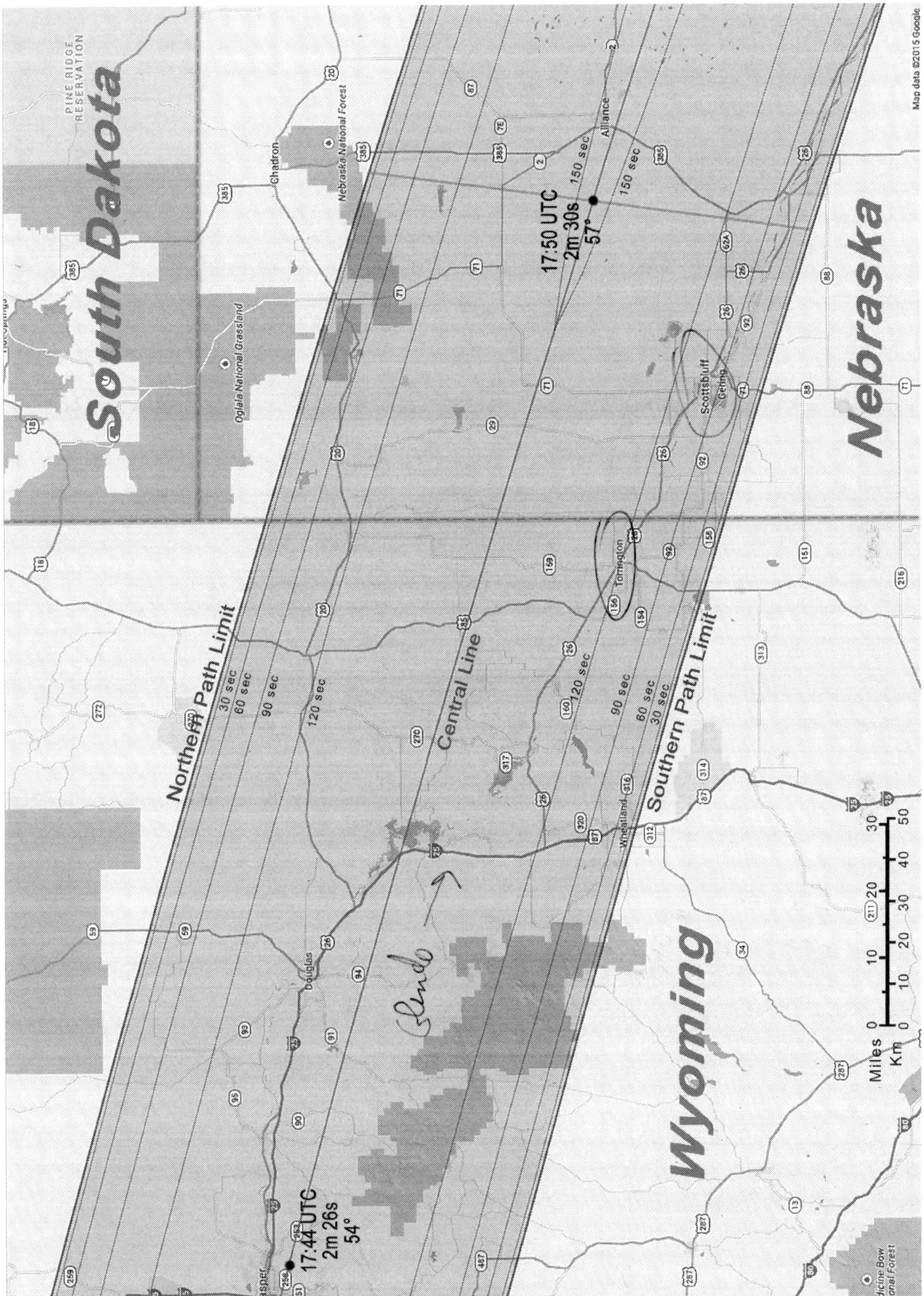

Figure 3–6: Detailed Path Map – Wyoming & Nebraska

Total Solar Eclipse of 2017 August 21 (Espenak & Anderson)

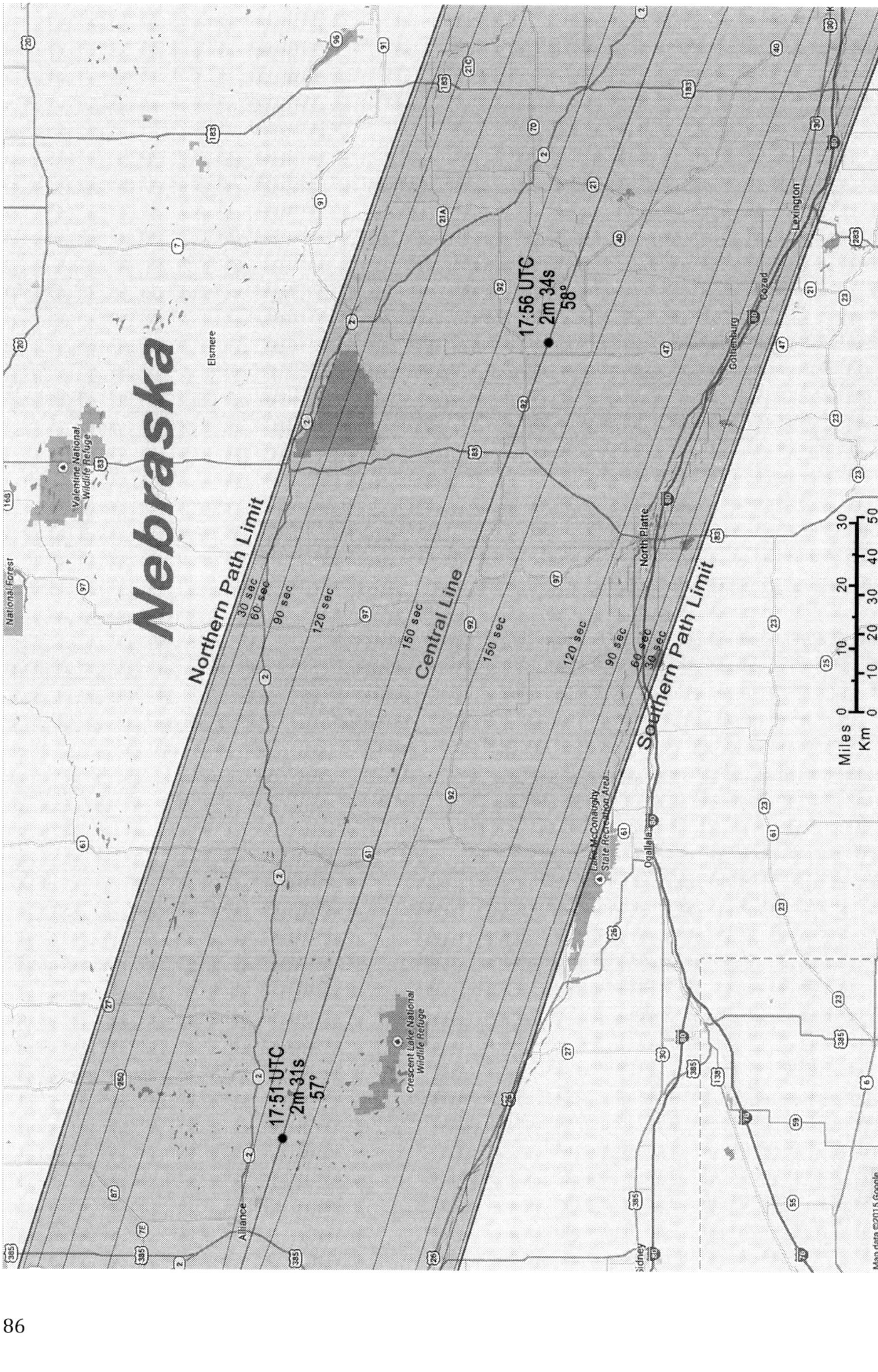

Figure 3–7: Detailed Path Map – Nebraska

Total Solar Eclipse of 2017 August 21 (Espenak & Anderson)

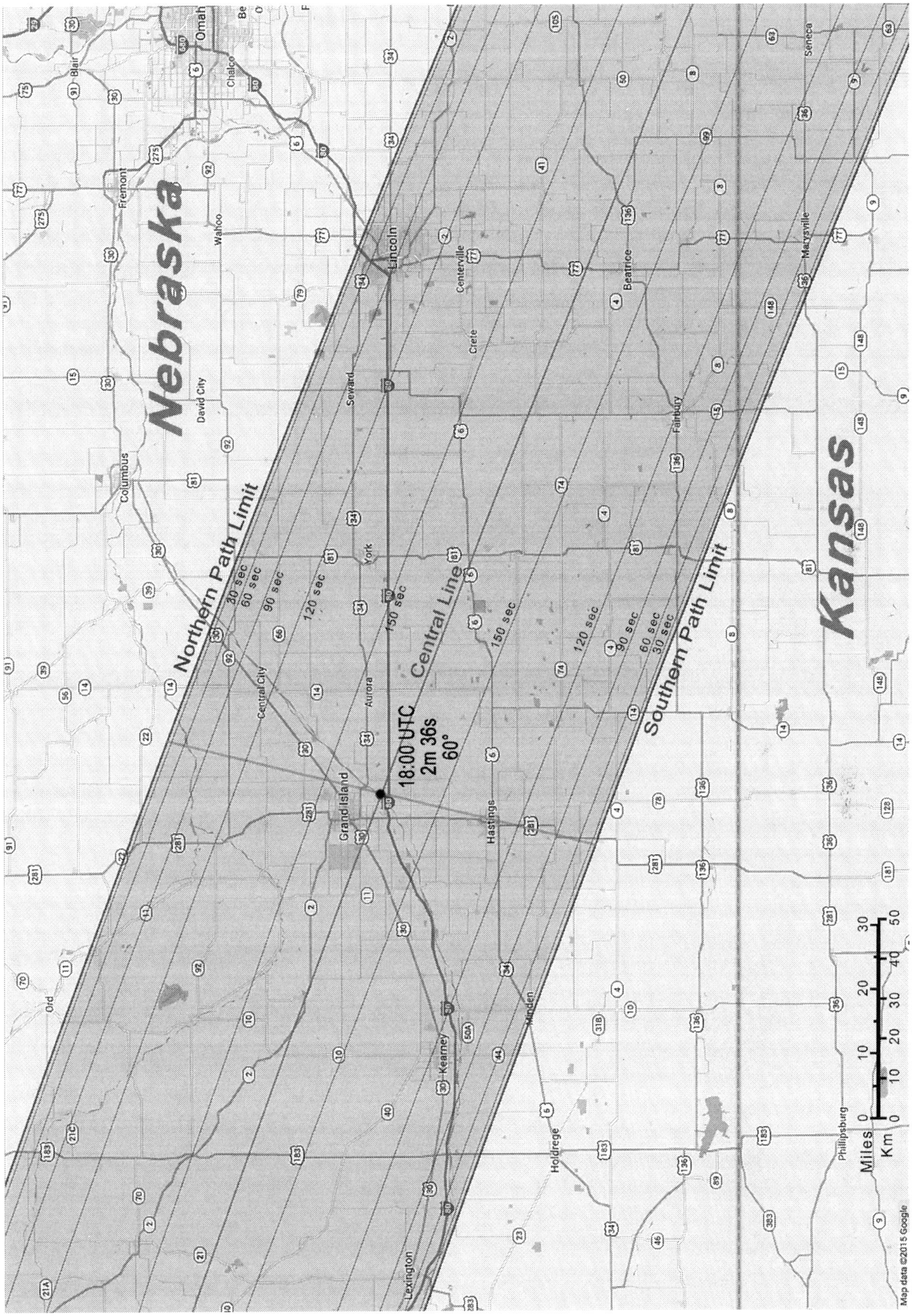

Figure 3–8: Detailed Path Map – Nebraska & Kansas   Total Solar Eclipse of 2017 August 21 (Espenak & Anderson)

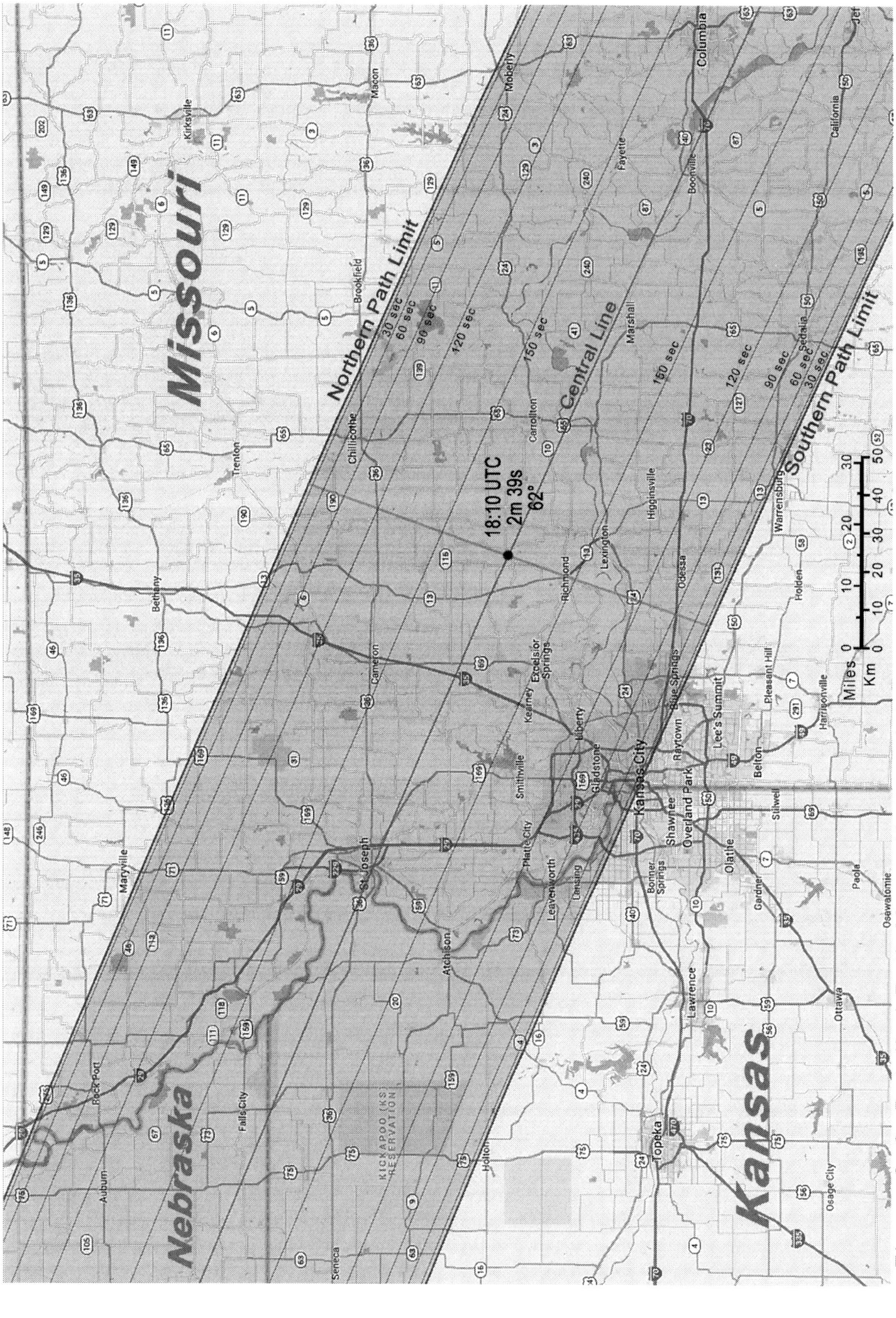

Figure 3–9: Detailed Path Map – Kansas & Missouri

Total Solar Eclipse of 2017 August 21 (Espenak & Anderson)

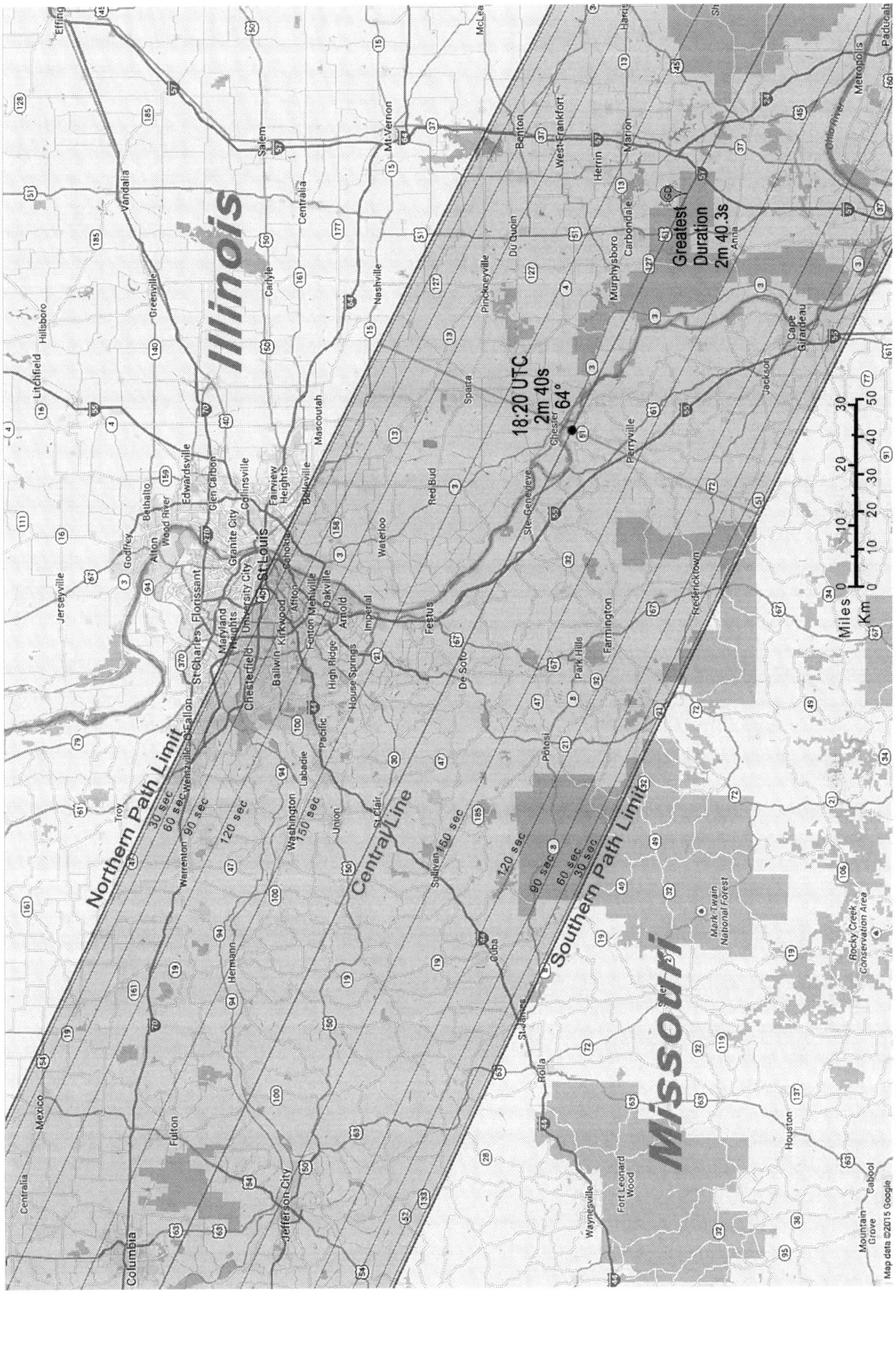

Figure 3–10: Detailed Path Map – Missouri & Illinois   Total Solar Eclipse of 2017 August 21 (Espenak & Anderson)

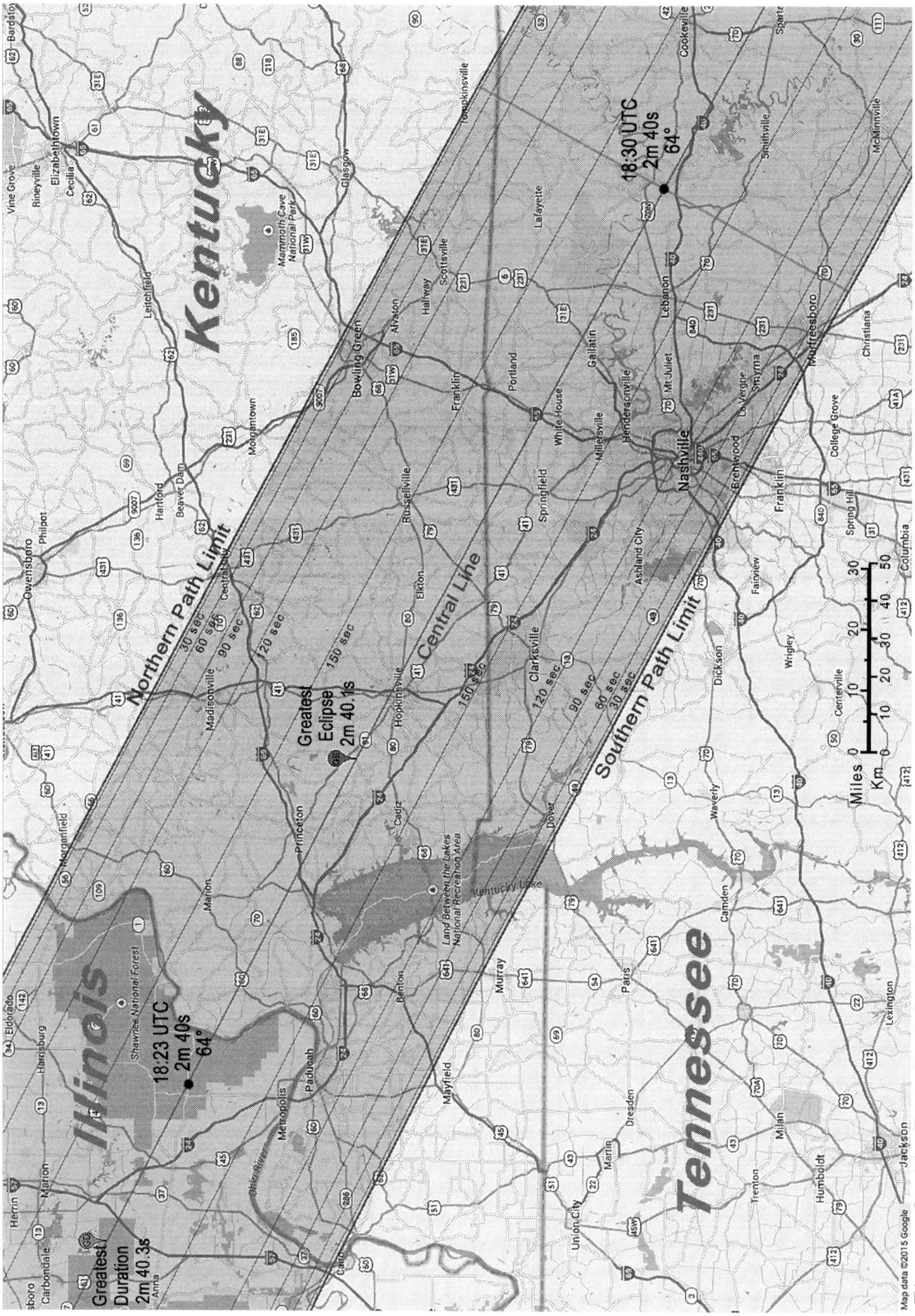

Figure 3–11: Detailed Path Map – Kentucky & Tennessee

Total Solar Eclipse of 2017 August 21 (Espenak & Anderson)

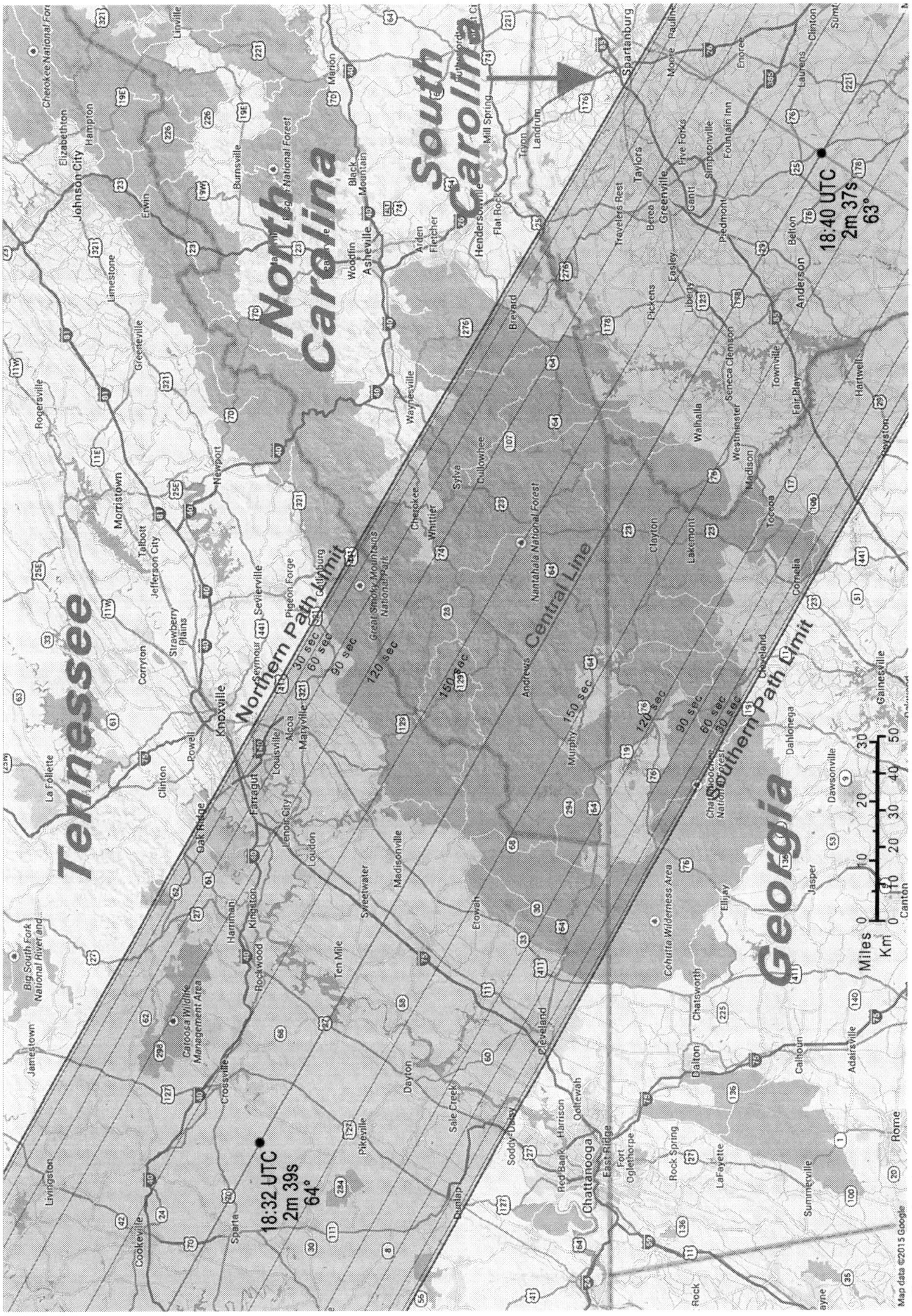

Figure 3–12: Detailed Path Map – Tennessee, Geogia & Carolinas Total Solar Eclipse of 2017 August 21 (Espenak & Anderson)

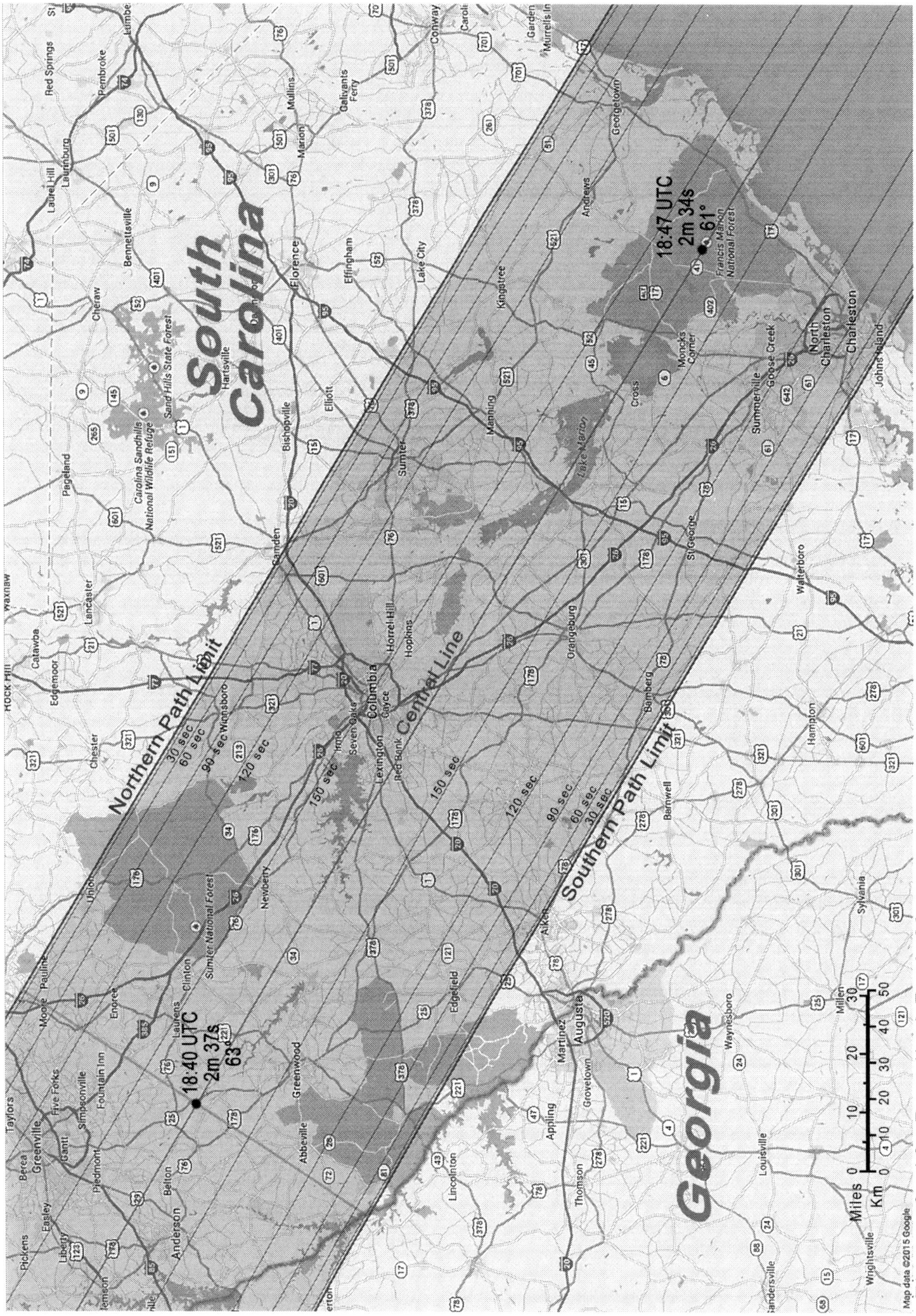

Figure 3–13: Detailed Path Map – South Carolina

Total Solar Eclipse of 2017 August 21 (Espenak & Anderson)

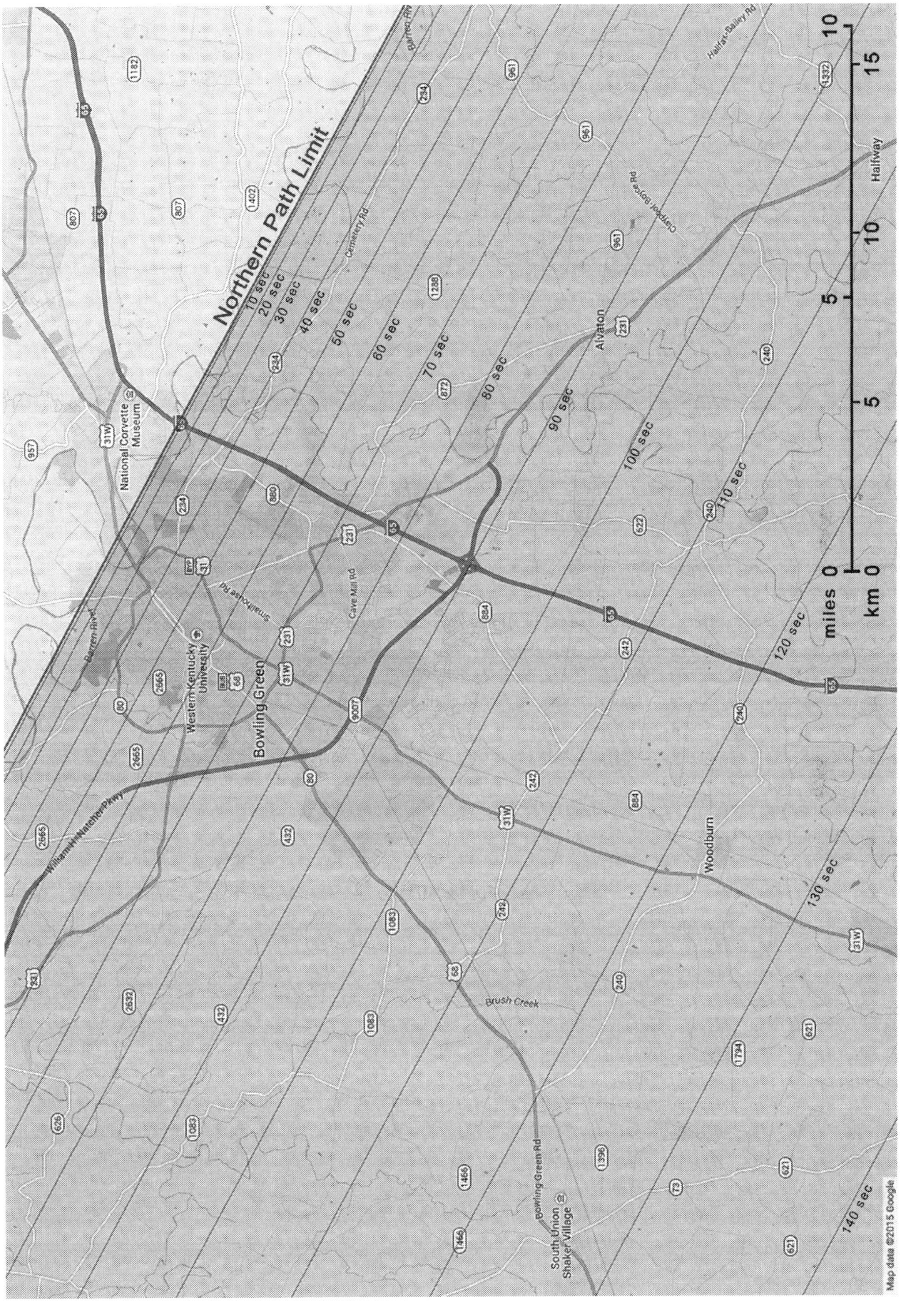

Figure 3–14: Detailed Path Map – Bowling Green, KY

Total Solar Eclipse of 2017 August 21 (Espenak & Anderson)

Total Solar Eclipse of 2017 August 21 (Espenak & Anderson)

Figure 3–15: Detailed Path Map – Charleston, SC

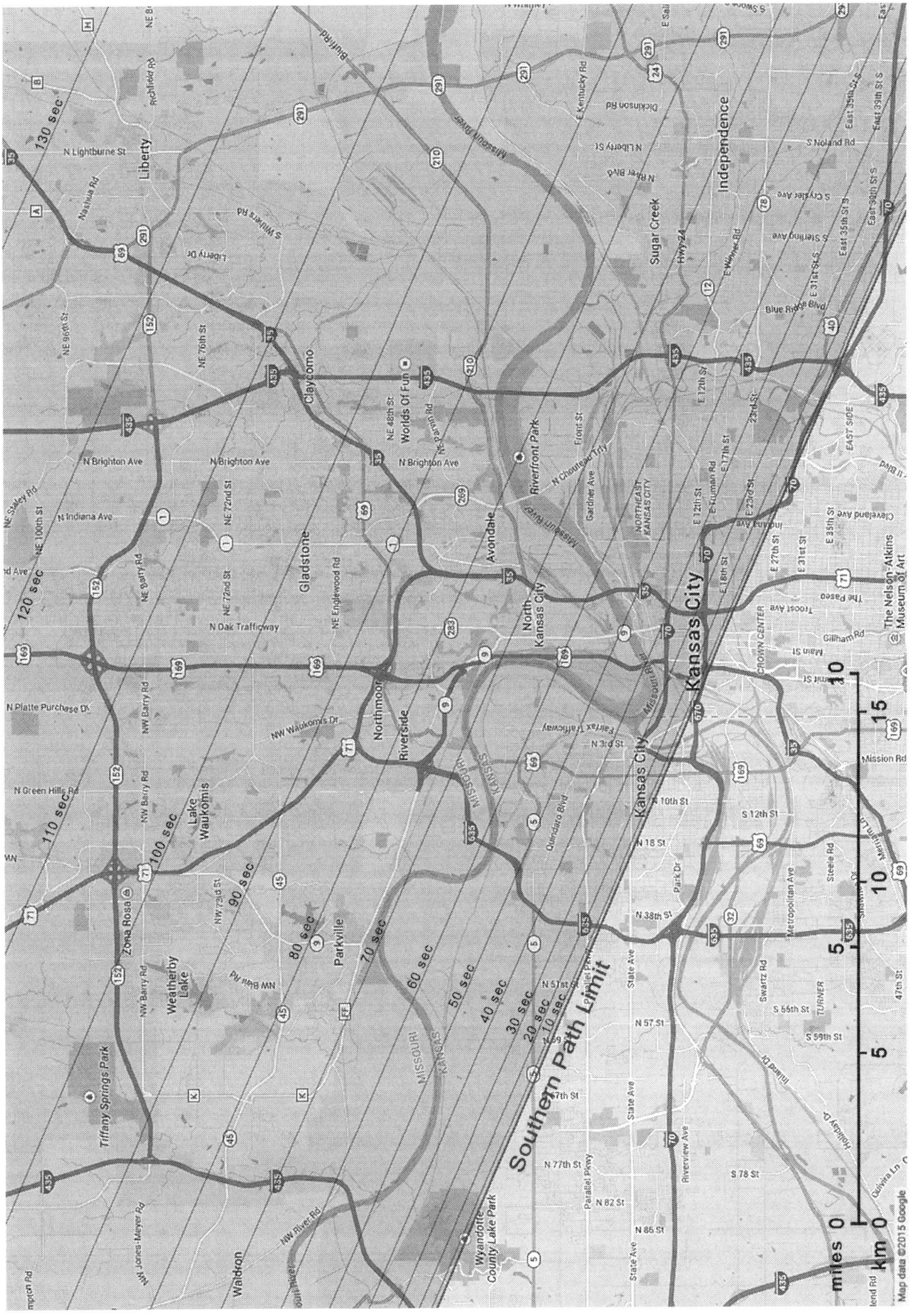

Figure 3–16: Detailed Path Map – Kansas City, KS

Total Solar Eclipse of 2017 August 21 (Espenak & Anderson)

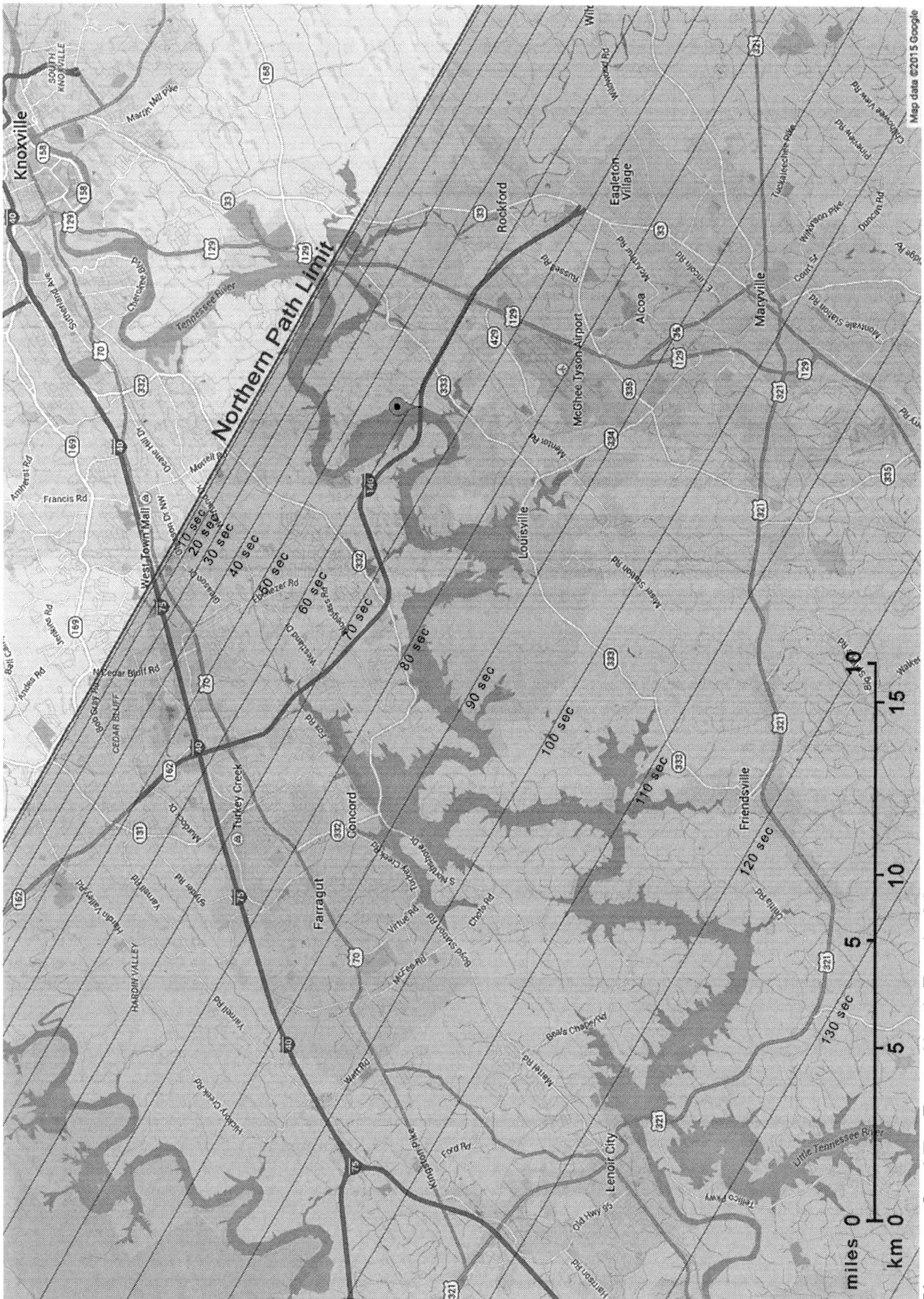

Total Solar Eclipse of 2017 August 21 (Espenak & Anderson)

Figure 3–17: Detailed Path Map – Knoxville, TN

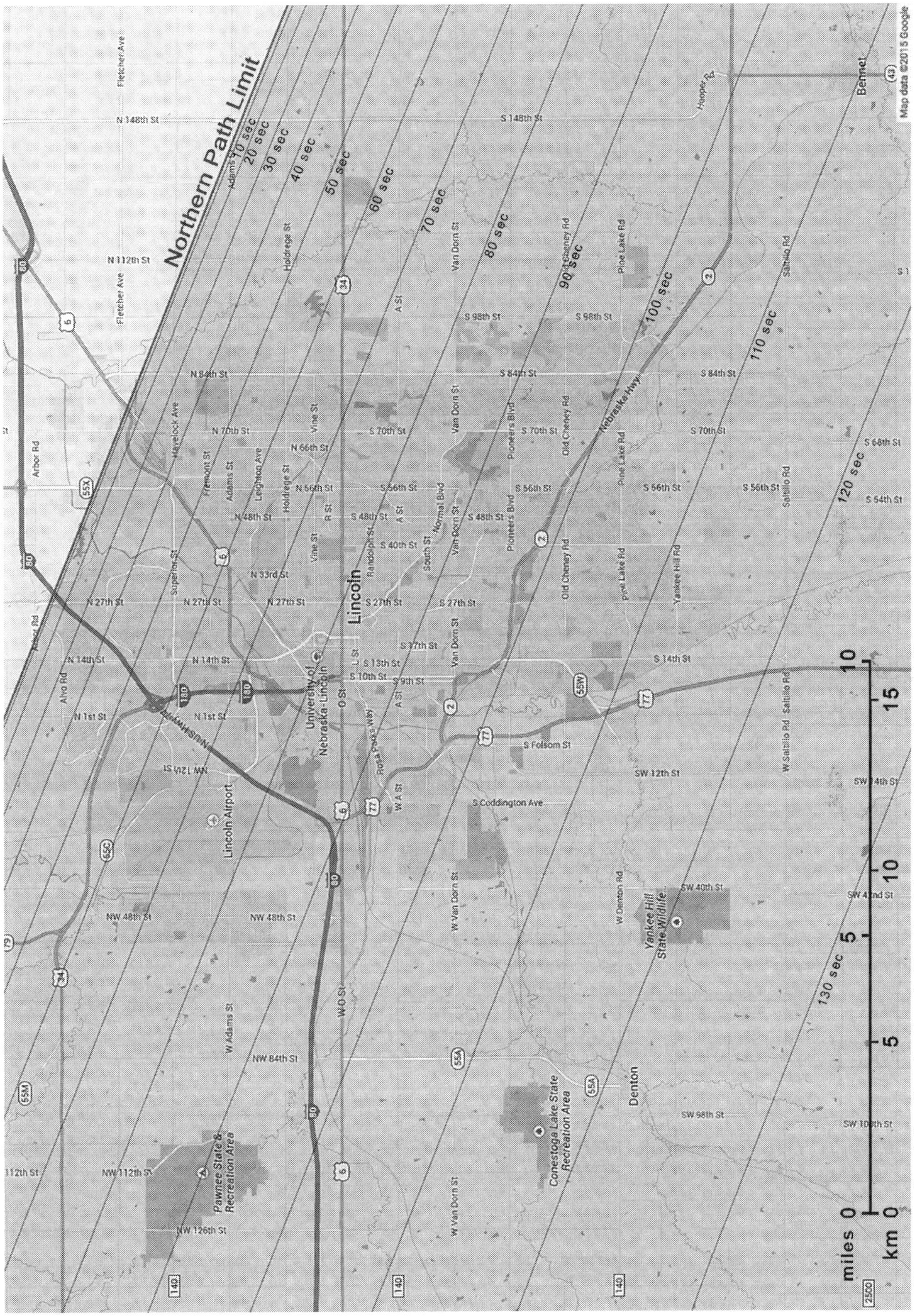

Figure 3–18: Detailed Path Map – Lincoln, NE

Total Solar Eclipse of 2017 August 21 (Espenak & Anderson)

Figure 3–19: Detailed Path Map – Nashville, TN

Total Solar Eclipse of 2017 August 21 (Espenak & Anderson)

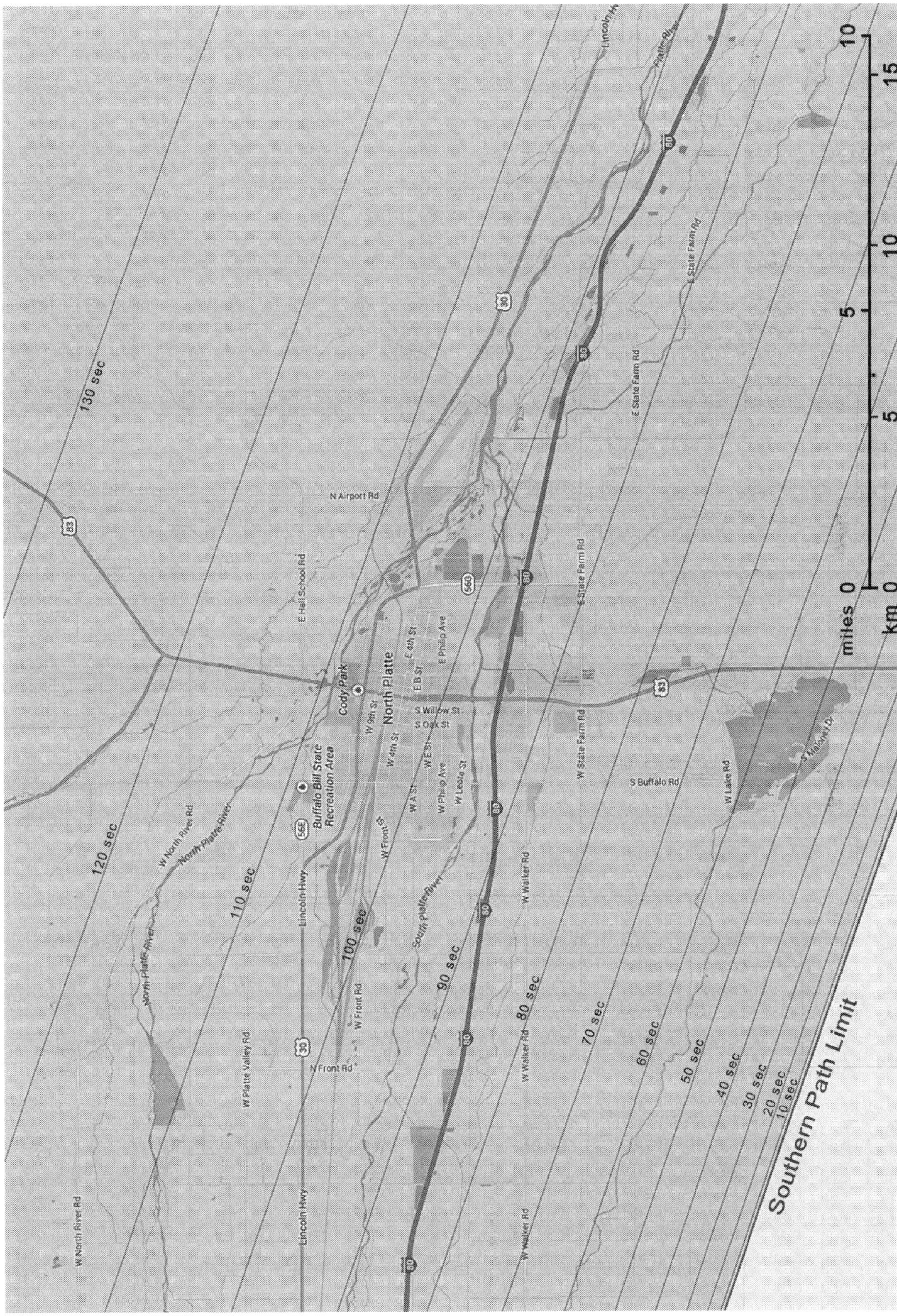

Figure 3–20: Detailed Path Map – North Platte, NE

Total Solar Eclipse of 2017 August 21 (Espenak & Anderson)

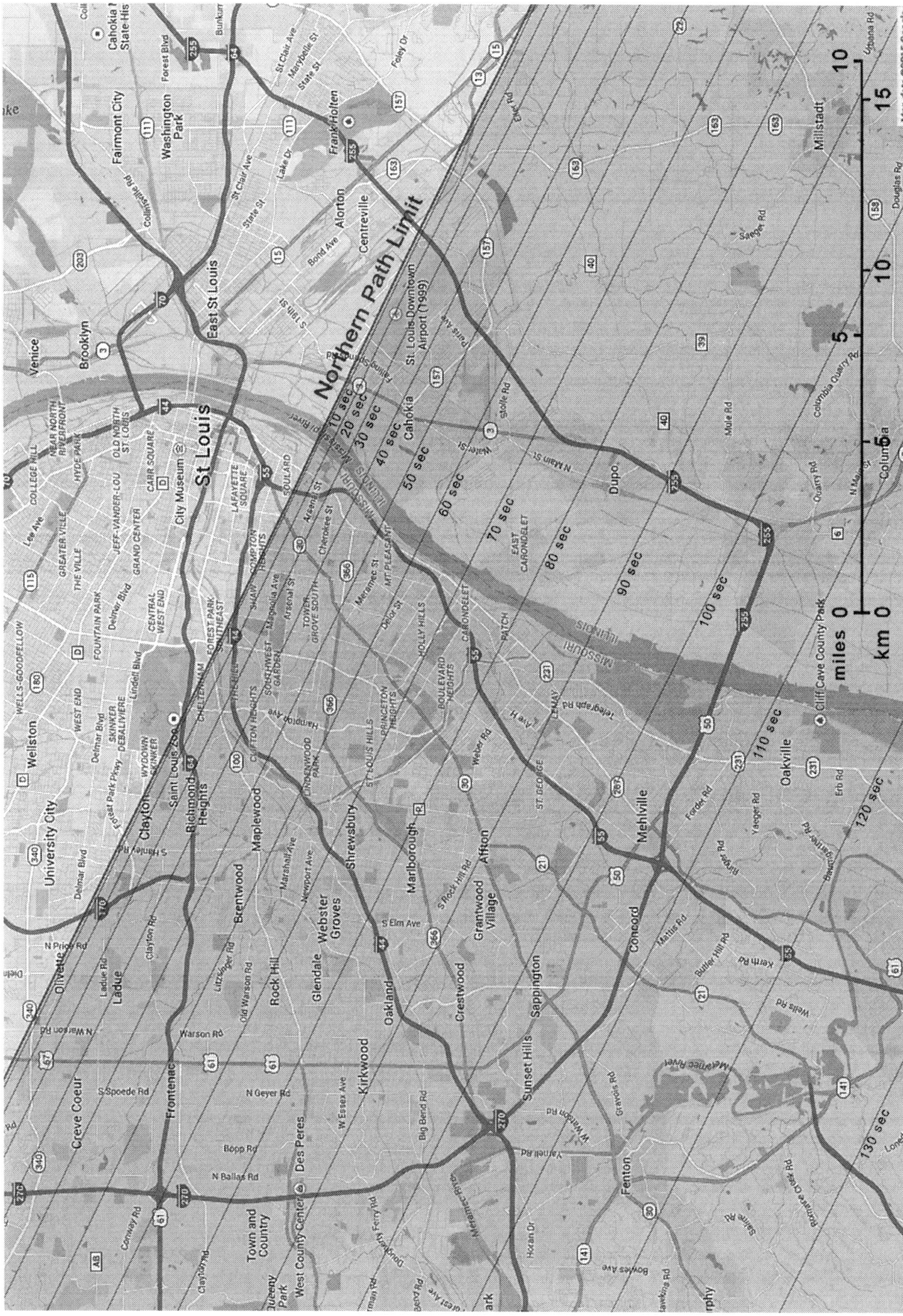

Total Solar Eclipse of 2017 August 21 (Espenak & Anderson)

Figure 3–21: Detailed Path Map – St. Louis, MO

Section 4: Observing and Photographing the Eclipse

4.1 Eye Safety and Solar Eclipses

A solar eclipse is probably the most spectacular astronomical event that many people will experience in their lives. There is a great deal of interest in watching eclipses, and thousands of astronomers (both amateur and professional) and other eclipse enthusiasts travel around the world to observe and photograph them.

A solar eclipse offers students a unique opportunity to see a natural phenomenon that illustrates the basic principles of mathematics and science taught through elementary and secondary school. Indeed, many scientists (including astronomers) have been inspired to study science as a result of seeing a total solar eclipse. Teachers can use eclipses to show how the laws of motion and the mathematics of orbits can predict the occurrence of eclipses. The use of pinhole cameras and telescopes or binoculars to observe an eclipse leads to an understanding of the optics of these devices. The rise and fall of environmental light levels during an eclipse illustrate the principles of radiometry and photometry, while biology classes can observe the associated behavior of plants and animals. It is also an opportunity for children of school age to contribute actively to scientific research—observations of contact timings at different locations along the eclipse path are useful in refining our knowledge of the orbital motions of the Moon and Earth, and sketches and photographs of the solar corona can be used to build a three-dimensional picture of the Sun's extended atmosphere during the eclipse.

Observing the partial phases of an eclipse, however, can be dangerous if the proper precautions are not taken. The solar radiation that reaches the surface of Earth ranges from ultraviolet (UV) at wavelengths longer than 290 nm, to radio waves in the meter range. The tissues in the eye transmit a substantial part of the radiation between 380–400 nm to the light-sensitive retina at the back of the eye. While environmental exposure to UV radiation is known to contribute to the accelerated aging of the outer layers of the eye and the development of cataracts, the primary concern over improper viewing of a partial solar eclipse is the development of "eclipse blindness" or retinal burns.

Photo 4–1 Two boys from a Zambian village use eclipse glasses to watch the partial phases of the total solar eclipse of 2001 June 21. © 2001 F. Espenak, www.MrEclipse.com

Exposure of the retina to intense visible light causes damage to its light-sensitive rod and cone cells. The light triggers a series of complex chemical reactions within the cells, which damages their ability to respond to a visual stimulus, and in extreme cases, can destroy them. The result is a loss of visual function, which may be either temporary or permanent depending on the severity of the damage. When a person looks repeatedly, or for a long time, at the Sun without proper eye protection, this photochemical retinal damage may be accompanied by a thermal injury—the high level of visible and near-infrared radiation causes heating that literally cooks the exposed tissue. This thermal injury or photocoagulation destroys the rods and cones, creating a small blind area. The danger to vision is significant because photic retinal injuries occur without any feeling of pain (the retina has no pain receptors), and the visual effects do not become apparent for at least several hours after the damage is done (Pitts 1993). Viewing the Sun through binoculars, a telescope, or other optical devices without proper protective filters can result in immediate thermal retinal injury because of the high irradiance level in the magnified image.

The only time that the Sun can be viewed safely with the naked eye is during a total eclipse, when the Moon completely covers the disk of the Sun. ***It is never safe to look at a partial or annular eclipse, or the partial phases of a total solar eclipse, without the proper equipment and techniques.*** Even when 99% of the Sun's surface (the photosphere) is obscured during the partial phases of a solar eclipse, the remaining crescent Sun is still intense enough to cause a retinal burn, even though illumination levels are comparable to twilight (Chou 1981 and 1996, and Marsh 1982). Failure to use proper observing methods may result in permanent eye damage and severe visual loss. This can have important adverse effects on career choices and earning potential, because it has been shown that most individuals who sustain eclipse-related eye injuries are children and young adults (Penner and McNair 1966, Chou and Krailo 1981, and Michaelides et al. 2001).

The same techniques for observing the Sun outside of eclipses are used to view and photograph annular solar eclipses and the partly eclipsed Sun (Sherrod 1981, Pasachoff 2012, Pasachoff and Covington 1993, and Reynolds and Sweetsir 1995). The safest and most inexpensive method is by projection. A pinhole or small opening is used to form an image of the Sun on a screen placed about a meter behind the opening. Multiple openings in perf board, a loosely woven straw hat, or even interlaced fingers can be used to cast a pattern of solar images on a screen. A similar effect is seen on the ground below a broad-leafed tree: the many "pinholes" formed by overlapping leaves creates hundreds of crescent-shaped images. Binoculars or a small telescope mounted on a tripod can also be used to project a magnified image of the Sun onto a white card. All of these methods can be used to provide a safe view of the partial phases of an eclipse to a group of observers, but care must be taken to ensure that no one looks through the device. The main advantage of the projection methods is that nobody is looking directly at the Sun. The disadvantage of the pinhole method is that the screen must be placed ~1 meter behind the opening to get a solar image that is large enough to be easily seen.

The Sun can only be viewed directly when filters specially designed to protect the eyes are used. Some of these filters have a thin layer of chromium alloy or aluminum deposited on their surfaces that attenuates both visible and near-infrared radiation. Other filters use a black polymer to accomplish the same thing. A safe solar filter should transmit less than 0.003% (density ~4.5) of visible light and no more than 0.5% (density ~2.3) of the near-infrared radiation from 780–1400 nm. (In addition to the term transmittance [in percent], the energy transmission of a filter can also be described by the term density [unit less] where density, d, is the common logarithm of the reciprocal of transmittance, t, or d=log10[1/ t]. A density of '0' corresponds to a transmittance of 100%; a density of '1' corresponds to a transmittance of 10%; a density of '2' corresponds to a transmittance of 1%, etc.). Figure 4–1 shows transmittance curves for a selection of safe solar filters.

One of the most widely available filters for safe solar viewing is shade number 14 welder's glass, which can be obtained from welding supply outlets. A popular inexpensive alternative is aluminized polyester that has been specially made for solar observation. (This material is commonly known as "mylar," although the registered trademark "Mylar®" belongs to DuPont, which does not manufacture this material for use as a solar filter. Note that "space blankets" and aluminized polyester film used in gardening are NOT suitable for this purpose!) Unlike the welding glass, aluminized polyester can be cut to fit any viewing device, and does not break when dropped. It has been pointed out that some aluminized polyester filters may have large (up to approximately 1 mm in size) defects in their aluminum coatings that may be hazardous. A microscopic analysis of examples of such defects shows that despite their appearance, the defects arise from a hole in one of the two aluminized polyester films used in the filter. There is no large opening completely devoid of the protective aluminum coating. While this is a quality control problem, the presence of a defect in the aluminum coating does not necessarily imply that the filter is hazardous. When in doubt, an aluminized polyester solar filter that has coating defects larger than 0.2 mm in size, or more than a single defect in any 5 mm circular zone of the filter, should not be used.

An alternative to aluminized polyester that has become quite popular is "black polymer" — first developed by Thousand Oaks Optical — in which carbon particles are suspended in a resin matrix. This material is somewhat stiffer than polyester film and requires a special holding cell if it is to be used at the front of binoculars, telephoto lenses, or telescopes. Intended mainly as a visual filter, the polymer gives a yellow-white image of the Sun (aluminized polyester produces a blue-white image). This type of filter may show significant variations in density of the tint across its extent; some areas may appear much lighter than others. Lighter areas of the filter transmit more infrared radiation than may be desirable. The advent of high resolution digital imaging in astronomy, especially for photographing the Sun, has increased the demand for solar filters of higher optical quality. Baader AstroSolar Safety Film, a metal-coated resin, can be used for both visual and photographic solar observations. A much thinner material, it has excellent optical quality and much less scattered light than polyester filters. The Baader material comes in two densities: one for visual use and a less dense version optimized for photography. Filters using optically flat glass substrates are available from several manufacturers, but are more expensive than polyester and polymer filters.

Figure 4–1: Spectral Response of Some Commonly Available Solar Filters

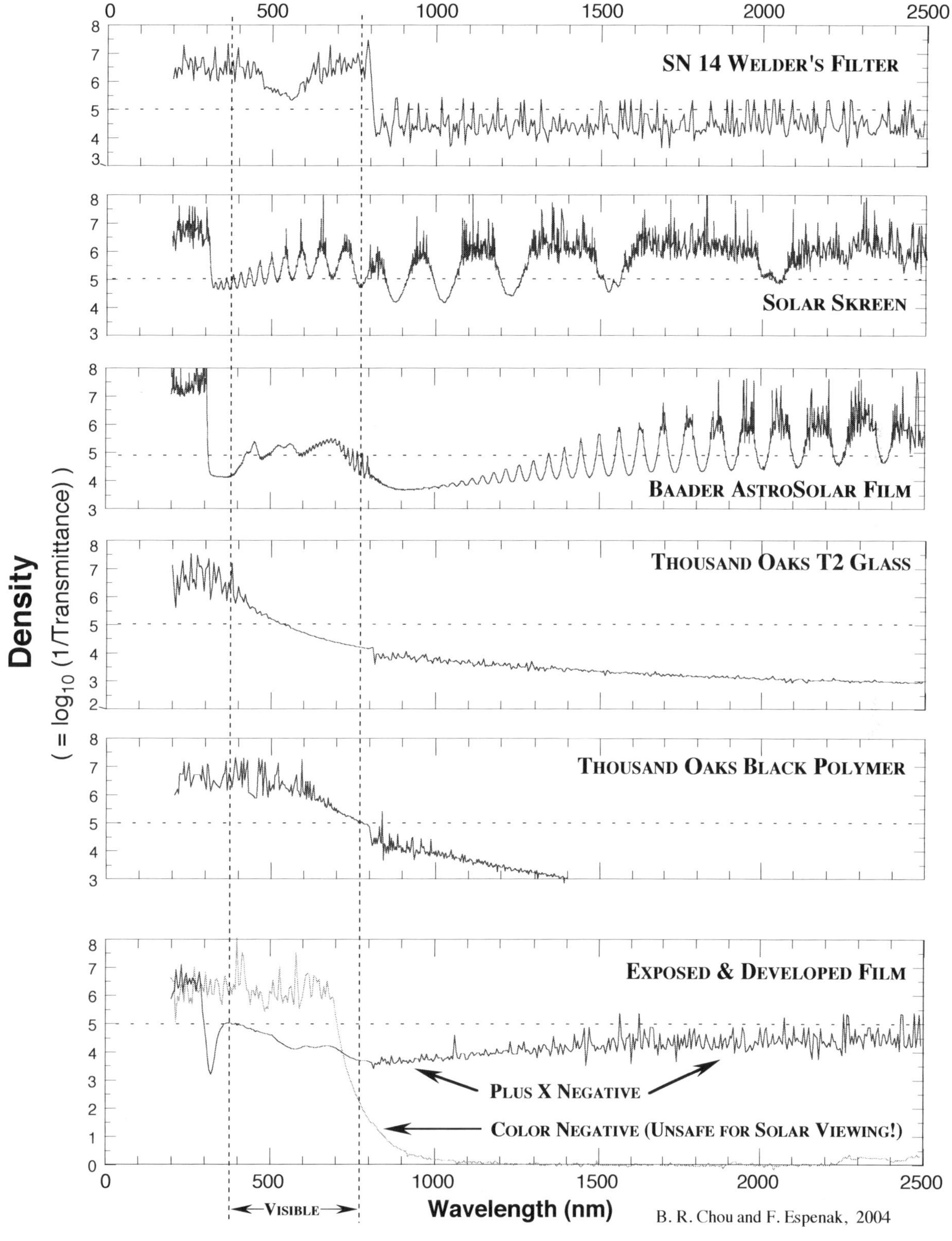

Many experienced solar observers use one or two layers of black-and-white film that has been fully exposed to light and developed to maximum density. Not all black-and-white films contain silver so care must be taken to use a silver-based emulsion. The metallic silver contained in the film acts as a protective filter; however, any black-and-white negative containing images is not suitable for this purpose. More recently, solar observers have used floppy disks and compact disks (CDs and CD-ROMs) as protective filters by covering the central openings and looking through the disk media. However, the optical quality of the solar image formed by a floppy disk or CD is relatively poor compared to aluminized polyester or welder's glass. Some CDs are made with very thin aluminum coatings that are not safe—if a lighted light bulb can be seen through the CD, it should not be used! No filter should be used with an optical device (e.g., binoculars, telescope, camera) unless it has been specifically designed for that purpose and is mounted at the front end. Some sources of solar filters are listed below.

Unsafe filters include color film, black-and-white film that contains no silver (i.e., chromogenic film), film negatives with images on them, smoked glass, sunglasses (single or multiple pairs), photographic neutral density filters and polarizing filters. Most of these transmit high levels of invisible infrared radiation, which can cause a thermal retinal burn (see Figure 4–1). The fact that the Sun appears dim, or that no discomfort is felt when looking at the Sun through the filter, is no guarantee that the eyes are safe.

Solar filters designed to thread into eyepieces that are often provided with inexpensive telescopes are also unsafe. These glass filters often crack unexpectedly from overheating when the telescope is pointed at the Sun, and retinal damage can occur faster than the observer can move the eye from the eyepiece. Avoid unnecessary risks. Local planetariums, science centers, or amateur astronomy clubs can provide additional information on how to observe the eclipse safely.

There are some concerns that ultraviolet-A (UVA) radiation (wavelengths from 315–380 nm) in sunlight may also adversely affect the retina (Del Priore 1999). While there is some experimental evidence for this, it only applies to the special case of aphakia, where the natural lens of the eye has been removed because of cataract or injury, and no UV-blocking spectacle, contact or intraocular lens has been fitted. In an intact normal human eye, UVA radiation does not reach the retina because it is absorbed by the crystalline lens. In aphakia, normal environmental exposure to solar UV radiation may indeed cause chronic retinal damage. The solar filter materials discussed in this article, however, attenuate solar UV radiation to a level well below the minimum permissible occupational exposure for UVA (ACGIH 2004), so an aphakic observer is at no additional risk of retinal damage when looking at the Sun through a proper solar filter.

In the days and weeks before a solar eclipse, there are often news stories and announcements in the media, warning about the dangers of looking at the eclipse. Unfortunately, despite the good intentions behind these messages, they frequently contain misinformation, and may be designed to scare people from viewing the eclipse at all. This tactic may backfire, however, particularly when the messages are intended for students. A student who heeds warnings from teachers and other authorities not to view the eclipse because of the danger to vision, and later learns that other students did see it safely, may feel cheated out of the experience. Having now learned that the authority figure was wrong on one occasion, how is this student going to react when other health-related advice about drugs, AIDS2, or smoking is given (Pasachoff 2001). Misinformation may be just as bad, if not worse, than no information.

Remember that the total phase of an eclipse can, and should, be seen without any filters, and certainly never by projection! It is completely safe to do so. Even after observing 14 solar eclipses, the author finds the naked-eye view of the totally eclipsed Sun awe-inspiring. The experience should be enjoyed by all.

Section. 4.1 was contributed by: *B. Ralph Chou, MSc, OD*
Professor Emeritus, School of Optometry & Vision Science
University of Waterloo
Waterloo, Ontario, Canada N2L 3G1

4.2 Sources for Solar Filters

The following is a brief list of sources for filters that are specifically designed for safe solar viewing with or without a telescope. The list is not meant to be exhaustive, but is a representative sample of sources for solar filters currently available in North America and Europe. For additional sources, see advertisements in *Astronomy* and or *Sky & Telescope* magazines.

Companies listed in ***bold*** have advertisements in the final section of this publication in the Solar Eclipse Marketplace. We encourage you to shop with them.

Sources in the USA:

American Paper Optics: www.3dglassesonline.com/
Astronomics: www.astronomics.com/
Celestron: www.celestron.com/
DayStar Filters: www.daystarfilters.com/
Lunt Solar Systems: luntsolarsystems.com/
Meade Instruments: www.meade.com/
OPT Telescopes: www.optcorp.com/
Orion Telescopes and Binoculars: www.telescope.com/
Rainbow Symphony: www.rainbowsymphony.com/
Seymour Solar: www.seymoursolar.com/
Spectrum Telescope: www.spectrumtelescope.com
Thousand Oaks Optical: www.thousandoaksoptical.com/

Sources in Canada:

Kendrick Astro Instruments: www.kendrickastro.com/
Khan Scope Centre: www.khanscope.com/
KW Telescope: www.kwtelescope.com/

Sources in Europe:

Baader Planetarium: www.baader-planetarium.com/
First Light Optics: www.firstlightoptics.com/solar-filters.html
Rother Valley Optics: www.rothervalleyoptics.co.uk/
Solar Scope: www.solarscope.co.uk/

Photo 4–2: A team of NASA scientists prepare for the total solar eclipse of 1995 October 23 from Dundlod, India.
© 1995 by F. Espenak, www.MrEclipse.com

4.3 Eclipse Photography

A solar eclipse can be safely photographed provided the above precautions are followed. Almost any kind of camera can be used to capture this rare event, but Single Lens Reflex (SLR) cameras offer interchangeable lenses and zooms. A lens with a fairly long focal length is recommended in order to produce as large an image of the Sun as possible. A standard 50 mm lens on a 35 mm film camera yields a minuscule 0.5 mm solar image, while a 200 mm telephoto or zoom lens produces a 1.9 mm image (Figure 4–2). A better choice would be one of the small, compact, catadioptric or mirror lenses that have become widely available in the past 20 years. The focal length of 500 mm is most common among such mirror lenses and yields a solar image of 4.6 mm.

With one solar radius of corona on either side, an eclipse view during totality will cover 9.2 mm. Adding a 2x tele-extender will produce a 1000 mm focal length, which doubles the Sun's diameter to 9.2 mm. Focal lengths in excess of 1000 mm usually fall within the realm of amateur telescopes.

Consumer digital cameras have become affordable in recent years and many of these may be used to photograph the eclipse. Most recommendations for 35 mm SLR cameras apply to digital SLR (DSLR) cameras as well. The primary difference is that the imaging chip in most DSLR cameras is only about 2/3 the area of a 35 mm film frame (check the camera's technical specifications). This means that the Sun's relative size will be 1.5 times larger in a DSLR camera so a shorter focal length lens can be used to achieve the same angular coverage compared to a 35 mm SLR camera. For example, a 500 mm lens on a digital camera produces the same relative image size as a 750 mm lens on a 35 mm camera (Figure 4–2). Another issue to consider is the lag time between digital frames required to write images to the DSLR's memory card. Better DSLRs have a buffer to temporarily store a burst of images before they are written to the card. It is also advisable to turn off the autofocus because it is not reliable under these conditions; focus the camera manually instead. Preparations must also be made for adequate battery power and space on the memory card.

Photo 4–3: Fred Espenak shoots the total solar eclipse of 1995 October 23 from Dundlod, India in the self-portrait using a 16mm Nikkor fisheye lens. © 1995 by F. Espenak, www.MrEclipse.com

Figure 4–2: Lens Focal Length Vs. Image Size for Eclipse Photography

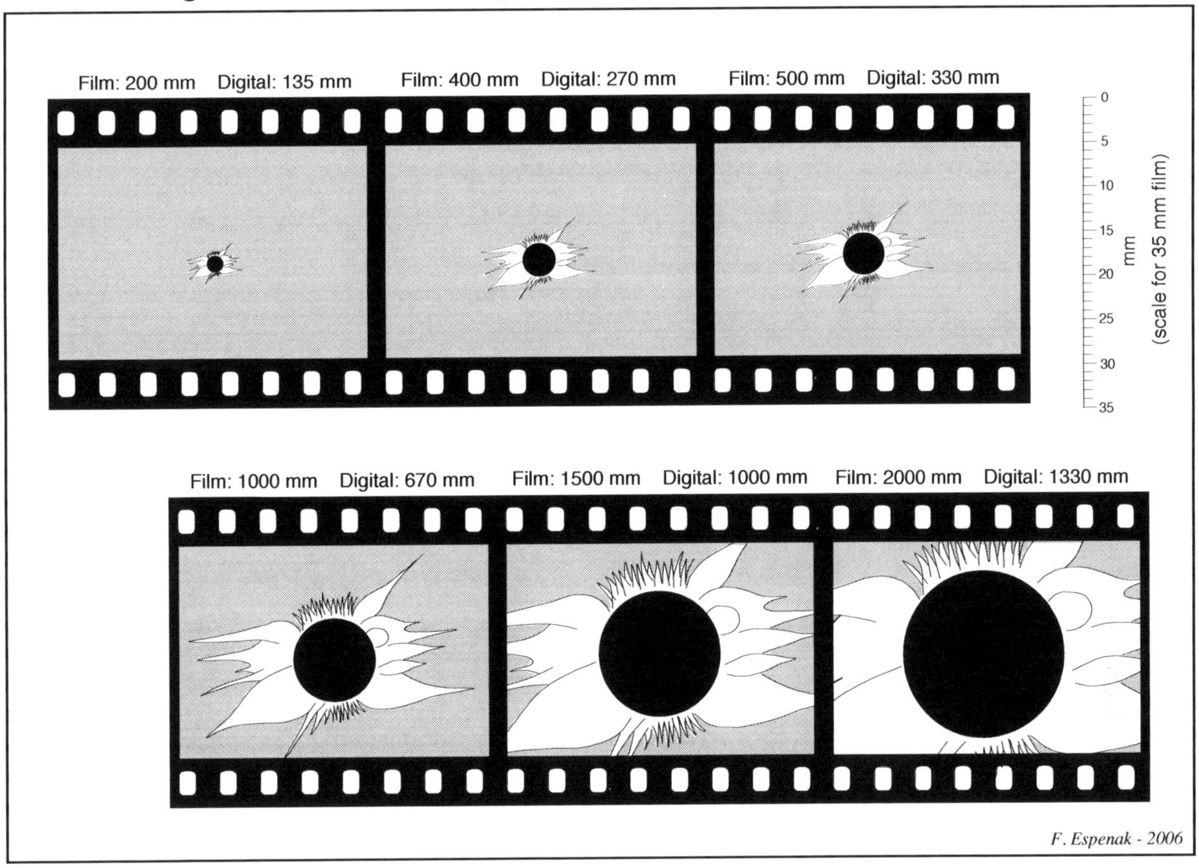

If full disk photography of an annular or partial eclipse is planned, the focal length of the optics must not exceed 2500 mm on 35 mm format (1700 mm on digital). Longer focal lengths permit photography of only a magnified portion of the Sun's disk. In order to photograph the Sun's corona during a total eclipse, the focal length should be no longer than about 1500 mm (1000 mm on digital); however, a shorter focal length of 1000 mm (700 mm digital) requires less critical framing and can capture some of the longer coronal streamers. Figure 4–2 shows the apparent size of the Sun (or Moon) and the outer corona in both film and digital formats for a range of lens focal lengths. For any particular focal length, the diameter of the Sun's image (on 35 mm film) is approximately equal to the focal length divided by 109 (Table 4–1).

Table 4–1: Field of View & Sun Size for Various Photographic Focal Lengths

Focal Length	Field of View (35mm)	Field of View (digital)	Size of Sun
14 mm	98° x 147°	65° x 98°	0.2 mm
20 mm	69° x 103°	46° x 69°	0.2 mm
28 mm	49° x 74°	33° x 49°	0.2 mm
35 mm	39° x 59°	26° x 39°	0.3 mm
50 mm	27° x 40°	18° x 28°	0.5 mm
105 mm	13° x 19°	9° x 13°	1.0 mm
200 mm	7° x 10°	5°x 7°	1.8 mm
400 mm	3.4° x 5.1°	2.3° x 3.4°	3.7 mm
500 mm	2.7° x 4.1°	1.8° x 2.8°	4.6 mm
1000 mm	1.4° x 2.1°	0.9° x 1,4°	9.2 mm
1500 mm	0.9° x 1.4°	0.6° x 0.9°	13.8 mm
2000 mm	0.7° x 1.0°	0.5° x 0.7°	18.4 mm

Image Size of Sun (mm) = Focal Length (mm) / 109

A solar filter must be used on the lens throughout the partial phases for both photography and safe viewing. Such filters are most easily obtained through manufacturers and dealers listed in *Sky & Telescope* and *Astronomy* magazines (see Section 4.2, ***Sources for Solar Filters***). These filters typically attenuate the Sun's visible and infrared energy by a factor of 100,000. The actual filter factor and choice of International Organization for Standardization (ISO) speed, however, will play critical roles in determining the correct photographic exposure. Almost any ISO can be used because the Sun gives off abundant light.

Table 4–2: Solar Eclipse Exposure Guide

ISO		***f/Number***								
25		**1.4**	**2**	**2.8**	**4**	**5.6**	**8**	**11**	**16**	**22**
50		**2**	**2.8**	**4**	**5.6**	**8**	**11**	**16**	**22**	**32**
100		**2.8**	**4**	**5.6**	**8**	**11**	**16**	**22**	**32**	**44**
200		**4**	**5.6**	**8**	**11**	**16**	**22**	**32**	**44**	**64**
400		**5.6**	**8**	**11**	**16**	**22**	**32**	**44**	**64**	**88**
800		**8**	**11**	**16**	**22**	**32**	**44**	**64**	**88**	**128**
1600		**11**	**16**	**22**	**32**	**44**	**64**	**88**	**128**	**176**
Subject	***Q***	***Shutter Speed***								
Solar Eclipse										
Partial[1] - 4.0 ND	**11**	—	—	—	1/4000	1/2000	1/1000	1/500	1/250	1/125
Partial[1] - 5.0 ND	**8**	1/4000	1/2000	1/1000	1/500	1/250	1/125	1/60	1/30	1/15
Baily's Beads[2]	**11**	—	—	—	1/4000	1/2000	1/1000	1/500	1/250	1/125
Chromosphere	**10**	—	—	1/4000	1/2000	1/1000	1/500	1/250	1/125	1/60
Prominences	**9**	—	1/4000	1/2000	1/1000	1/500	1/250	1/125	1/60	1/30
Corona - 0.1 Rs	**7**	1/2000	1/1000	1/500	1/250	1/125	1/60	1/30	1/15	1/8
Corona - 0.2 Rs[3]	**5**	1/500	1/250	1/125	1/60	1/30	1/15	1/8	1/4	1/2
Corona - 0.5 Rs	**3**	1/125	1/60	1/30	1/15	1/8	1/4	1/2	1 sec	2 sec
Corona - 1.0 Rs	**1**	1/30	1/15	1/8	1/4	1/2	1 sec	2 sec	4 sec	8 sec
Corona - 2.0 Rs	**0**	1/15	1/8	1/4	1/2	1 sec	2 sec	4 sec	8 sec	15 sec
Corona - 4.0 Rs	**-1**	1/8	1/4	1/2	1 sec	2 sec	4 sec	8 sec	15 sec	30 sec
Corona - 8.0 Rs	**-3**	1/2	1 sec	2 sec	4 sec	8 sec	15 sec	30 sec	1 min	2 min

Exposure Formula: $t = f^2 / (I \times 2^Q)$ where: t = exposure time (sec)
f = f/number or focal ratio
I = ISO film speed
Q = brightness exponent

Abbreviations: ND = Neutral Density Filter.
Rs = Solar Radii.

Notes: [1] Exposures for partial phases are also good for annular eclipses.
[2] Baily's Beads are extremely bright and change rapidly.
[3] This exposure also recommended for the 'Diamond Ring' effect.

F. Espenak - 2006 Oct

The easiest method for determining the correct exposure is accomplished by running a calibration test on the uneclipsed Sun. Shoot a roll of film of the mid-day Sun at a fixed aperture (f/8 to f/16) using every shutter speed from 1/1000 to 1/4 second. After the film is developed, note the best exposures and use them to photograph all the partial phases. With a digital camera, the process is even easier: shoot a range of different exposures and use the camera's histogram display to evaluate the best exposure. The Sun's surface brightness remains constant throughout the eclipse, so no exposure compensation is needed except for the narrow crescent phases, which require two more stops due to solar limb darkening. Bracketing by several stops is also necessary if haze or clouds interfere on eclipse day.

Certainly the most spectacular and awe-inspiring phase of an eclipse is totality. For a few brief minutes or seconds, the Sun's pearly white corona, red prominences, and chromosphere are visible. The great challenge is to obtain a set of photographs that captures these fleeting phenomena. The most important point to remember is that during the total phase, all solar filters must be removed. The corona has a surface brightness a million times fainter than the photosphere, so photographs of the corona must be made without a filter. Furthermore, it is completely safe to view the totally eclipsed Sun directly with the naked eye. No filters are needed, and in fact, they would only hinder the view. The average brightness of the corona varies inversely with the distance from the Sun's limb. The inner corona is far brighter than the outer corona so no single exposure can capture its full dynamic range. The best strategy is to choose one aperture or f/number and bracket the exposures over a range of shutter speeds (e.g., 1/1000 to 1 second). Rehearsing this sequence is highly recommended because great excitement accompanies totality and there is little time to think.

Photo 4–4: Fine details in the solar corona are captured during the total solar eclipse of 2006 March 29 from Jalu, Libya.
© 2006 F. Espenak, www.MrEclipse.com

Exposure times for various combinations of ISO speeds, apertures (f/number) and solar features (chromosphere, prominences, inner, middle, and outer corona) are summarized in Table 4–2. The table was developed from eclipse photographs made by F. Espenak, as well as from photographs published in Sky and Telescope. To use the table, first select the ISO speed in the upper left column. Next, move to the right to the desired aperture or f/number for the chosen ISO speed. The shutter speeds in that column may be used as starting points for photographing various features and phenomena tabulated in the 'Subject' column at the far left. For example, to photograph prominences using ISO 400 at f/16, the table recommends an exposure of 1/1000 second. Alternatively, the recommended shutter speed can be calculated using the 'Q' factors tabulated along with the exposure formula at the bottom of Table 4–2. Keep in mind that these exposures are based on a clear sky and a corona of average brightness. The exposures should be bracketed one or more stops to take into account the actual sky conditions and the variable nature of these phenomena.

Point-and-shoot cameras, and even cell phone cameras, are excellent for capturing the quickly changing light in the seconds before and during totality. If possible, use a tripod or brace with the camera on a wall or fence because slow shutter speeds may be needed. In addition, disable or turn off the camera's electronic flash so that it does not interfere with anyone else's view of the eclipse. If the flash cannot be turned off, cover it with black tape. Another eclipse effect that is easily captured with point-and-shoot and cell phone cameras should not be overlooked. Use a straw hat or a kitchen sieve and allow its shadow to fall on a piece of white cardboard placed several feet away. The small holes act like pinhole cameras and each one projects its own image of the eclipsed Sun. The effect can also be duplicated by forming a small aperture with the fingers of one's hands and watching the ground below. The pinhole camera effect becomes more prominent with increasing eclipse magnitude. Virtually any camera can be used to photograph the phenomenon, but they must have their flashes turned off because this would otherwise obliterate the pinhole images.

Much more on solar eclipse photography can be found on *MrEclipse.com* at:

www.mreclipse.com/SEphoto/SEphoto.html

For more information on eye safety eclipse observations and eclipse photography, see *Further Reading on Eclipses* in the *Bibliography* section.

4.4 Sky at Totality

As the partial phases progress, the temperature drops noticeably. This can affect the focus of cameras and telescopes, which should be checked as totality approaches. The total phase of an eclipse is accompanied by the onset of a rapidly darkening sky whose appearance resembles evening twilight about half an hour after sunset. The effect presents an excellent opportunity to view planets and bright stars in the daytime sky. Aside from the sheer novelty of it, such observations are useful in gauging the apparent sky brightness and transparency during totality.

During the 2017 eclipse, the Sun will be in Leo. Four of the naked-eye planets and a number of bright stars will be above the horizon within the total eclipse path. Figure 4–3 depicts the appearance of the sky during totality as seen from the central line at 18:00:00 UT1. This corresponds to eastern Nebraska near Grand Island, but the view is similar from other states along the eclipse path.

The four planets[1] lie almost symmetrically east and west of the Sun in a string spanning 90°. The most conspicuous of the planets will be Venus (m_v[2] = –4.0) located 34° west of the Sun. Having reached greatest western elongation on June 3, Venus's apparent disk is 13 arc-seconds and shrinking as it approaches superior conjunction in 2018 Jan 09.

Jupiter is also quite bright (m_v = –1.8) but is 51° east of the Sun having passed through opposition 4½ months earlier on April 07. Just 8° to the Sun's west, Mars will be much more challenging (m_v = +1.8) having recently passed through solar conjunction on July 27. Finally, little Mercury will be the faintest (m_v = +3.3) making it unlikely to spot 10° east of the Sun. The swiftest planet races to solar conjunction 5 days after the eclipse on August 26.

The bright star Regulus (m_v = +1.36) lies 1.2° east of the Sun, so its image will likely be captured in longer exposures of the corona during totality. A number of other bright stars will be scattered around the sky and may become visible during the eerie twilight of totality. They include Castor (m_v = +1.94) and Pollux (m_v = +1.14) located 35° to 39° west, and Procyon (m_v = +0.40) 36° southwest, Sirius (m_v = –1.44) 57° southwest, Spica (m_v = +1.05) 55° to the east and Arcturus (m_v = –0.05) 61° to the northeast. Star visibility requires a very dark and cloud free sky during the total phase.

Table 4–3 contains a geocentric ephemeris (using Bretagnon and Simon 1986) and gives the apparent positions of the naked eye planets during the eclipse. Delta is the distance of the planet from Earth (in Astronomical Units), *App. Mag.* is the apparent visual magnitude of the planet, and *Solar Elong* gives the elongation or angle between the Sun and planet.

---

[1] Saturn is not visible because it is below the horizon along the entire path of totality through the USA.
[2] The apparent visual magnitude (m_v) of a celestial body is a measure of its brightness in the visible spectrum as seen by an observer on Earth, adjusted to the value it would have in the absence of the atmosphere. The brighter the object appears, the lower the value of its apparent magnitude.

Table 4–3: Geocentric Solar System Ephemeris for 2017 August 21 at 18:00 UT1

Planet	RA	Dec	Delta	App. Mag.	Apparent Diameter	Phase	Elong
Sun	10h04m00s	+11°52'05"	1.01154	-26.7	1897.4	-	-
Mercury	10h32m33s	+04°24'05"	0.61997	3.3	10.8	0.05	10.3E
Venus	07h46m46s	+21°22'05"	1.28264	-4.0	13.0	0.81	34.1W
Mars	09h34m29s	+16°07'06"	2.64974	1.8	3.5	1.00	8.3W
Jupiter	13h15m56s	-07°05'10"	6.02251	-1.8	32.7	0.99	51.3E
Saturn	17h21m46s	-22°40'03"	9.63385	0.4	17.3	1.00	112.3E

Figure 4–3: Sky During Totality As Seen From Central Line At 18:00:00 UT1

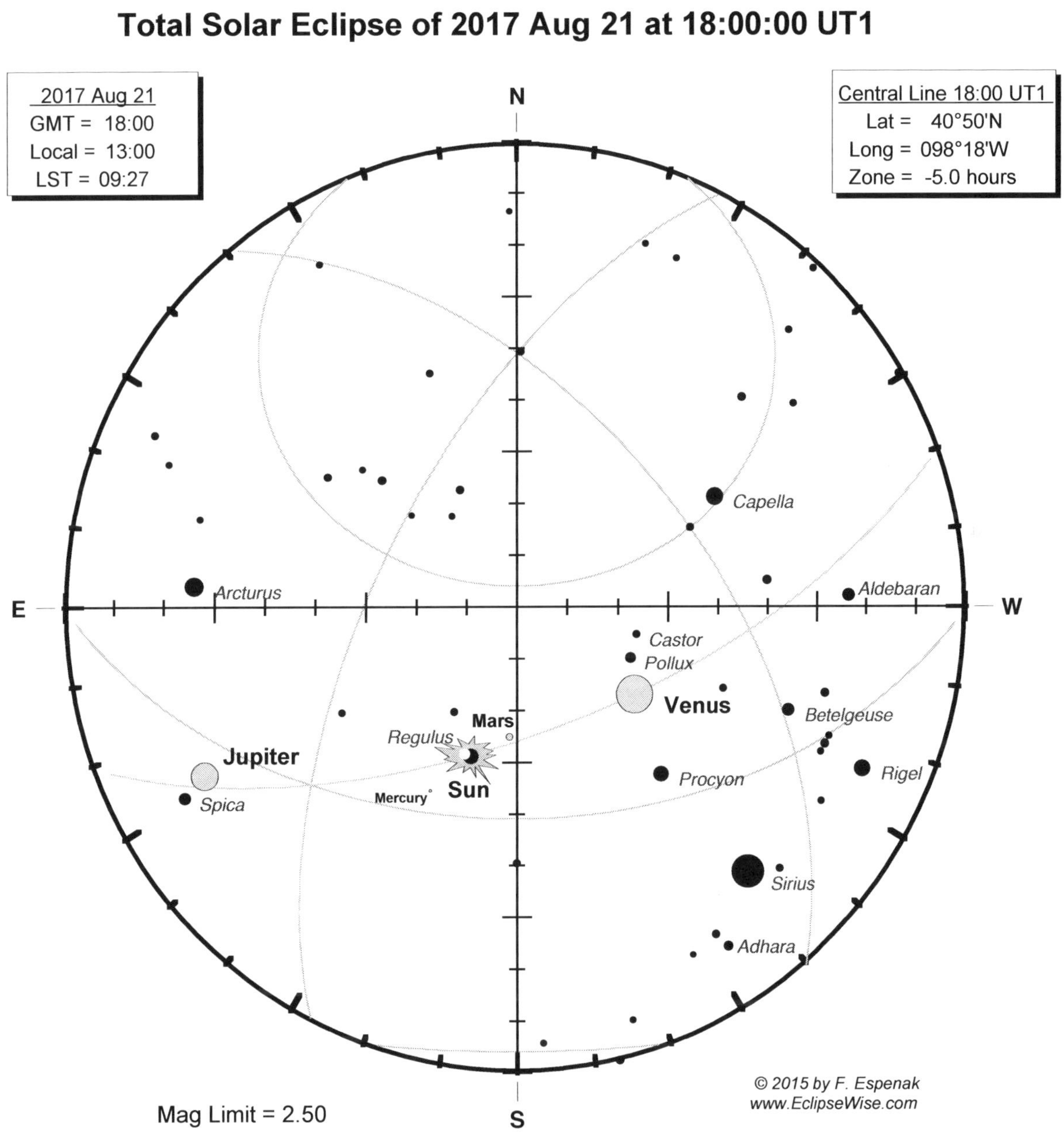

4.5 Contact Timings from the Path Limits

Precise timings of beading phenomena made near the northern and southern limits of the umbral path (i.e., the graze zones), may be useful in determining the diameter of the Sun relative to the Moon at the time of the eclipse. Such measurements are essential to an ongoing project to try to detect changes in the solar diameter. For more information on how to time and report beading phenomena, visit the IOTA webpage for the 2017 solar eclipse at:

http://occultations.org/eclipse2017

4.6 Eclipse Path on Google Maps

The 2017 eclipse path is also plotted on interactive Google Maps on the EclipseWise.com web site. The northern and southern path limits of an eclipse path are plotted in blue and the central lines are red. The yellow lines crossing the path indicate the position of maximum eclipse at 10-minute intervals. The four-way toggle arrows (upper left corner) is for navigating around the map. The zoom bar (left edge) is used to change the magnification. The two buttons (top right) turn on the map view or satellite view.

The green marker labeled GE is the point of Greatest Eclipse. The magenta marker labeled GD is the point of Greatest Duration. This is the location where the total eclipse lasts the longest along the entire path.

Clicking anywhere on a map marks a position and calculates the eclipse times at that location. Moving the cursor over a marker reveals the eclipse circumstances for that position. The marker predictions can also be viewed in a new window via the Eclipse Times Popup button. The information in the popup window can be selected, copied, and pasted into a word processor. All markers can be removed using the Clear Markers button above. Choosing the Large Map check box produces a bigger map (for users with large monitors and fast Internet connections).

The Google Map of the 2017 total solar eclipse is located at:

eclipsewise.com/solar/SEgmap/2001-2100/SE2017Aug21Tgmap.html

Photo 4–5: Pat and Fred Espenak gaze at the Sun's corona during the total solar eclipse of 2008 August 1 from Jinta, China. © 2008 F. Espenak, www.MrEclipse.com

Section 5: Climatology Along the Umbral Path

5.1 Introduction

When August 21 looms on the calendar, the minds of hopeful eclipse watchers will be much engaged with the weather. Forecasts in the days before an eclipse can give good short-range advice on the best places to await the shadow passage, but for longer-range planning, we must exploit the climatological statistics from years past. A major reservation that affects all of the data in this study is cloud statistics, both surface and satellite, which should be used to compare one site with another and not to decide on the absolute probability of seeing the eclipse. That latter probability is best derived from sunshine data, but even then, only imperfectly.

August is a very good month for viewing an eclipse across the USA. The Moon's shadow crosses the continent at a time of year when the peak of the thunderstorm season has passed, and sunshine is generous, particularly west of the Missouri River. The cloud-cover pattern is relatively simple. The Rocky Mountain and Pacific States have a mixed cloud climatology that is dictated by the up-and-down character of the terrain. The plains have a simple cloud climatology that reflects the daily formation of convective clouds and the development of afternoon thundershowers. The states east of the Missouri and Mississippi Rivers have the heaviest average cloud because of a ready supply of moisture from both the Gulf of Mexico and the Atlantic Ocean and the influence of the jumbled mountains of the Appalachians. Nevertheless, across the continent there are good spots and poorer spots in all regions of America.

The best weather prospects are in the protected mountain valleys and interior plateau in parts of Oregon, Idaho, and Wyoming. The Front Range of Western Nebraska is also a promising spot, before the eclipse track gets too far away from the Rockies. Mountains are efficient cloud makers and efficient cloud eaters depending on whether the wind flow is uphill or downhill, and so weather conditions can change greatly over very short distances. In contrast, without a convoluted topography to contend with, the Great Plains offers a more even, long-term cloud climatology, though on a daily scale, the development and movement of thunderstorms is just as complex a problem as the mountain flows. And while the rough terrain of the Appalachians might be expected to bring the same variable benefits as in the Rockies, the lower mountain peaks and the narrow valleys in that range are not so capable of eroding passing weather systems, which tend to be more extensive and deep than those crossing the western mountains.

Figure 5–1: Oregon Topography Along the Eclipse Track

5.2 Oregon

Pacific weather systems that cross the coast of North America are obligated to rise over the complex mountain terrain that defines the western third of the continent. On the windward side of each mountain range, air is compelled to rise, cooling and forming clouds in otherwise clear skies or thickening clouds that are already present in passing weather disturbances. In contrast, on the downwind side of mountain ridges, air descends, warming and drying, and dissipating clouds and precipitation. On top of this up-and-down roller coaster ride is superimposed the daily formation of convective clouds on the mountain slopes, promoted by the dark forest slopes that absorb sunlight and blossom into afternoon buildups.

Oregon topography is characterized by two north-south mountain chains, the 3,000-foot-high Coast Range and the 10,000-foot-high Cascade Range, and an eastern highland, the Columbia Plateau, which dissolves into a rough jumble of small mountain ranges and river basins toward the Idaho border. Between the Coast and Cascade Ranges, the terrain descends into the Willamette Valley; on the Plateau, the terrain is lowest against the Cascade Range and rises slowly toward the Idaho border, where it encounters the Ochoco Mountains (Figure 5–1).

The central line of the Moon's umbral shadow first touches the west coast of the United States just before 10:16 PDT on the south side of the community of Lincoln Beach, Oregon. More precisely, it comes ashore 100 yards south of a small public park called Fishing Rock, a scenic, rocky point accessed by a rough trail that begins on a suburban street amid a cluster of single-family homes and continues through a dense wood before opening toward the ocean. It's a fitting place to start the eclipse, since the area was also in the path of the last total eclipse through mainland USA in 1979.

Table 5–1: August Cloud-Cover Statistics for Oregon

Oregon		Percent Frequency of Sky Condition							
Airport Observing Site	Nearby communities	Clear	Few	Scattered	Broken	Overcast	Thin Fog	Thick Fog	Average Cloud Cover (%)
Coast									
Southwest Oregon Regional	North Bend	19.8	14.2	4	13.4	48.4	0.2	0	63
Umpqua River CGS	Winchester Bay	0	30.5	20.9	48.7	0	0	0	52
Newport Municipal *	Newport	18.4	15.6	0	18.4	32.8	14.4	0.3	64
Astoria	Astoria	7.7	9.6	5	17.1	60.4	0.2	0	78
Willamette Valley									
Corvallis Muni *	Corvallis	27.6	15.5	0.9	21.6	34.5	0	0	54
Mahlon Sweet Field	Eugene	36.9	13.7	5.8	14.1	29.5	0	0	45
McNary Field *	Salem	36.7	13.9	4.4	13	32	0	0	46
Aurora State *	Aurora	38.1	11.7	2.1	16.7	31.4	0	0	47
Troutdale	Troutdale	21	16.8	1.7	17.9	42.5	0.2	0	60
Portland International	Portland	22.4	12.5	7.8	18.2	39.1	0	0	59
Columbia Plateau									
Klamath Falls Airport	Klamath Falls	42.2	36.1	0.7	16.2	4.9	0	0	24
Roberts Field *	Redmond	50.4	18.1	7.1	15.9	8.3	0	0.3	27
Burns Municipal Airport	Burns	57.9	14.9	4.1	16.9	6.2	0	0	24
Eastern Oregon Regional	Pendleton	47.3	17	8	17.3	10.4	0	0	30
Baker Muni *	Baker City	38.3	29.8	2.3	16.8	12.9	0	0	32
Ontario Muni *	Ontario	77.4	6.5	0	4.3	11.8	0	0	16
* = within total eclipse track									

Table 5–1: August cloud-cover statistics at eclipse time for Oregon derived from surface observations at selected locations along the eclipse path. Data: National Climate Data Center.

Even in summer, the Oregon coast is a cloudy region (Table 5–1), lying within the main track of Pacific weather systems moving onto the continent and subject to the cloud-making influence of the Coast Range, which lies barely a mile or two inland. Observations from weather stations (mostly airports) record an average cloud amount of 64% at Newport and a range from 52% to 78% at other locations along the coast. Cloud statistics based on satellite observations (Figure 5–2) confirm the high cloud average along the Pacific coast. The satellite graphs reveal that afternoon cloudiness along the coast is much lower than in the morning, but the eclipse comes at mid-morning when the cloud-clearing influence of the Sun is just beginning to be felt. Cooling associated with the oncoming lunar shadow may bring an increase in cloudiness though it will depend on the character of the day's weather.

The cool offshore waters along the northwest Pacific coast are good fog makers, and though the fog tends to remain offshore, tendrils of mist are commonly found drifting across the beaches. The misty air is drawn inland by the land warming during the day and, though it tends to be relatively transparent, the moist and salty air is not kind to cameras and telescopes. In Photo 5–1, taken at eclipse time in 2013, a transparent haze can be seen in the background, partly a wind-blown spray and partly the remnants of an earlier fog. The dress of the beachgoers testifies to the unpleasant temperature on the coast and of an onshore wind that drives stinging sand against unprotected ankles. Of course, an offshore wind will dissipate the fog and warm the air, so attention to the eclipse-day forecast will tell whether an inland location might be better. Table 5–1 shows that at Newport Airport, ~15% of August days have fog at eclipse time. Presumably the frequency is higher right on the beach, as the airport lies 700 yards inland.

Photo 5–1: A view from the beach at Lincoln City on August 21, 2013.

Inland from Lincoln Beach, the coastal plain quickly transforms to the rough, forested terrain of the Coast Range. Eclipse-viewing options are limited on the coast mountains, in part because the westerly flows from the Pacific bring a lot of cloud when forced to rise over the terrain and partly because visibility is limited on the heavily treed and twisty highway. However, once over the Coast Range and into the Willamette Valley, the eclipse adventurer's chances improve, though only by a modest amount. The Coast Range is not very high and its ability to dry the air limited. Nevertheless, cloud measurements at Corvallis, Salem, and Aurora show a 15% to 20% reduction in average cloud cover compared to that at Newport. Here again, the afternoon Sun is very efficient at breaking up the morning cloud cover, but the late-morning hour of the eclipse suggests that this may be limited.

A far better eclipse-watching location — perhaps the best anywhere in the USA — is on the east side of the Cascade Mountains, which, at 10,000 feet, are high enough to make a major dent in the cloud-cover statistics. After crossing this range, the eclipse track drops sharply into "The Basin," a low valley centered on Madras, OR; this valley is the southernmost extent of the Columbia Basin. While there are no cloud observations at Madras, airport observations at Redmond, at the south limit of the umbral shadow, show an average cloud cover of 24% at eclipse time. Over 50% of the reports at Redmond show "clear" skies and 75% have scattered cloud or less. Satellite observations indicate that Madras has the distinction of having the least August cloudiness anywhere along the eclipse track, and just as promising, has almost no difference between morning and afternoon cloud amounts.

Of course, if large-scale frontal systems are moving across the mountains on eclipse day, the impact of terrain will be much less dramatic, even though descending winds at some leeside locations will be able to erode holes in an otherwise extensive cloud cover.

Once past Madras, the lunar shadow moves onto the Ochoco Mountains, where the terrain becomes rougher and the average elevation begins to climb, eventually crossing the peaks of the Columbia Plateau. The Ochoco Mountains and the Plateau are not particularly high ranges, but lofty enough that the average cloud amount increases by ~10%, in step with the terrain. Most of the cloud in eastern Oregon will be convective — showers and thundershowers — and so are susceptible to dissipating as cooler temperatures arrive with the

Figure 5–2: Graphs of Average Cloud Cover Along the Eclipse Path

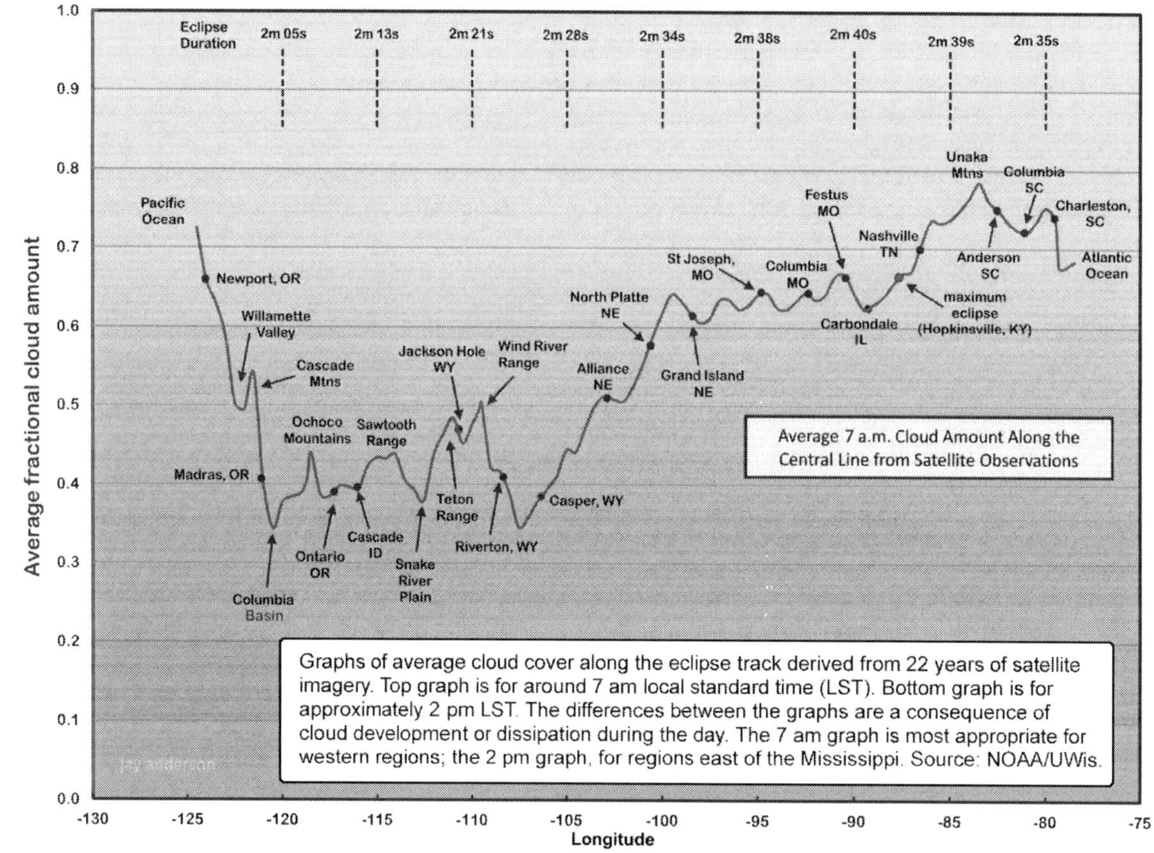

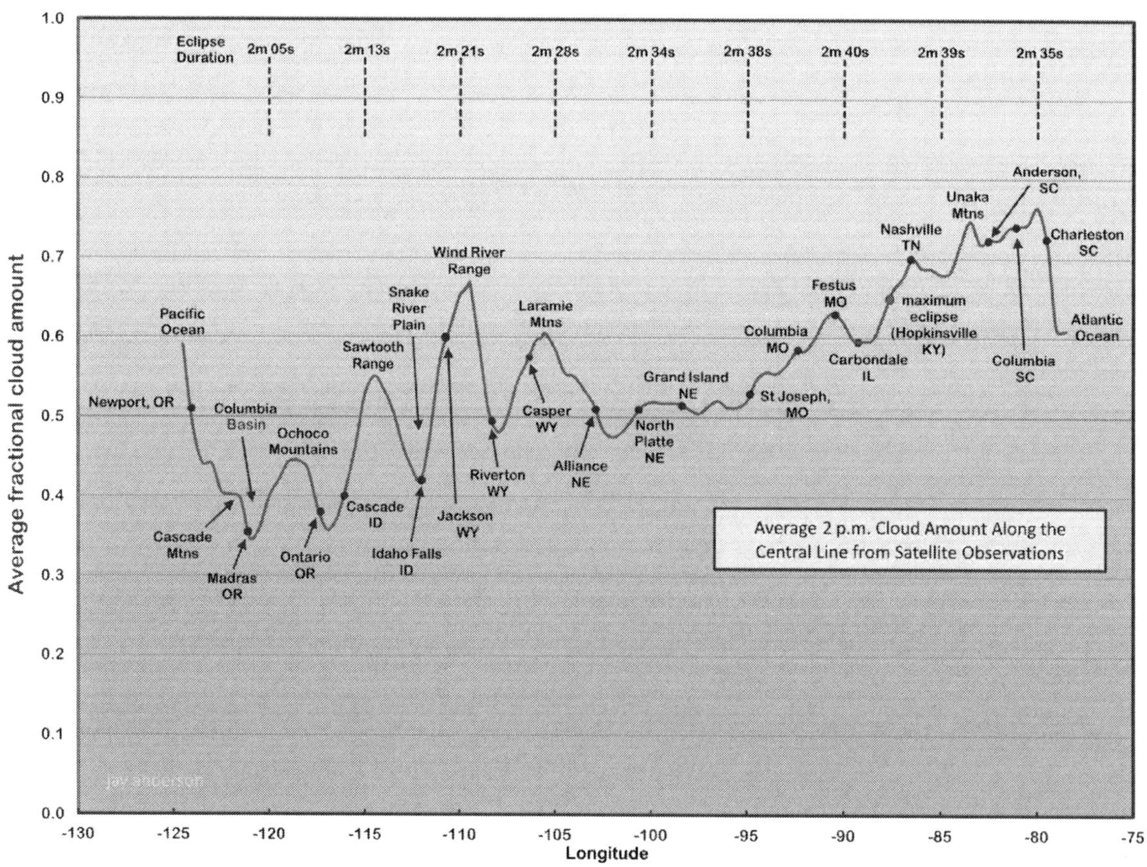

eclipse, provided the buildups are not too large. Nevertheless, if the day is even half promising, these mountains are a good area for eclipse chasing, as Highway 26 follows the track of the umbral shadow all the way to Idaho.

Close to the Idaho border, the eclipse track descends from the Columbia Plateau onto the northwest corner of the Snake River Plain, reaching Interstate 84 and the Idaho border at Ontario, OR. As the track descends to lower elevations, the average cloud cover also retreats to lower values, reaching a mean of ~30% at the Idaho border. Airport observations at Ontario show an average cloudiness of only 16% (with ~75% clear skies!), but the site has only provided a small number of observations over the past years and the values must be viewed with caution. The surface statistics cannot be discounted, however, as satellite observations are also kind to the region, showing average cloud amounts that rival the low values around Madras.

5.3 Idaho

Idaho has one of the rockiest landscapes in the Union, but by good fortune, a significant portion of the track crosses the lowlands of the Snake River Plain, a 100-mile-wide, 375-mile-long depression in the Rocky Mountains that stretches from Ontario, on the Oregon border, in a broad arc that terminates against the Teton Mountains near Idaho Falls (Figure 5–3). On the west side of Idaho, near Payette and Council, the track barely intercepts the low ground, so that the north limit moves across the narrow Long Valley of the Payette River, while the south limit intercepts westernmost portion of the Snake River Plain. Immediately afterward, the track crosses the rugged and cloudy 10,000-foot peaks of the Sawtooth Range. Before it leaves the state, it drops once more down to the lower elevations of the Snake River, passing over Idaho Falls before reaching the Wyoming border.

Figure 5–3: Idaho Topography Along the Eclipse Track

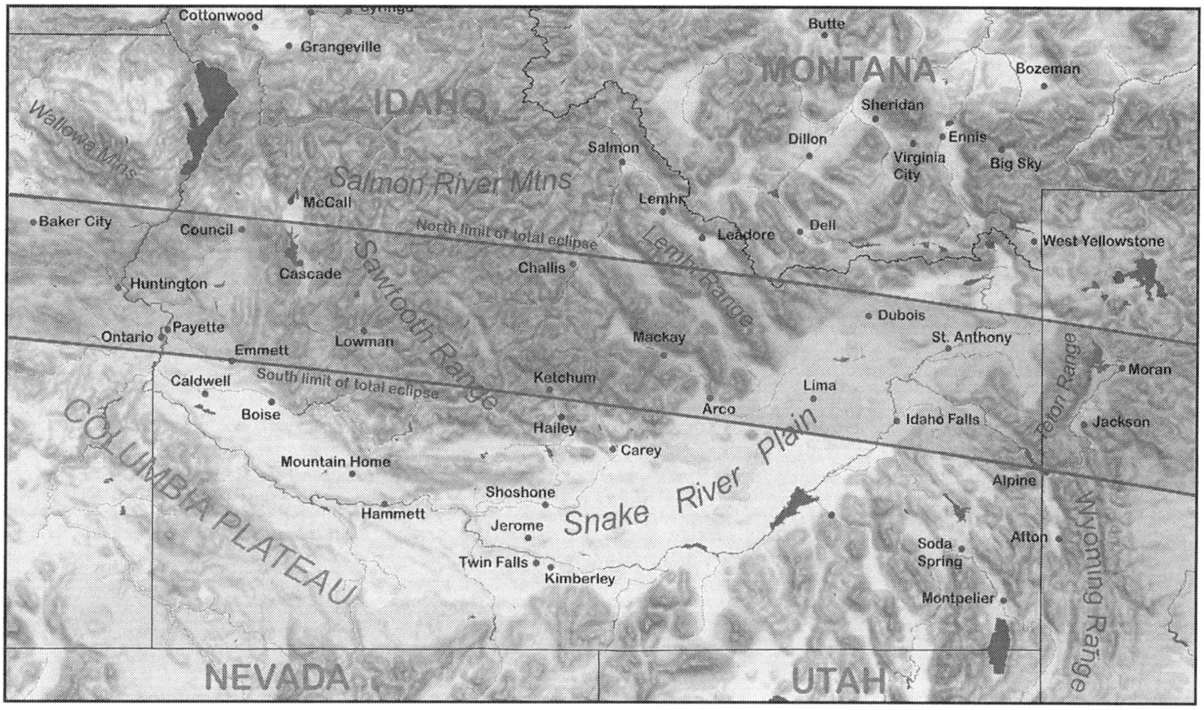

The lowest cloud averages in Idaho can be found near the Oregon border, where the track descends from the Ochoco Mountains onto the Snake River Plain. Surface-based observations show a 16% average cloudiness at Ontario, OR (though suspect because of a short observing record), and 27% at Boise. At the eastern end of the plain, where the eclipse track descends from the Sawtooth Range, average cloudiness measured at airport weather stations is ~5% higher – around 30% at both Idaho Falls and Pocatello. Satellite observations of cloud cover mimic the pattern of observations from the ground, but with higher overall values: an average of around 38% near Ontario, climbing to almost 60% over the Sawtooth Range, and then declining to 44% at Idaho Falls.

Satellite observations show that the cloud cover on the Sawtooth Range climbs dramatically during the day, by almost 20%, as the slopes warm and convective clouds build. Clouds also build in the afternoon in the eastern Snake River Plain, near Idaho Falls, but the increase is less than 5% according to space-based observations. Since the lunar shadow passes just before noon, the increase in valley cloud is likely to be small and overcome by eclipse cooling, but the cloud buildups in the mountains may not be so responsive.

Table 5–2: August Cloud-Cover Statistics for Idaho

Idaho		Percent Frequency of Sky Condition								
Airport Observing Site	Nearby communities	Clear	Few	Scattered	Broken	Overcast	Thin Fog	Thick Fog	Average Cloud Cover (%)	Percent of Possible Sunshine
Boise Municipal	Boise	50.3	18.9	6.1	15.2	9.6	0	0	27.1	85
McCall Municipal	McCall	35.4	24.2	0	32.3	7.1	1	0	37.1	
Friedman Memorial	Hailey	39.8	35.3	0.3	17.1	6.7	0.3	0.5	26.5	
Lemhi County	Salmon	34.7	35.3	0	20.5	9.1	0.3	0	30.9	
Pocatello Municipal	Pocatello	43.7	21	7.5	18.1	9.7	0	0	30.4	81
Idaho Falls Regional *	Idaho Falls	42.5	24.6	1.1	17.7	12.8	0.2	1.1	32.3	
Allen H. Tigert	Soda Springs	38.4	39.3	0	17.9	3.6	0	0.9	24.6	
* = within total eclipse track										

The best appraisal for eclipse visibility is the direct measurement of sunshine itself, but the instrumentation to take such recordings are only found at a few stations across the USA. In Idaho, there are two such stations along the eclipse track, at Boise and at Pocatello. Both indicate that August sunshine averages over 80% of the maximum possible, a number that is comparable to some of the sunniest locations on the globe. No suitable measurements are available for Oregon, but based on the cloud statistics, the percent of possible sunshine is similar and possibly even larger. Because the sunshine measurements take place across the whole of the day and the eclipse is in the less-cloudy morning hours, the sunshine measurements are likely a little pessimistic.

The broad channel of the Snake River Plain provides a conduit for Pacific moisture to pass through the larger part of the Rocky Mountains, skirting the higher elevations that would wring moisture from the air. This has a small impact in the valley itself, but when the moist Pacific air encounters the barrier formed by the Teton Range, it is forced upward, turning into clouds and precipitation and giving the eastern end of the plain a heavier precipitation climatology than might otherwise be expected. August rainfall at Boise averages just 0.2 inches; that at Driggs, at the foot of the Tetons, averages 1.1 inch. While the effect on the cloud climatology is small, eclipse chasers looking for the best weather prospects should stay back from the Tetons unless the forecast for the day is favorable.

5.4 Wyoming

In Wyoming, the eclipse track crosses the Continental Divide and enters a new climatological regime – one not dictated solely by the flow of westerly winds across the Rockies but also by exposure to the moisture of the Great Plains to the east. Wyoming is basin-and-range country: a complex jumble of mountain ranges separated by flat dryland or desert basins. Since the mountains in western Wyoming lie perpendicular to the prevailing westerly flows, they too create a barrier that forces the air upward to deposit its moisture on the westerly slopes. This leaves the central and eastern parts of the state with a semiarid climatology that is shaped more by flows from the east than the west. August is the driest month of the year, and while cloud conditions for the eclipse may not be as favorable as those in Idaho and Oregon, the climatology is still one of generous sunshine.

After crossing the Tetons, the cross-mountain flow drops sharply down into Jackson Hole, the valley that contains the resorts at Jackson and Teton Village. This is a prime eclipse-watching site, in part for the grandeur of the Teton Mountains and in part for the high frequency of sunny conditions brought on by the leeside protection from those mountains. It's a chancy location for the eclipse, as mountain atmosphere in Wyoming responds very strongly to daytime heating - more than any other part of the eclipse track. In Figure 5-2, afternoon cloudiness in the Teton and Wind River Ranges increases by as much as 20% between 7 am and 2 pm The eclipse comes at 11:30 am local time – late enough for that some cloud development may have already started, but perhaps early enough that eclipse cooling will cause it to dissipate.

Figure 5–4: Wyoming Topography Along the Eclipse Track

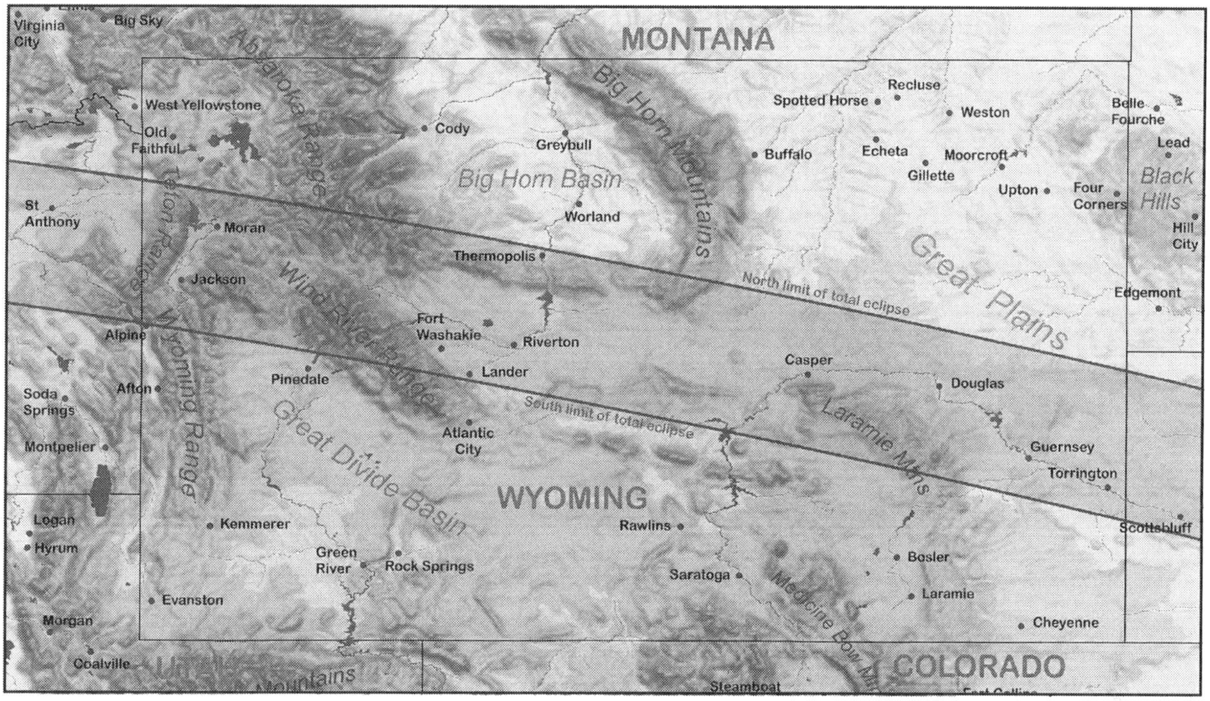

As might be expected, observation sites on the peaks above Teton Village (reached by chairlift) are cloudier than locations in the 10-mile-wide valley below and it is not uncommon for convective clouds to bring broken clouds to the heights while sunshine dapples the valley below (Photo 5–2). When thunderstorms threaten, the chairlifts are closed, and watchers must remain on the peaks, in the limited shelter available until the threat has passed.

Observations at Jackson Hole Airport show an average daily cloudiness of 34%. And though the valley is protected on all sides by mountains, its altitude gives it a relatively high average August precipitation for Wyoming: nearly 1⅓ inches of rain spread across 9 days of the month.

Photo 5–2: A view across Jackson Hole from the heights of the Tetons, showing the tendency for the eastern part of the valley to be sunny while the slopes sprout convective clouds. On this day, the showers and thundershowers were caused by a passing upper disturbance and moved off to leave bright sunshine in the early afternoon.

The central axis of the eclipse lies north of Jackson – right through the middle of the Jackson Hole airport – where the valley opens, helping to create the descending air flow that dries out the atmosphere. Because of the larger size of the valley, the central line is probably sunnier than at Jackson itself, a few miles away. From this location, there is a grand view of the Tetons to the west, and open terrain east. If the day should prove cloudy, Highway 26 leads to the east side of the Wind River Range near Riverton where skies have a tendency to be sunnier than at Jackson Hole. Riverton is well shielded from Pacific moisture, but lies in a long east-west gap in the terrain that allows Great Plains moisture to reach deep into Wyoming on easterly winds. Fortunately, the altitude of Riverton is itself a barrier to the Plains moisture – a barrier that becomes less and less effective as the elevation along the track descends past Casper to Nebraska.

Table 5–3: August Cloud-Cover Statistics for Wyoming

Wyoming		Percent Frequency of Sky Condition								
Airport Observing Site	Nearby communities	Clear	Few	Scattered	Broken	Overcast	Thin Fog	Thick Fog	Average Cloud Cover (%)	Percent of Possible Sunshine
Jackson Hole Airport *	Jackson Hole/Teton Village	25.7	42.5	0	21.6	9.9	0	0.3	33.7	
Hunt Field Airport *	Lander/Boulder Flats	21.6	32.3	10.3	22.5	13.4	0	0	40.5	75
Riverton Regional *	Riverton	29.5	40.5	0	21.6	8.4	0	0	31.6	
Worland Municipal Airport	Worland/ McNutt	35.1	37	0	18.2	9.2	0	0.5	29.7	
Natrona County *	Casper	25.5	26.7	9.5	20.3	17.9	0	0	42.1	
Gillette-Campbell County Arpt	Gillette	29.5	19.3	17.7	24.3	8.9	0	0.2	38.8	

* = within total eclipse track

Satellite observations of central line cloud cover (Figure 5–2) shows that cloudiness increases significantly in the lee of the Rockies between morning and afternoon – a change that is partly mirrored by the surface observations, which show a smaller increase through the region. However, the drying effect of the Wind River Range is still obvious in the graphs, with a minimum in the cloud cover centered on Riverton. Toward Casper, the satellite-measured cloud cover increases again, but this is probably due to the presence of the Laramie Mountains to the south of the city. The satellite footprint – the area over which it measures cloud cover – is large enough that both the city and the mountains are included in the same field of view. Surface measurements of cloudiness at Casper show an average amount of just over 40% (Table 5–3), which, by comparing with the sunshine measurement at Lander, probably means a percent of possible sunshine around 75%. Riverton, with nearly 10% less cloud than Lander and Casper, may have a sunshine frequency that reaches 80% or more.

August is thunderstorm season in the mountains, brought on by the summer heat, instability, available moisture, and upslope winds from the east. At the mid and upper levels of the atmosphere, the westerly flow prevails and so the air is dry and cool from its passage over the Rockies unless a large-scale weather system is crossing the mountains. Lower down in the atmosphere, moist air is often available, carried onto the Great Plains and up into eastern Wyoming from its source region on the Gulf of Mexico. The combination of low-level moisture and cool, dry air aloft creates instability in the atmosphere, and if other conditions contribute, afternoon's will blossom into thunderstorms.

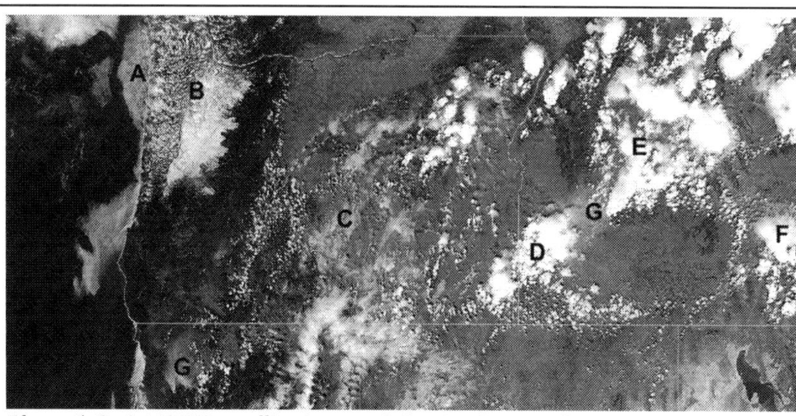

Photo 5-3: A MODIS satellite image of the Northwest taken on August 12, 2013. Low cloud and fog is visible offshore at A. At B, a small disturbance is crossing the Willamette Valley; behind the system, small cumulus clouds have formed over the dark forests of the Coast Range. At C, a mixture of cumulus and thin mid cloud can be seen over the Columbia Basin. Deep convective clouds - probably thunderstorms, have formed over the Columbia Plateau at D and are moving over the eastern parts of the Snake River Plain, but the central part of the plain, immediately to the east, remains sunny. Similar storms have formed over the Sawtooth Range at E and against the Tetons at F. At G, in northern California and again in central Idaho, forest fire smoke gives a fuzzy gray tone to the landscape. Image: NASA/SSEC/U. of Wisconsin-Madison

In Wyoming, thunderstorms are most frequent on the High Plains east of Casper and least frequent between Casper and Riverton, though the frequency depends very much on terrain and is highly variable through the state. Along the eclipse track, one of the most prolific thunderstorm generators is the Laramie Mountains, only 9 miles south of Casper. Typically, small convective clouds will form on the mountains in the late morning, growing into large towering cumulus and thunderstorms by mid-afternoon. At that point, the storms may move off of the mountains, tracking east and northeastward onto the plains of Nebraska. Such storms may dissipate as they leave the mountains, or blossom into large severe-storm systems that continue into late evening as they move eastward onto the Great Plains.

The mountain storms feed off of moisture drawn in from the plains, a flow that is promoted by the dark, forested slopes of the mountains that absorb radiation from the Sun and give energy to the forming clouds. The approaching lunar shadow will cut off the solar energy that feeds these buildups, and since the time of the shadow passage is still before noon, any small but growing clouds will lose their energy source and settle back to await the return of the sunlight. It is a common feature of eclipses, that convective clouds, if not too large, cannot sustain themselves through the hour or two of cooling that comes with the onrushing shadow. Such cloud buildups should disappear between first and second contact.

Casper records an average of 7 thunderstorm days and Lander 6.5 in August. The state is not noted for severe convective weather, but still has a significant historical record of hail damage and tornadoes, particularly on the Front Range east of Casper. June is the peak month for tornadoes in the state, with an average of 42 in the month; August is much quieter with just under 10. An average of about 29 large hail events occur in the summer, concentrated in the southeast near Laramie. In the area of the eclipse track, severe hail events occur about once per season, from May until September.

One feature of the climatology of Wyoming, Idaho, and even Oregon, is the occurrence of forest fires and the presence of smoke. The dry summer season almost always sprouts forest fires in the western mountains, from the Mexican border to Northern Canada (Photo 5-3). When winds are right, the smoke will be carried over the eclipse track, putting a haze across the sky and possibly masking the dimmest parts of the corona. As a rule, the reduction in visibility is only serious if the fire is relatively close, in which case it may be avoided by moving (Photo 5-4).

Photo 5-4: forest fire smoke along the Idaho-Oregon border, viewed from the Snake River Plain.

Wyoming provides a generous amount of space for eclipse viewing. Highway 20, from Casper to Riverton, follows the shadow axis for a hundred miles, passing through an open, high-plains countryside. West of Riverton, another 75 miles of highway takes the traveller to the shoulders of the Wind River Range, all within the depths of the shadow path and all with a favorable cloud climatology. East of Casper, for 70 miles, Interstate 25 never deviates from the central axis of the path by more than 10 miles. The generally favorable weather, ease of movement, and the wide-open skies make Casper a very enticing home base from which to plan an eclipse expedition.

5.5 Nebraska and Kansas

Nebraska and Kansas are states of transition, where the elevation lowers steadily as the eclipse track moves from the High Plains of central and western regions eastward to the Central Lowlands that straddle the Missouri River (Figure 5-5). The route passes over undulating hills, grassy dunes, occasional buttes and escarpments, rugged Badlands and the rich ecology of intermingled lakes and wetlands. The lunar shadow moves over a gradually descending prairie topography – a countryside of rolling, open hills with none of the complex cloud-making terrain of the Mountain States.

Figure 5–5: Nebraska and Kansas Topography Along the Eclipse Track

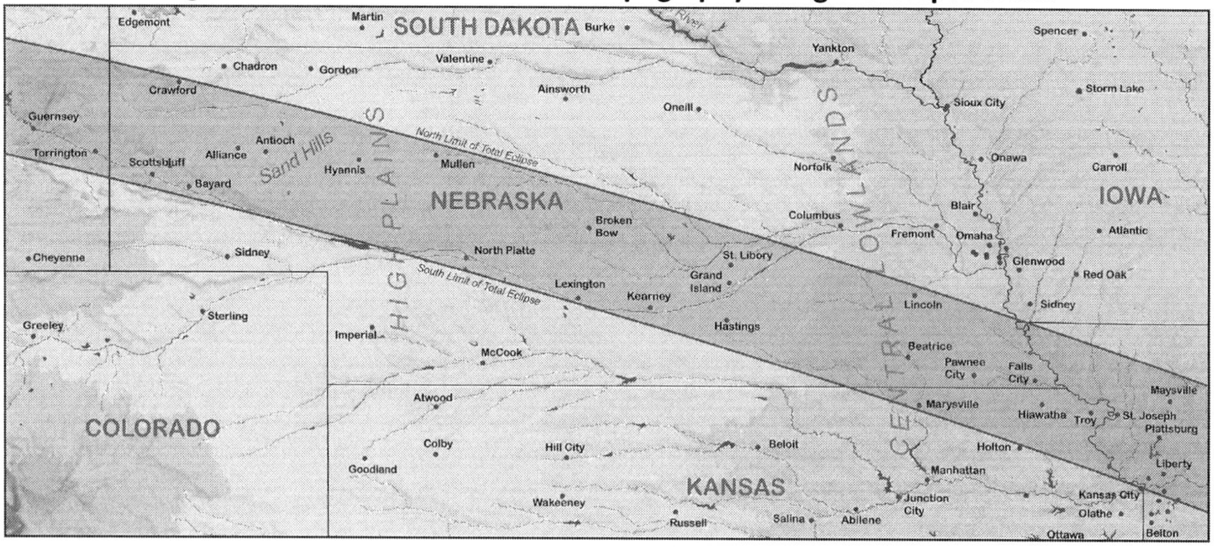

The descent from High Plains to Central Lowlands is matched by a steady increase in atmospheric moisture, which often has its origin over the Gulf of Mexico. Without any topographic barriers to block the winds, the Great Plains and the valleys of the Missouri and Mississippi form an atmospheric highway. Northerly flows are cool and dry, southerlies are humid and hot, westerlies are warm and dry, and easterlies usually mean a low-pressure system has just passed. In middle and upper layers, the westerly flows carry the cool, dried-out remnants of Pacific air across the mountains and onto the Plains. If these westerlies cross a humid, low-level flow of Gulf air, the combination creates the instability that helps initiate the thunderstorms for which the Prairies are so well known.

Interestingly, however, cloudiness tends to decrease during the afternoon across the Great Plains (Figure 5–2) rather than increase, as it does in the mountains. Morning clouds are broken up by solar heating, and so across Nebraska and Kansas cloudiness decreases by ~10% between 7 am and 2 pm The Moon's shadow passes across the Plains just before 1 pm CDT, well timed to take advantage of the diurnal trend.

Table 5–4: August Cloud-Cover Statistics for Nebraska and Kansas

Nebraska		Percent Frequency of Sky Condition								
Airport Observing Site	Nearby communities	Clear	Few	Scattered	Broken	Overcast	Thin Fog	Thick Fog	Average Cloud Cover (%)	Percent of Possible Sunshine
Western Nebraska Regional *	Scottsbluff/Minatare	25.8	22.5	11.6	24.6	15.6	0	0	43	
Alliance Muni *	Alliance	32.6	32.8	0	22.5	12.0	0	0	35	
Hooker County Airport *	Mullen	21.9	32.1	0	20.1	25.6	0.2	0.0	47	
North Platte Regional Arpt *	North Platte									75
Broken Bow Muni *	Broken Bow	20.1	32.4	0	27.7	19.3	0.5	0	47	
Kearney Regional Airport *	Kearney/Gibbon	28.0	27.7	0	12.6	31.5	0.2	0	46	
Hastings *	Hastings	20.6	31.9	0	32.3	15.2	0	0	45	
Central Nebraska Regional Arpt *	Grand Island	25.5	14.6	14.0	23.4	22.5	0	0	49	
Lincoln Municipal *	Lincoln	21.9	18.5	7.7	24.2	27.5	0.2	0	53	70
Beatrice Municipal *	Beatrice	16.4	31.1	0	33.0	19.4	0	0	50	
Columbus Muni	Columbus	17.4	33.7	0	32.0	16.1	0.6	0.2	47	
Brenner Field Airport *	Falls City	20.9	27.9	0.0	36.4	14.8	0	0	48	
Sidney Municipal/ Lloyd W. Carr	Sidney	20.3	20.0	22.2	22.6	14.0	0.9	0	45	
Chadron Municipal	Chadron	19.5	41.5	0	25.7	13.2	0	0	40	
Kansas										
Forbes Field	Topeka	18.5	32.7	0	29.6	19.2	0	0	47	70
Johnson County Executive Airport	Olathe	15.7	34.7	0	31.4	17.1	0.8	0.2	48	
* = within total eclipse track										

In southeastern Nebraska and northeast Kansas, where the eclipse crosses the Central Lowlands along the Missouri River, the terrain is wide open to sub-tropical moisture reaching northward from the Gulf of Mexico and also subject to extensive evapotranspiration of soil moisture from fields of grain, sorghum, and corn planted on the Prairies. Over the High Plains (toward the Wyoming border), the influx of low-level moisture is held at bay by the rising terrain, limiting the moisture supply for afternoon convective storms on the western side of Nebraska. A consequence of this gradual increase in moisture is reflected in the average August precipitation along the track, which ranges from a low of ~0.3 inches in central Wyoming to 4 inches along the Kansas-Missouri border.

The increase in humidity toward the southeast along the track results in a small, slow increase in average cloud amount (Table 5–4) across Nebraska, from around 35% to 45% in the west near Alliance, to 50% to 55% in the east around Lincoln and Beatrice. Satellite measurements of average cloudiness along the central line across the state (Figure 5–2, afternoon graph) show a similar trend, rising sharply from ~50% near Alliance to nearly 65% at North Platte, whereafter it remains nearly constant. Sunshine measurements (Table 5–4) along the track are not so pessimistic, showing that North Platte, in central Nebraska, receives 75% of the maximum possible sunshine, while Lincoln and Topeka are content with an average of 70%.

Nebraska is a part of "tornado alley," but by the date of the eclipse, the threat of tornado-producing thunderstorms has declined markedly, from the June peak of 21 to an August average of just 2. Nevertheless, eclipse seekers will still have to contend with the convective climatology of the plains, which tends to be a little less predictable than the large-scale weather systems of other seasons. Thunderstorm frequency increases toward the southeast, into northeast Kansas, where growing storms can tap into the Gulf moisture more readily.

Nebraska favors the well-prepared expedition, as highways in the state permit rapid movement to clearer skies, which are often not too distant when thunderstorms are responsible. Good weather information will be essential – satellite images work particularly well in the short range, but a forecast from the day before will probably prove more convenient.

A good home base for eclipse expeditions is Alliance, NE, or nearby Antioch, both located in the best weather zone on the plains. Highway 2 from Alliance offers 200 miles of in-the-track travel toward the east if cloud avoidance is necessary and a westerly escape can go all the way to Casper and still stay in the umbral region. In central and eastern Nebraska, Interstate 80 allows rapid relocation, cutting diagonally across the track, from Lincoln to Lexington, a distance of 150 miles. Nebraska and northeast Kansas don't offer any terrain for cloud dissipation, so flight is the best option if a new viewing location is wanted.

5.6 Missouri and Illinois

The portion of the eclipse track through Missouri and Illinois could very well be called the "river" section, as it follows the Missouri River to the Mississippi River, and then passes over the lower Ohio River on its path across Illinois. Because so much of the track is along or across these rivers, it's generally a lowland transit along the central line, except in southeast Missouri. There, the south side of the track skirts the Ozark Plateau and then passes over the 1700-foot-high St. Francois Mountains before descending onto the floodplain of the Mississippi and Ohio Rivers, where elevations are about 1300 feet lower (Figure 5–6).

Table 5–5: August Cloud-Cover Statistics for Missouri and Illinois

Missouri		Percent Frequency of Sky Condition								
Airport Observing Site	Nearby communities	Clear	Few	Scattered	Broken	Overcast	Thin Fog	Thick Fog	Average Cloud Cover (%)	Percent of Possible Sunshine
Rosecrans Memorial Arpt *	St Joseph	20.3	23.4	2.7	29.8	23.8	0	0	52	
Kansas City International *	Kansas City/Platte	21.8	19.3	12.1	25.9	20.9	0	0	49	67
Columbia Regional *	Columbia/Ashland	13.2	24.1	11.2	24	27.4	0	0	55	64
Jefferson City Mem *	Jefferson City	14.6	33.4	0	30.7	21.3	0	0	51	
St Louis International Apt	St Louis	11	21.4	13.9	30.8	22.9	0	0	56	
Spirit of St Louis Airport *	St Louis	14.8	32.3	3.9	28.6	20.3	0	0	49	65
Cape Girardeau Muni *	Cape Girardeau/Jackson	11	34.6	4.6	30.9	18.6	0	0.3	50	
Illinois										
St Louis Downtown Apt/Cahokia *	St Louis	8.4	31.5	0	41	19.1	0	0	56	
Abraham Lincoln Capital Arpt	Springfield	11.9	17.4	15.6	29	25.9	0.2	0	58	70
Scott AFB/MidAmerica	Scott AFB/Shilo	7.9	44.2	0	31.4	16.6	0	0	48	
Southern Illinois *	De Soto/Murphysboro	11.8	38.4	0	33.8	15.5	0	0.5	48	
Cairo	Cairo									75
Williamson County Regional *	Marion/Herrin/Carterville	5.7	35.5	0	41	17.8	0	0	55	

* = within total eclipse track

Figure 5–6: Missouri and Illinois Topography Along the Eclipse Track

Average August afternoon cloud cover across Missouri and Illinois is nearly constant, at around 65% in the satellite-based statistics (Figure 5–2, afternoon), a feature that is mirrored by weather-station observations from the ground. From the satellite viewpoint, the highest cloud levels on the central line occur south of St. Louis, near Festus, which is probably showing the influence of the nearby St. Francois Mountains. August cloud is typically convective cloud, usually building in the afternoon and settling overnight, though this is the general pattern and often interrupted by stronger weather systems that keep thunder and lightning going all through the night.

Table 5–5 reveals that the average cloudiness is close to 50% across lowland Missouri but the percent of possible sunshine is still a promising 65% or thereabouts. Satellite and surface-station observations do not distinguish between opaque and transparent cloud, and so the relatively large sunshine readings are not surprising – some portion of it will come from semi-transparent clouds through which the Sun can shine and record its presence on the instruments. Just remember that cloud statistics – surface and satellite – should be used to compare one site with another and not to decide on the absolute probability of seeing the eclipse. That latter probability is best derived from the sunshine data, but even then, only imperfectly.

After crossing the St Francois Mountains, the track descends to the Mississippi and Ohio valleys and across the Big Muddy Watershed, where the average cloud cover declines once again. It's only a ~5% decline according to Figure 5–2, but it comes at a propitious place: near Carbondale, IL. Carbondale could justifiably be called America's "Eclipse City," for it lies at the point where the eclipse tracks of 2017 and 2024 cross (Figure 1–3).

During the summer months, hot and muggy weather from the Gulf of Mexico is entrenched over Missouri and Illinois. Afternoon temperatures average close to 90 degrees and relative humidities hover near 50%. Such energy-sapping weather often persists for days, but in late August, relief comes irregularly when cool and dry air masses push southward from Canada and the Upper Midwest to bring a break in the sultry weather. For those not familiar with Midwest weather, the humid conditions will prove a challenge. These conditions are also important elements in thunderstorm development, though by August, the frequency of severe storms at least is much less than at its peak in May and June; tornadoes are quite uncommon in the month.

5.7 Kentucky and Tennessee

It is early afternoon – around 1:30 pm – when the track reaches Kentucky and afternoon clouds, if there are any, will be building in the sky. Afternoons in Kentucky and much of Tennessee are no cloudier than the early morning according to observations from the weather satellites. What changes is the type of cloud: convective clouds build in the afternoon as the ground heats and warm thermals begin to rise. In contrast to eclipse sites in western states, the cloudiness measured in states east of the Mississippi shows a higher frequency of overcast conditions - just short of 20% of recordings in much of Tennessee and Kentucky. Compare this to just 10% to 15% in Wyoming and to 5% to 10% on the Columbia Plateau in Oregon.

Figure 5–7: Kentucky and Tennessee Topography Along the Eclipse Track

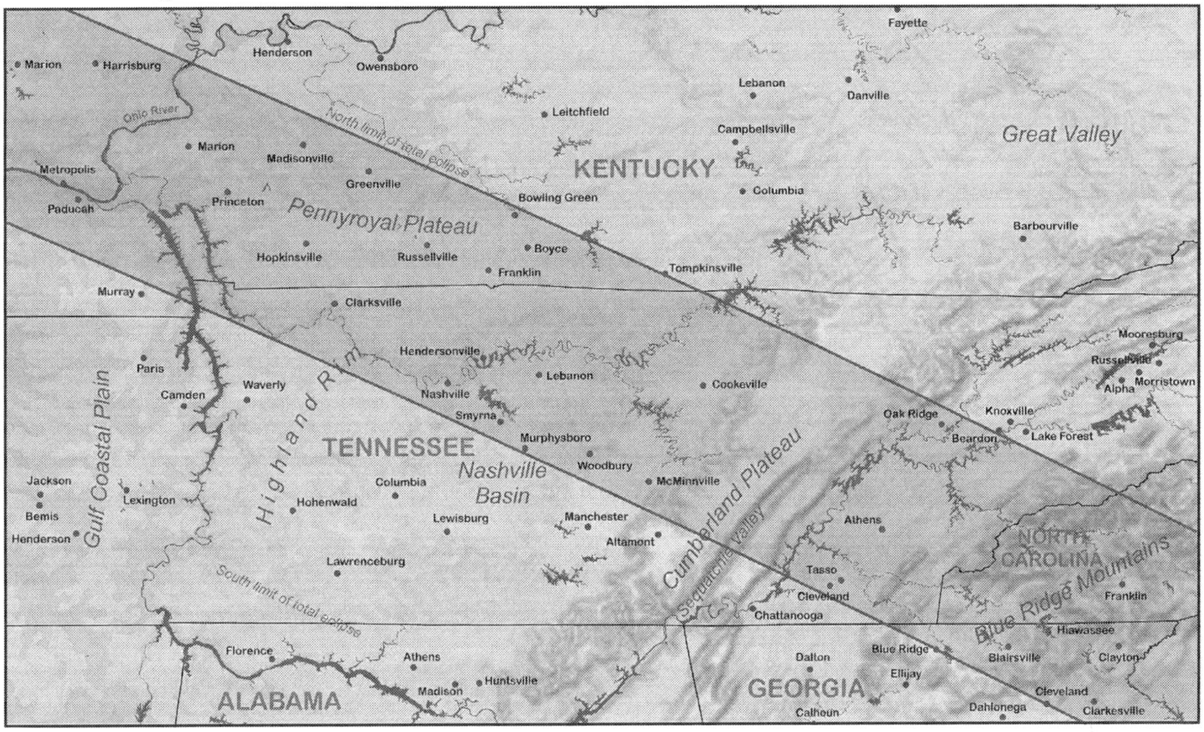

The terrain across western Kentucky is relatively flat – a landscape of small, rolling hills that make up the Pennyroyal (Pennyrile) Plateau. Off the Interstates, roads are tree lined, limiting the view, but accompanied by generous openings that look across the many farms in the region. As the track enters Tennessee, the land begins to rise, crossing the low hills of the Highland Rim before dropping into the Nashville Basin.

Table 5–6: August Cloud-Cover Statistics for Kentucky and Tennessee

Kentucky		Percent Frequency of Sky Condition								
Airport Observing Site	Nearby communities	Clear	Few	Scattered	Broken	Overcast	Thin Fog	Thick Fog	Average Cloud Cover (%)	Percent of Possible Sunshine
Barkley Regional Airport *	Paducah	8.2	20.9	16.8	34.7	19.4	0	0	57	71
Owensboro-Daviess County	Owensboro	11.3	37.9	0	31.6	19.2	0	0	50	
Bowling Green/Warren County *	Bowling Green	6.1	33.7	0	40.4	19.6	0	0.2	57	
Tennessee										
Nashville International Arpt	Nashville	4.9	19.1	23.8	32.9	19.2	0	0	58	63
Smyrna Arpt	Smyrna	1.1	46.5	0	40.1	12.3	0	0	51	
Chattanooga Metropolitan Arpt	Chattanooga	7.9	23.7	37.2	24.8	6.4	0	0	46	
* = within total eclipse track										

Beyond Nashville, the track climbs the higher Cumberland Plateau, and then promptly sinks into the Tennessee Valley. In the satellite record, none of these ups and downs in the terrain have much influence on the cloud cover, which is high and climbing slowly, formed by the generous sub-tropical humidity of the area and a steady rise in the land's elevation toward the Appalachians. The surface observations are a little less daunting, showing a 12% decrease in average cloud cover at Chattanooga compared to the 58% at Nashville. Across both states, broken cloudiness is the dominant afternoon sky condition, while clear skies are uncommon - only 5% to 10% of the observations at eclipse time.

On the border with Georgia and the Carolinas, the track reaches the Blue Ridge Mountains, a section of the Appalachians, and both the terrain and average cloud cover climb sharply. This region has the highest cloud cover along the track, reaching nearly 80% in the satellite measurements. In the Moon's course across Kentucky and Tennessee, the percent of possible sunshine declines from a tolerable 71% at Paducah to 63% at Nashville, and presumably a fair bit lower over the Appalachians, though there are no measurements to show that.

5.8 Georgia, North Carolina, and South Carolina

It's now mid afternoon when the shadow passes, and convective cloud buildups will be increasing, replacing the more sedate cloudiness of the morning hours. Even so, the cooling that comes with the eclipse shadow should have a dramatic effect in dissipating the cloudiness provided the buildups haven't reached the rain-producing stage.

Figure 5–8: North and South Carolina Topography Along the Eclipse Track

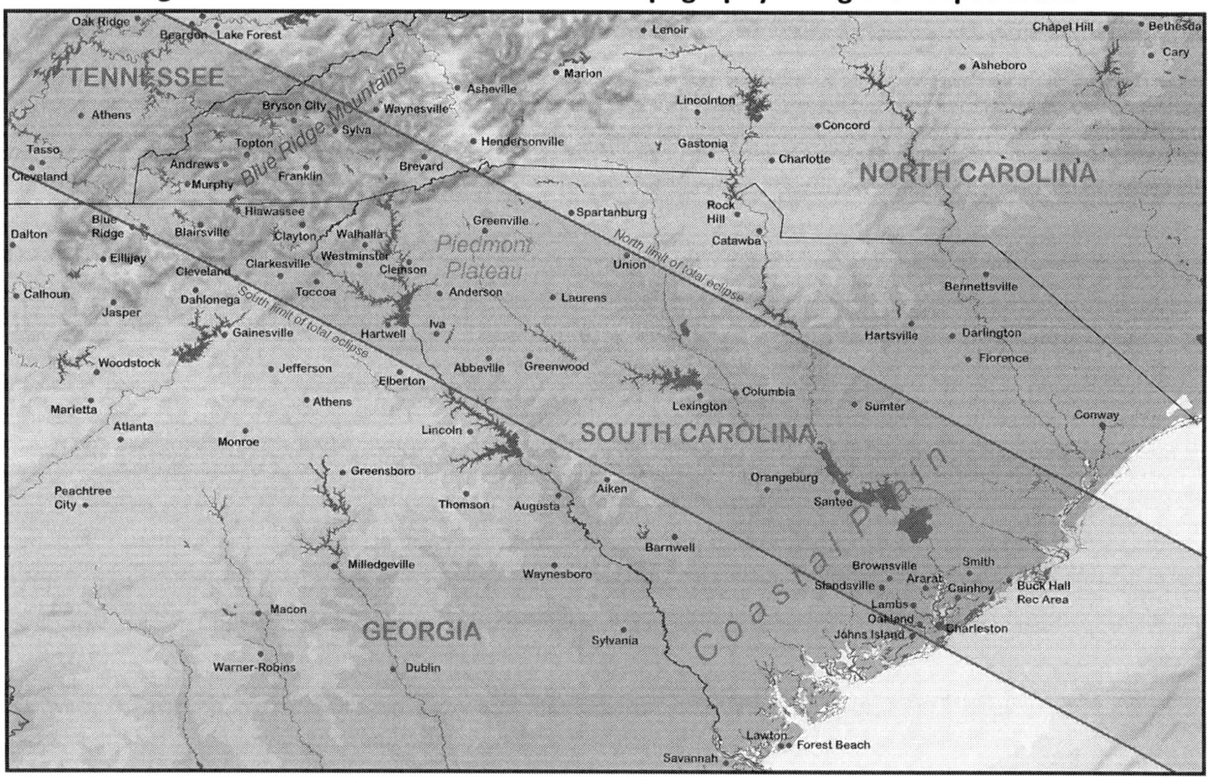

Once over the Blue Ridge Mountains, the track passes over a landscape that is dropping sharply in elevation toward the Atlantic coast (Figure 5-8). Alas, it has little impact on the cloud cover, as this part of America is now supplied by moisture from the warm waters along the coast and moist air is available in abundance no matter which way the wind blows.

Average cloudiness does not decline on the east side of the Appalachians according to both satellite and surface observations, at least until the Atlantic coast is reached. Eclipse observations will be a challenge. In Table 5-7, Anderson, SC, has an observed cloudiness of only 50% on average, but the calculation is affected by a strange absence of "scattered" cloud reports and a high frequency in the "few" category in the official record. In all likelihood, the value there will be similar to those elsewhere in the state.

Table 5–7: August Cloud-Cover Statistics for Georgia, North Carolina and South Carolina

North Carolina										
Asheville Municipal	Asheville	3.6	23.4	15.2	32.7	25.2	0	0	61	54
Georgia										
Athens-Ben Epps Airport	Athens	3.4	21.5	20.2	36.5	18.3	0	0	59	
August Cloud-Cover Statistics										
South Carolina		Percent Frequency of Sky Condition								
Airport Observing Site	Nearby communities	Clear	Few	Scattered	Broken	Overcast	Thin Fog	Thick Fog	Average Cloud Cover (%)	Percent of Possible Sunshine
Anderson Regional Arpt *	Anderson	4.9	44.5	0	34.4	15.8	0	0.3	50	
Greenville Spartanburg *	Greenville	4	20.3	17.6	37.2	20.8	0	0	61	61
Columbia Metropolitan *	Columbia	2.8	20.9	19.1	34.7	22.5	0	0	61	66
Charleston AFB/International *	Charleston	2.8	13.8	15.4	39.2	29	0	0	69	64
Beaufort MCAS	Beaufort	0.6	14.7	14.1	53.4	17.1	0	0	67	

The best observing site in the Carolinas is likely to be right on the coast. Since convective clouds build from the heating of the land, ocean side viewpoints are able to take advantage of lower cloud levels in the cooler air that flows inland from the offshore waters. A site could be selected in Charleston, though the city is well off the central line and the time penalty would be significant. Better is a seaside park such as Buck Hall Recreation Area, about 2 ½ miles from the central line. Buck Hall is a small park however and cannot accommodate a large crowd. Access to the waterfront is limited by marshy offshore islands and private homes along the coast, but it should be sufficient to be within a mile or two of the Atlantic to take advantage of the cloud-suppressing winds from the east.

South Carolina is no stranger to Atlantic hurricanes and so the possibility of one of these storms in the days surrounding the eclipse, though very small, cannot be discounted. Storm statistics for the Carolina coast reveal that the eclipse comes at a time when the frequency of tropical storms is on the rise toward its mid-September peak. Since 1867, 43 hurricanes have made landfall over South Carolina. Of these, 5 occurred within the four-week period centered on the date of the eclipse. Talk about hurricanes with any long-term South Carolina resident and the names Hugo (1989), Charley (2004), and perhaps Gracie (1959) and Hazel (1954) will come quickly into the conversation. Of these, Hugo, at category 5, was the strongest, bringing wind speeds of 140 mph when it made landfall right over the eclipse track on September 22.

5.9 In Retrospect

Climatology favors the Mountain States along the eclipse track, but only in locations within the valleys or atop plateaus that are protected by higher terrain, usually to their west. In particular, the Columbia Basin near Madras, OR; the Snake River Plain both near Ontario, OR, and Idaho Falls, ID; and possibly the Front Range of the Rockies west of Casper, WY. On the Great Plains, western Nebraska leads the choices, particularly near Alliance, NE. East of the Missouri River, it's difficult to identify a "best" location because it will depend on how quickly the afternoon cloudiness builds. However, sites in and near the Appalachians tend to have the highest average cloudiness, building strongly in the afternoons from Nashville eastward, only improving on the shores of the Atlantic near Charleston, SC.

While the focus in this study has been on afternoon cloudiness because it dominates the statistics in most regions, the eclipse track is also subject to migrating weather systems that bring much heavier and prolonged cloudiness. These are well forecast by numerical models and can be avoided for the most part by selecting a viewing site a few days ahead based on the forecast for eclipse day. There will almost certainly be one or two such systems somewhere along the track on eclipse day, so the determined eclipse chaser should be prepared to move to sunnier prospects.

5.10 Getting a Forecast Ahead of Eclipse Day

Weather forecasts are ubiquitous across the USA, but there is often little consistency from one media version to another until the eclipse is only a few days away. To sort out the conflicting predictions that may arise before August 21, it helps to take a look at the actual output from the numerical models provided by the National Weather Service. There are several sources for this output, but here are a few that are most useful:

1. The National Center for Environmental Prediction (NCEP)
   Web site: *http://mag.ncep.noaa.gov/model-guidance-model-area.php*

The NCEP is the primary site for original output from the model – they are the folks that design and run the computer. However, it is a site more sculpted for a knowledgeable user rather than for a casual weather forecast. If you find your way here, select either the GFS model (long range, out to 15 days) or the NAM model (short range, out to 84 hours). Choose NAMER (North America) for the domain and you will be presented with a menu of weather elements. Cloudiness is not one of them, but you can have a look at the relative humidity (which is related to cloud cover) at three different levels in the atmosphere: 500_RH_ht for high levels, 700_RH_ht for mid levels, and 850_RH_ht for low levels. Run the animation and you can watch weather systems develop and move.

2. College of DuPage
   Web site: *http://weather.cod.edu*

The COD is a go-to site for storm chasers, but they also have a display of weather elements from several numerical models. Click on "Weather Analysis Tools" and select "Numerical Models." Nine tabs will give you a choice of models, but the one you will find most useful is under the "NAM" button. Once you have selected this tab, a menu will drop down on the left side. You will find cloud-cover forecasts under the "Precipitation Products." The GFS model is also available, but has no cloud-cover forecast; you will have to rely on relative humidity. Once you have selected a product, a time bar will appear near the top of the display and you can move a slider along to see what problems the models may be forecasting for you on eclipse day.

The COD also has an excellent satellite and radar image display site.

3. The National Weather Service
   Web site: *http://www.weather.gov*

The NWS is the U.S. government agency responsible for providing forecasts and supporting commercial meteorological companies. You will find forecasts for the nation, radar, satellite images, and other information. Much of what you find will also appear on commercial providers.

4. Many commercial providers – Accuweather, Wunderground, The Weather Network – rework the information from the NWS's models or occasionally, run their own models. It helps to look at several model forecasts to see how much consistency there is from one to another, a measure of the models' reliability. Most private companies will not identify the model source of their data, but much, probably nearly all of it, comes from the NCEP. Nevertheless, they may use proprietary algorithms to predict secondary elements such as cloud cover and so are worth a look. Presentation of the weather elements is usually better done by private forecast companies.

Section 6: A Travelogue Along the Umbral Path

6.1 Introduction

This is the story of an eclipse odyssey undertaken by two Canadians. Stan Runge and Jay Anderson left Winnipeg in August 2013 to travel the length of the 2017 eclipse path, a kind of four-wheeled "walkabout" that took them from the west coast to the east. These are some of their observations.

6.2 Oregon

The Moon's shadow glides onto the North American continent over Oregon, an ephemeral darkness that slips inland at a speed of 2400 mph on August 21. The central axis of the shadow meets the coast at Fishing Point, a largely unknown local landmark whose name tells more about its function than the small signage that announces the path that leads toward the water. It's a quiet local attraction, thrust into national prominence by this accident of solar system geometry. It's not Fishing Point National Park or Fishing Point State Park, just "Fishing Point," though it could perhaps be more aptly named "fishing point."

Fishing Point lies on the south side of Lincoln Beach, which in turn lies a little south of Lincoln City. It's a rough grass-and-boulder promontory, not very flat and margined by smooth, timeworn stones that tumble down a fifty-foot cliff to the water. Reaching the water to fish is not a casual saunter, but a cautious climb over car sized, smooth sided rocks that ends in a few rough perches extending into the Pacific. On the bluff above, there is room for perhaps fifty or a hundred watchers, with some risk of being tumbled down the sides of the outcrop, as there are no railings to constrain the unwary or unfortunate. It is not impossible that cameras and sprawling tripods on eclipse day may reduce space to a precious commodity that will have to be defended from a press of other eclipse watchers. Fortunately, the Sun will be high (39° at totality) as the Moon begins its interruption, and nearby streets will offer adequate observing, though at a considerable loss of ambience. Fishing Point is a unique vantage point from which to view the eclipse, but one too easily crowded.

Photo 6–1: The view from Fishing Point, OR. Beyond the sandy edge, the promontory drops sharply to a rocky beach below.

The bluff is approached by a rough trail through a small forest, ambling back and forth across tree roots and overhanging branches for a few hundred feet – a bit of a gloomy start for a day that will celebrate the Sun and Moon. Cameras and tripods will make it through the brush just fine, but heavier or larger equipment will be a bit of a slog. Signs at the entrance gate warn of "Unfenced Cliff Ahead," and also, thankfully, no "Commercial Activities without a Permit."

The Point is situated in the midst of a subdivision of single-family bungalows and two-story houses, a tiny oasis of nature within the overly humanized landscape of this part of Oregon's coast. A small parking lot for perhaps 30 cars lies across the street from the Point's entrance, and a gated community of upscale houses marks its southern perimeter. The private houses form a barrier against any approach to the water line, and so those eclipse seekers who cannot find space on the Point will have to watch from surrounding streets unless fortunate enough to find friends along this stretch of coast. The unimpeded view to the west across the water afforded by Fishing Point will give the first view of the approaching shadow and a panorama of the sunlit horizon at totality.

Photo 6–2: Lincoln City beach.

The central axis of an eclipse is only a symbolic point, and there are many other places to settle on August 21 without sacrificing more than a few seconds of total eclipse time. Lincoln City is one of these, boasting Pacific coast beaches and a summer holiday culture, though the cold offshore waters are most suited for the young and adventurous or those clad in a wetsuit. It was a windy day when we visited, and though there was a cloud free sky overhead, onshore winds blowing from the cold waters made for a chilly day. Of more concern was the blowing sand and the salt laden air, which quickly sullied our camera optics, making lens caps essential and photography cautious.

South of Fishing Point lies the town of Newport, a larger resort town than Lincoln City and endowed with a few advantages that may make it a more convenient site than those farther north. Newport has South Beach State Park and Yaquina Bay State Recreation Site to accommodate the crowds, both affording a great view to the west to greet the oncoming shadow. The eastern horizon is blocked by the nearby Coast Mountains, but at eclipse time, the Sun will be well above the slopes. Newport is a lively town, with many restaurants, sea front bars, souvenir shops, and stores that specialize in seafront vacations.

The Coast Range lurks over the narrow coastal plain and though it is not a high one by topographic standards, it is the first to intercept the Pacific winds, and so hits way above its weight class as far as climate is concerned. Eastward, on Highway 20 from Newport, takes us through dense coniferous forests that almost engulf the twisting, climbing road, blocking the sky and presenting few opportunities to view the eclipse. Nevertheless, there is a certain attraction to the forest, as we imagine a photograph of the Sun in eclipse viewed through a gap in the trees, framed by a press of dark branches against a darkened sky – perhaps a gamble in view of the high average precipitation on the slopes, but a prize winner if it happens.

Mountain forest gives way rather abruptly to an open landscape of small towns and farms as we cross onto the east side of the Coast Range into the Willamette Valley. The openness would be welcome on eclipse day, allowing a view of the evolution and devolution of the Moon onto the Sun, but comes with a less photogenic surrounding. The valley is narrow, squeezed for space between the substantial heights of the Cascade Mountains to the east and the lower Coast Range to the west. One almost wishes for a low altitude eclipse so that the solar/lunar disks could be photographed hanging above the mountain terrain. The forested Cascades impose the same eclipse viewing limitation as the Coast Range, though the scenery is more picturesque because of the more dramatic surrounding peaks. The climb over the range on Highway 20 presents a number of attractive photographic and observational possibilities, as there are many spots where a traveler could stop to view the Sun (and scenic views), even though the overall trip is through crowding forests and over mountain passes. The Pacific side of the route is a steady upward climb to 4300 feet, sneaking between 12,000-foot peaks, but the ascent is not mirrored by the descent, which comes out on the elevated Columbia Basin at an altitude of 2100 feet.

Photo 6–3: A view of Madras from the north. The central line of the eclipse is behind the photographer. Madras's climatology reveals that August has sunny skies more than 80 percent of the time.

It is strikingly apparent that the Columbia Basin, protected by mountain peaks that defend against moisture bearing winds from the Pacific, does not share the moist climate of the west sides of the Coast and Cascade Mountains. Instead, the treed slopes give way to an open plain of towns and farms, and vegetation changes from lush forest to dry grasses. The flat, slightly rolling landscape in the Columbia Basin lacks the rugged character of the mountains, but the bland terrain has its benefits, providing wide, open skies and a climatology that promises one of the best chances in America of seeing the eclipse in a cloud free sky.

We stopped at Madras to talk to the city managers and tell them about the eclipse and, in particular, that their small city is located where climatology promises the sunniest skies along the whole of the track. Though it was news at the time, the community is now very much aware of what is coming. Madras is a little south of the central line; the line itself overlooks flat farm fields with more than enough space to accommodate all of the travelers who might arrive on August 21. Some of the residents were a little guarded when we pulled up to talk, but quickly grew interested with the news of an impending eclipse. "Guess I'll have to invite the relatives," was one of the most frequent reactions to the news.

As the lunar shadow speeds eastward, the terrain gradually rises and roughens, becoming mountainous toward the Idaho border. This is the Columbia Plateau, a landscape of small towns and weather beaten, round topped hills: mountains that qualify for a senior's discount. It's an attractive part of Oregon, sparsely treed, so that the intricacies of the geology are revealed to the traveller. Highway 26 wanders back and forth within the shadow path, mostly on the south side, following an assortment of river valleys toward the Idaho border. Here and there, small communities beckon to the motorist, but the most beguiling is the tiny mining town of Mitchell, which entrances the sightseer with a facade of old buildings and necessary supplies. The mountains have closed in a bit, but not enough to impede the view to the Sun, which stands high up on eclipse day. The Little Pine Cafe and Lodge tempted, both for lunch and as an eclipse viewing site, but don't try to rent one of the few rooms on the floors above the cafe for eclipse day. "We rented those a couple of years ago," we are told.

Eastern Oregon holds a number of tempting state and national parks, but along the eclipse track, the John Day Fossil Beds are the most beguiling. There are three separate locations associated with the John Day beds, but when highway signs announced the Thomas Condon Paleontology Center at the junction with Highway 19, we turned left and took a 2-mile side trip to visit the center and take in its impressive fossil collection. It's impossible to drive those two miles without stopping to absorb the astonishing sight of 40 million years of geology revealed in bands of olivine tinted volcanic ash, tilted by eons of tectonic forces. The Condon Center is not quite on the axis of the shadow path, but it's only 3½ miles short and the time penalty is less than a second. The central line itself passes over a narrow, flat valley of scrubby rangeland, compressed on each side by low hills and more distant peaks that also show the layered paleontology of the Plateau. While they had no plans for the eclipse at the time, the Center would be a world-class location from which to view an eclipse and incidentally take in some of the numerous hikes through the valley.

Photo 6–4: The Columbia Plateau near the Thomas Condon Paleontology Center. The layers of ancient volcanic ash take on magnificent blue-green shades in this view from just south of the Center.

An hour east of the Paleontology Center, Highway 26, still within the eclipse shadow, begins to ascend to the summit of the Columbia Plateau, a short 35-mile, up-and-down ride between Prairie City and Unity. It's not particularly rugged – a rise of only 1400 feet or so – and at the higher elevations, the pine forests lend a freshening smell to the air. Clouds were growing when we passed, advertising the tendency of the altitude and the dark, forested landscape to promote the development of afternoon showers.

6.3 Idaho

After Unity, the highway drops steadily toward the Idaho border, into the western section of the Snake River Plain, a descent of about 2300 feet. The Snake River bottomlands are a dramatic lowering in the Rocky Mountain landscape, covering the lower third of Idaho between the Oregon and Wyoming borders. Once in Idaho, highways divert southward, out of the eclipse track, to get around the towering Sawtooth Range, which plants itself in the middle of the track like a cat plants itself in a lap. This leaves a small area along the Oregon-Idaho border to watch the eclipse, but it's an area with excellent weather prospects, as the dry landscape confirms. We followed the Snake River Plain, past Boise, making a large loop through open farmland to eventually reconnect with the eclipse track near Idaho Falls. On the way, we dropped in on the Craters of the Moon National Monument for an exploration of not-very-ancient lava flows, a welcome break along the highway. The extensive network of irrigated farms through the Snake River Plain testifies to the sparse rainfall this deep in the Rocky Mountains and to the very good eclipse viewing prospects in the area. Idaho Falls is in the eclipse track, on the south side, and it is only a short 22-mile drive to the middle of the shadow, past Rexburg to the hamlet of Thornton, if an extra few seconds of totality is on your agenda.

The Snake River Valley comes to an abrupt end against the Teton Range, but the encounter is a spectacular one with some of the most attractive scenery along the track. The impressive Tetons, seen from the west side, immediately invite the question, "How do I frame these against an eclipsed Sun?" It doesn't seem possible with a Sun that stands 50° high at totality, well above the surrounding scenery. A little exploration in Google Earth© points to a few routes heading east from Alta, ID, but almost certainly, someone will watch the eclipse from atop the Tetons. Nevertheless, a hike into the backcountry, either from the east or the west, would be rewarding for the serious photographer.

Photo 6–5: A view across the Snake River Plain. The grassland bears witness to the light precipitation on the Plain, but when irrigation is applied, the land becomes productive. Scattered cumulus clouds are typical of the weather along the Snake River; higher cloud marks a passing upper-level disturbance.

6.4 Wyoming

The Tetons are not only a boundary to the Snake River, but their ramparts also interrupt highway travel, which must divert southward to find a route around the range and into Jackson Hole. We chose the route through Tetonia, Driggs, and Victor, a comfortable, open road past farms and small towns, until we turned onto Highway 33 at Victor and headed across the Tetons. It was a spectacular ride, both for scenery and for the twisting mountain road that climbs upward to a difficult pass through the south side of the mountains.

Jackson Hole, named after its topographic appearance of a "hole" in the Rockies – much smaller than the Snake River Plain – presents a number of gorgeous viewing opportunities. The two communities, Jackson and Teton Village, put complete tourist resources at your fingertips, and we were lucky to find an open motel in Jackson during our journey. Jackson lies about 9 miles from the central line and Teton Village, 2 miles. Where Jackson has a "tucked-in" feeling, squeezed into a narrow valley, Teton Village lies up against the mountains on its west side but is open toward the east. One of the best experiences in the area is the aerial Tramway that carries visitors 4100 feet upward to the 10,000-foot summit of Rendezvous Mountain – an apt name for a place from which to watch a solar-lunar assignation.

Photo 6–6: Looking across the Tetons from the summit of Rendezvous Mountain. Clouds build on the hills while the "hole" below remains sunny. Two hours later, skies were completely clear, but only after the passage of a few thunderstorms.

Alas, as an eclipse site, the Tram has a few shortcomings. We found that the mountains were quick to form cloud, much earlier than the valley below, and when the clouds built into thunderstorms, the lift was suspended. There was scant shelter at the summit from the rain and lightning. The storms passed quickly and the summit was then washed in bright sunshine for the day, exposing the familiar peaks of the Tetons to the north – a world-class vista for a day with an eclipse. The thunderstorms would have been easily predictable, probably a day or more ahead, so we decided that a prudent approach would be to choose whether to Tram or not according to the forecast.

Photo 6-7: The Grand Tetons from the valley bottom in Jackson Hole. This location, near the airport, may be the most scenic vista along the entire eclipse track.

We were anxious to explore the terrain north from Jackson, as this would be our main "escape" route if the forecast for Jackson proved to be unfavorable. While Jackson itself is fairly confined by mountains, immediately northward, the "Hole" opens into a broader mountain meadow. This area, adjacent to the airport, gives an extended view to the east and a superb view of the best Teton peaks to the west, especially of 13,770-foot Grand Teton. The whole of the 20-mile route from the airport to Moran offers many scenic panoramas to garnish the spectacle of an eclipse. From Moran, Highway 26 and then 20 heads eastward, remaining in the path of the lunar shadow for over 300 miles, past Dubois and Riverton to Casper and a little beyond. And, even better, much of the highway route is close to the central line of the eclipse. This is prime Sun chasing country, naturally sunny because of its location to the lee of the Rockies, but with good fast transportation if clouds have to be avoided. By taking advantage of secondary highways, an avid eclipse chaser could go all the way to Grand Island, NE, without leaving the umbra's footprint – a distance of 650 miles as the shadow flies.

Riverton, along with Madras and Idaho Falls, enjoys very favorable cloud climatology, with a wide-open setting of high country rangeland and low, rolling hills. If scenery is not important, central and eastern Wyoming provide an unencumbered vista in which to enjoy the whole of the eclipse: incoming shadow, totality with all its phenomena, and departing shadow. Casper has the advantage of being the biggest city on the central line, which bisects the southern suburbs, and delivers a nearly 2 ½-minute eclipse. We were particularly captivated by the Fort Caspar Museum (yes, the spelling is correct), a recreation of an 1865 military post, which is situated just ¼ mile north of the eclipse axis. It's a historical spot on which to view the August 21st treat, but the city has a wider ranging celebration in mind, planning for a four-day "Wyoming Eclipse Fest." The Astronomical League will hold "ASTROCON 2017" from the 16th to the 19th, leaving time for eclipse-o-philes to scatter to their individual observing places. Casper is indeed a good starting point.

6.5 Nebraska

From Casper, Highway 26 (and for part of the journey, Interstate 25) first skirts the Laramie Mountains and then descends gradually onto the plains of western Nebraska, arriving on the Great Plains at Scottsbuff. It's a high desert environment in Wyoming, but by the time we arrived at the Nebraska border, it was clear that we'd entered the prairies, as stony vistas turned to grassland and irrigated farms. The shadow path crosses the Nebraska Sandhills, a vaguely mysterious landscape of gently rolling, grass covered hills and intervening wetlands, occasionally unclothed to reveal the underlying dunes.

Nearly astride the central line lies Alliance, NE, and nearby, Carhenge, a whimsical replica of its English namesake but constructed with the decaying remains of Detroit iron. Carhenge consists of 38 automobiles arranged in a circle with a 1962 Cadillac for a heelstone. It's a bit down-and-out these days, but the folks at the souvenir shop were aware of their coming date with the Moon when we stopped in. Carhenge lies a little over five miles from the eclipse axis, but if you like to mix your eclipses with whimsy and fun, this could be the best place for it in America. The weather chances aren't so bad either.

Photo 6–8: Carhenge, north of Alliance, NE. The "monument" is a bit tacky, but good fun for eclipse watchers.

Nebraska and Kansas offer wide-open skies for the eclipse, easy travel to avoid cloud, and space for everyone to settle in and wait for the spectacle. It's not as flat as most folks would think – in fact, the two states are better characterized as rolling – but flatness rules for those who drive Interstate 80 through the Platte River Valley. Perhaps unfortunately, I-80 runs for 226 miles within the eclipse track from North Platte to Lincoln – a good route to take if a quick change in position is needed, but not one with much variety in the terrain. The highway meets the central line near Grand Island.

We were delighted to drive unawares into the Nebraska State Fair in Grand Island, sampling a touch of mid-west energy and hospitality. In 2017, the State Fair celebrates Nebraska's sesquicentennial, and it will be underway before and after the day of the eclipse. Quite a celebration! Quite an omen!

6.6 Kansas and Missouri

Kansas gets a 40-mile section of the eclipse track over its northeast corner, crossing the twisty Missouri River three times near St. Joseph, MO. Riverboat cruise, anyone? One of the crossings takes place right at St. Joseph, but the crossing point is less than enticing: the Pony Express Bridges that span the border between Kansas and Missouri. Just upstream however, and within sight of the bridges, we were pleased to find the Remington Nature Center of St. Joseph – a science museum with a big parking lot, cool, tree-lined paths, and grassy open spaces: tailor made for an eclipse site.

6.7 Illinois

The route from Missouri to Illinois could be described as the "half-way" part of the track, not so much for its position within the track, but because the umbral shadow cuts half-way across Kansas City and half-way across St. Louis. We are also getting pretty close to the halfway point of the path, where the eclipse reaches a maximum, but first we had a stop at Carbondale. We dropped in to talk to the city managers and tell them about what's coming in August 2017. "You could call yourself's eclipse city," I tell them, "since you are at the crossing point of the 2017 and 2024 eclipses. They were intrigued with this information, as all civic Chambers of Commerce would be, and so you will find Carbondale billed as "America's Eclipse Crossroad." It's not bad for cloudiness either, though western regions take top billing for that statistic.

Photo 6–9: Downtown Carbondale, on South Illinois Avenue, where totality can be found in both 2017 and 2024. Continue south on South Illinois Ave. for 6 miles to reach the central line of the 2017 eclipse.

Carbondale is actually 6 miles north of the middle of the 2017 track, but it provides open spaces and a proud little-town-America ambience for the eclipse visitor. The actual central line, which lies farther down South Illinois Avenue into the Illinois countryside, is tree-lined and closed in. Carbondale is a seven-year-long celebration, and if you see the eclipse there this time, you are almost obligated to return in 2024.

6.8 Kentucky

Perhaps the town even more eager for this eclipse is Hopkinsville, Kentucky, the largest community in proximity to the point of Greatest Eclipse and a little farther down the track from Carbondale. The point of Greatest Eclipse actually depends on some technical definitions and astronomical constants used in the calculation[1]. From a purely practical standpoint, the location of "Maximum Duration" (where totality lasts within 1 second of its maximum possible value) falls within zone from western Missouri to eastern Tennessee. Nevertheless, Hopkinsville has taken the claim and has big plans for eclipse day.

Photo 6–10: A view to the central line claimed by Hopkinsville. Not much to look at.

This is all too big for the farmer who owns the Greatest Eclipse site. We spent some time driving backcountry roads to find that point, eventually creeping down Stone Quarry Road to a view across the sorghum fields. We stopped beside the owner, who paused in his mowing to talk. "I know why you're here," he said. "You've had others come by to look for the eclipse track?" I questioned. "Every week." he replied, "Dozens."

"You'll have quite a crowd here on eclipse day." I offered.

[1] The point of *Greatest Eclipse* (defined as the location intercepted by the Moon's shadow axis when it makes its closest pass to Earth's center) lies about 12 miles northwest of Hopkinsville, KY. The point of *Greatest Duration* (defined as the location where totality has its absolute maximum duration) is located 12 miles southeast of Carbondale, IL. The truth of the matter is that the difference in duration between these two points, separated by 90 miles, is a scant 0.1 seconds. So any community claiming bragging rights to *Maximum Eclipse* is splitting hairs of the most academic nature. Better to worry about the weather on eclipse day rather than whether you get the very longest possible duration of totality. For more, see Section 1.9.

"Maybe I'll let my grandkids set up a lemonade stand. Except the bureaucrats will want them to get a license."

The central line is a quiet country backwater, Amish country, with horse drawn buggies and picture-book farms and homes. The eclipse crowds will be totally out of character and probably crowded together. Under ordinary circumstances, it might be a peaceful place to watch the sky, but all bets are off for August 21, 2017.

6.9 Tennessee

Across Kentucky and much of Tennessee, the picturesque, rolling landscape offers a multitude of attractive eclipse watching sites. Back roads twist and turn, mostly at angles across the eclipse track, but Interstate 24 west into Nashville and Interstate 40 leading eastward remain under the lunar shadow for most of the length of Tennessee. Climatological observations show that the cloudiness in this part of the USA is higher than farther west, and so means of making a quick relocation to a better observing site becomes important for those determined to catch the spectacle.

Past Nashville, the countryside becomes more heavily forested and the topography adopts a noticeably rougher tone. Hills get bigger and more frequent, and before long, we are immersed in the Ridge-and-Valley Appalachians for which eastern Tennessee is famous. It's a tempting area, as small parks and wandering single lane roads beckon with a promise of quiet streamside resting places and a photographer's cornucopia. We cannot resist, and take time to sit beside a babbling brook, a travelogue cliché, but welcome after a week on the road. This countryside and the South Carolina Appalachians just ahead do not seem to be places for an eclipse however, as the view to the sky is restricted by the forest and besides, the mountains have a well deserved reputation for cloudiness. Notwithstanding, this part of our trip along the eclipse track is one of the most pleasant and easiest on the eyeballs. Maybe that's a good enough reason to visit on August 21 – after all, there are scenic overlooks along the way that provide generous views of the sky.

Photo 6–11: The magnificent forests and mountain valleys of the Appalachians – a bit of a challenge for eclipse viewing.

6.10 Georgia and the Carolinas

As we cross into North Carolina, touching a piece of Georgia along the way, the downward trend of the terrain accelerates our approach to the Atlantic. The dense forests of the mountain slopes give way to a more open vista of fields and tree-lined roads. We stop at Anderson (of course) admiring the mix of (for us) strange trees and impossibly green fields and woodlots. The landscape flattens, though still rolls, and the humid and pleasantly warm atmosphere speaks of a more subtropical climate than on the plains a few days before.

Getting about on eclipse day doesn't present a problem in South Carolina, as highways form a network across the state, a benefit that results from its relatively long history of settlement. But South Carolina is a cloudy state, and so we liked Santee as a pre-eclipse staging point; it lies next to Lake Marion, which would help subdue the normal convective clouds that build most days in August. Santee lies very nearly on the central axis of the shadow, and it has highways heading to all compass points, particularly nearby Interstate 26, which runs more-or-less up and down the shadow path, from Charleston to Columbia and up to Greenville.

Charleston was a treat – justifiably famous, immersed in history, and preserving an old world charm. It's pretty close to the southern limit, however, so the eclipse duration is a minute less than on the central line to the north. Still, we wondered if the view from Fort Sumter or some other suitably famous location in the harbor might not be worth the loss. The public piers jutting out into Charleston Harbor provide room for thousands and deliver a great ambience for an eclipse. Watching an eclipse in the midst of a large crowd can be an exhilarating experience as everyone suddenly becomes connected by the celestial show above.

We scouted northward from Charleston to find a spot to watch near the central line and near the Atlantic coast, since cloud cover tends to build inland when the afternoon Sun warms the ground. Sites near the water take advantage of the cool air that is drawn inland by the local sea breezes. There wasn't much available – the highway north is tree-lined and private property lined, leaving us with few options until we found Buck Hall Recreation Area within the Francis Marion National Forest. It satisfied all of the requirements for a great eclipse watching spot: a waterfront view, pleasant surroundings, proximity to the central line (just over 2 miles to the north), and a collection of birds to watch as the Sun is covered. It also has resident alligators, so swimming is not recommended.

Photo 6–12: The view north from Buck Hall Recreation Area. The Recreation Area extends only a short distance in this direction before it gives way to private homes.

Alas, Buck Hall is a small park, with ~20 camping spots and parking for 30 cars (plus another 20 for boat trailers). The site looks out across the small marshy islands that line this part of the coast, with private homes to the north and more-or-less inaccessible forest immediately to the south. All we can say is book early.

At Buck Hall, we enjoyed a conversation with a charter boat operator, who was more than a little pleased to hear of the coming eclipse. "I'll have to tell the guys in the (charter boat) group," he said. "We can offer rides to the central line and a little fishing along the way."

North of Buck Hall lies the small town of McClellanville. There's open space there for eclipse viewing and it's close enough to the coast to take advantage of the cloud suppressing afternoon breezes.

6.11 Final Thoughts

Our trip along the track revealed a myriad of places to watch the eclipse and a population that was interested and even excited about the prospect. Many had stories to tell of a previous eclipse, watched while young with a father or grandfather. This eclipse will create the same generational links, giving rise to stories to be told in later decades when other eclipses cross America[2].

[2] During the 21st century, the paths of 8 total solar eclipses cross parts of the continental USA in 2017, 2024, 2044, 2045, 2052, 2078, 2079, and 2099. See Section 7–11 for more information.

Section 7: Eclipse Resources

7.1 EclipseWise.com Web Site

For many years the NASA Eclipse Web Site was the leading Internet resource for predictions and information on eclipses of the Sun and Moon. The webmaster of the site (Fred Espenak) has now retired it as a location for new information but it remains an archival site for eclipse predictions prior to 2015. All future predictions for upcoming eclipses will be posted on the new web site *EclipseWise.com*.

EclipseWise.com has individual web pages, maps and diagrams for every solar and lunar eclipse from 2000 BCE to 3000 CE. This amounts to 11,898 solar eclipses and 12,064 lunar eclipses. Much of the design, layout and graphics were inspired by the recent publications *Thousand Year Canon of Solar Eclipses 1501 to 2500* and the *Thousand Year Canon of Lunar Eclipses 1501 to 2500*. (See: *astropixels.com/pubs/index.html)*

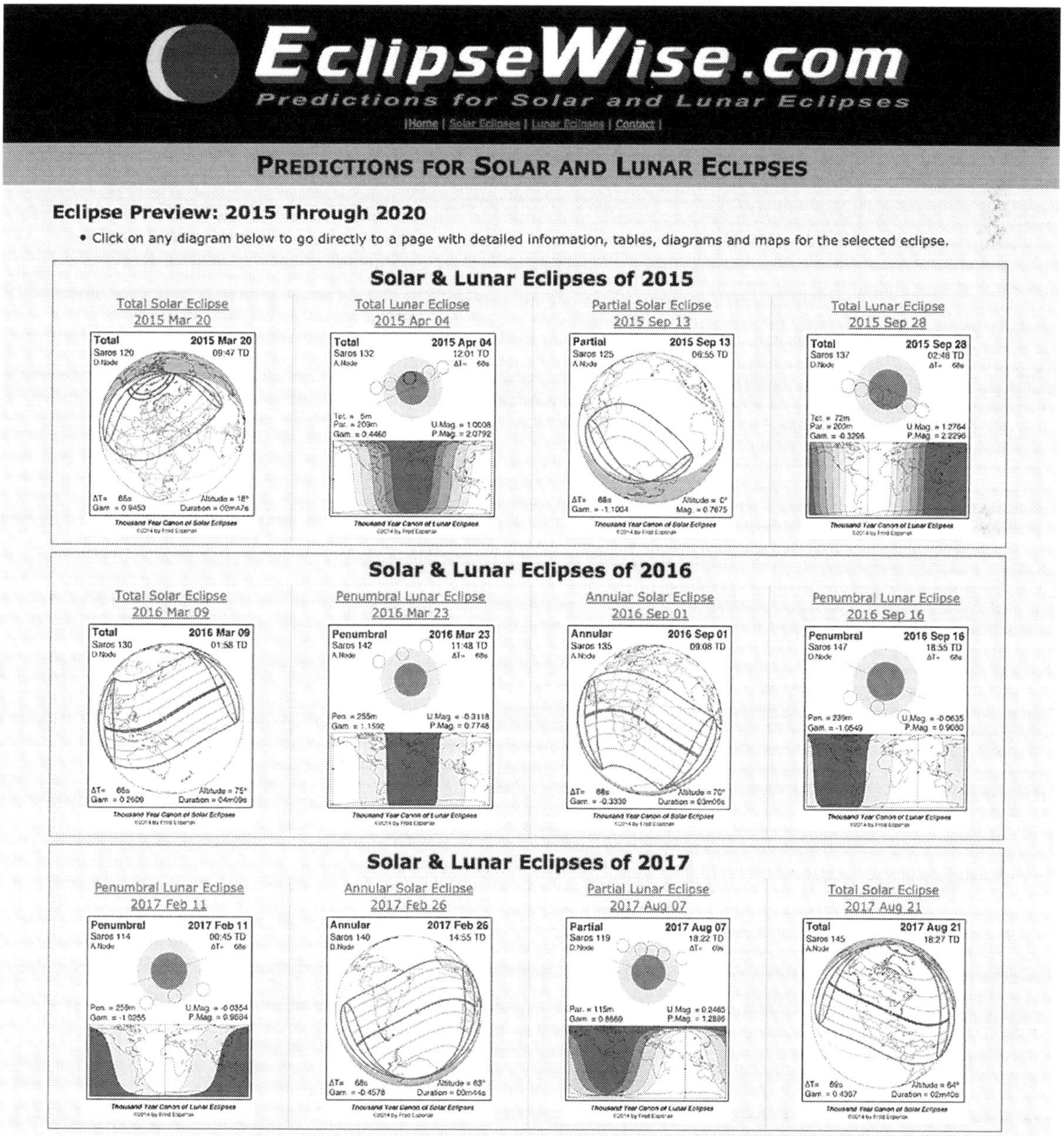

The Home Page of EclipseWise.com features a graphical interface to quickly preview upcoming eclipses with links to pages containing more detailed information.

The graphical user interface used by *EclipseWise.com* offers an intuitive way of accessing eclipse predictions. For example, the home page presents a concise preview of all upcoming solar and lunar eclipses over several years. Each small eclipse diagram gives a quick preview of an eclipse and links to a dedicated page for that particular eclipse.

The main or top pages of EclipseWise.com are:

> Home Page (both solar and lunar eclipses): *eclipsewise.com/eclipse.html*
> Solar Eclipses Page: *eclipsewise.com/solar/solar.html*
> Lunar Eclipses Page: *eclipsewise.com/lunar/lunar.html*

7.2 Road Atlas for the Total Solar Eclipse of 2017

Road Atlas for the Total Solar Eclipse of 2017 contains a comprehensive series of 40 full color maps of the path of totality. The large scale (1 inch ≈ 11 mile) is twice that of the *Eclipse Bulletin* and shows both major and minor roads, towns and cities, rivers, lakes, parks, and mountain ranges. The path of totality is depicted as a lightly shaded region with gray lines marking the duration of the total eclipse in 20 second steps. This makes it easy to estimate the duration of totality from any location in the eclipse path.

Armed with this atlas, the latest weather forecasts, and mobility strategy offers the best strategy for witnessing the total eclipse in clear skies. (See: *astropixels.com/pubs/Atlas2017.html)*

7.3 Web Sites on the 2017 Eclipse

Web Site: EclipseWise.com/solar/SEnews/TSE2017/TSE2017.html

EclipseWise.com has a series of pages and resources devoted to the 2017 eclipse. The main page (shown above) provides links to detailed eclipse path maps, tables of local eclipse circumstances for hundreds of cities, weather prospects along the eclipse path, and more. The link to an interactive Google Map with the eclipse path plotted on it allows the user to zoom into an part of the path. Click on any point on the map to display the eclipse circumstances and duration of totality at that location.

Other features include information on eye safety, eclipse photography, the sky during totality and additional data tables about the eclipse path. This web site will continue to add features as the eclipse approaches.

Web Site: GreatAmericanEclipse.com

GreatAmericanEclipse.com is published by Michael Zeiler and Polly White to educate the public on how to witness nature's greatest spectacle, a total eclipse of the Sun. Contributions in the form of guest blog posts, recommended viewing sites for a state, photographs of past solar eclipses, scans of historic drawings and maps, and similar items are welcome. Email GreatAmericanEclipse@gmail.com with text and images and your contributions will be considered for insertion if they are relevant, of high quality, and authored by you or in the public domain.

Web Site: Eclipse2017.org

Eclipse2017.org has been established to provide information to news media, elected officials, school administrators and teachers, and the general public on how to view the total solar eclipse of 2017. Maintained by Dan McGlaun (a veteran of eleven total solar eclipses), this site will continue to grow additional information and late breaking news of events surrounding the eclipse.

7.4 Additional Web Sites On Solar Eclipses

The following list contains a number of notable web sites on solar eclipses.

Being in the Shadow (Dr. Kate Russo): *www.beingintheshadow.com/*
Eclipse-Chasers.com (Bill Kramer): *www.eclipse-chasers.com/Map.html*
Eclipse-Maps.com (Michael Zeiler): *www.eclipse-maps.com/Eclipse-Maps/Welcome.html*
EclipseGuy (David Makepeace): eclipseguy.com/
Eclipser (Jay Anderson's Eclipse Meteorology): *eclipser.ca/*
EclipseWise.com (Fred Espenak): *www.eclipsewise.com/eclipse.html*
Eclipse Photography (Miloslav Druckmüller): *www.zam.fme.vutbr.cz/~druck/eclipse/Index.htm*
Front Page Science (Michael Bakich): *fpsci.com/start.html*
Hermit Eclipse: *moonblink.info/Eclipse*
IAU Working Group on Eclipses: sites.*williams.edu/iau-eclipses/*
Interactive Eclipse Maps (Xavier Jubier): *xjubier.free.fr/en/site_pages/SolarEclipsesGoogleMaps.html*
International Occultation Timing Association (IOTA): *www.lunar-occultations.com/iota/iotandx.htm*
MrEclipse.com (Fred Espenak): *www.mreclipse.com/MrEclipse.html*
NASA Eclipse Web Site (Fred Espenak): *eclipse.gsfc.nasa.gov/eclipse.html*
Shadow & Substance (Larry Koehn): *shadowandsubstance.com/*
Solar Eclipse Mailing List (SEML): *groups.yahoo.com/neo/groups/SEML/info*
Solar Eclipses (Date and Time): *www.timeanddate.com/eclipse/total-solar-eclipse.html*
The Clock Tower (Sheridan Williams): *www.clock-tower.com/*
UMBRAPHILLIA (Dr. Glenn Schneider):
nicmosis.as.arizona.edu:8000/ECLIPSE_WEB/UMBRAPHILE/UMBRAPHILE.html
USNO Eclipse Portal (USNO): *astro.ukho.gov.uk/eclbin/query_usno.cgi*
Waiting For The Shadow (Joe Cali): *joe-cali.com/eclipses/*
Williams College Solar Eclipse Expeditions (Jay M. Pasachoff): *web.williams.edu/Astronomy/eclipse/* and *totalsolareclipse.org*

7.5 IAU Working Group on Eclipses

As of the 2015 reorganization of Divisions and Commissions of the International Astronomical Union, functions of both the original Working Group on Solar Eclipses of the International Astronomical Union of the Solar Division and the Program Group on Public Education at the Times of Eclipses and Transits have been combined in an Inter-Divisional Working Group.

Professional scientists are asked to send descriptions of their eclipse plans to the Working Group on Eclipses of the International Astronomical Union (IAU), so they can keep a list of observations planned. Send such descriptions, even in preliminary form, to:

International Astronomical Union/
Working Group on Eclipses
Prof. Jay M. Pasachoff, Chair
Williams College–Hopkins Observatory
Williamstown, MA 01267, USA

Fax: (413) 597-3200
E-mail: eclipse@williams.edu
Web Site: *www.eclipses.info*

The members of the Working Group on Eclipses of the Solar Division of the IAU are: Jay M. Pasachoff (USA), Chair, Iraida S. Kim (Russia), Hiroki Kurokawa (Japan), Jagdev Singh (India), Vojtech Rusin (Slovakia), Zhongquan Qu (Chin), Ralph Chou (Canada), Fred Espenak (USA), Jay Anderson (Canada), Glenn Schneider (USA), Michael Gill (UK), Michael Zeiler (USA), Bill Kramer (USA), Xavier Jubier (France) and Shadia Habbal (USA).

To ensure that astronomers and public health authorities have access to information on safe viewing practices, the IAU's Commission on Education and Development, set up a Program Group on Public Education at the Times of Eclipses and Transits, now combined with the Working Group on Solar Eclipses. Under Prof. Pasachoff, the Commission assembled information on safe methods of observing solar eclipses, eclipse-related eye injuries, and samples of educational materials on solar eclipses. (see *www.eclipses.info*).

For more information, contact Prof. Pasachoff. Further information on safe solar filters can be obtained by contacting Dr. B. Ralph Chou (bchou@uwaterloo.ca).

7.6 AAS Eclipse 2017 Workshops

American Astronomical Society is organizing a series of workshops on the 2017 eclipse. The main goal of these workshops is to pursue the execution of the tasks needed to best prepare the public and the nation for this event. The activities identified by the tasks are driven by the collective and contagious enthusiasm of amateurs and professionals alike. We encourage whoever is interested in getting involved or to just learn more about the Eclipse 2017, to either contact us and/or attend the upcoming workshop.

Dr. Shadia Habbal
E-mail: eclipse2017@ifa.hawaii.edu
Web Site: *https://aas.org/education/outreach/eclipse-2017*

7.7 Solar Eclipse Mailing List

The Solar Eclipse Mailing List (SEML) is an electronic news group dedicated to solar eclipses. Published by British eclipse chaser Michael Gill (eclipsechaser@yahoo.com), it serves as a forum for discussing anything about eclipses and facilitates interaction between both the professional and amateur communities. The SEML is hosted at URL *groups.yahoo.com/group/SEML/.*

7.8 International Solar Eclipse Conference in 2018

Over the last several decades, there have been dramatic changes in solar eclipse traveling. Solar Eclipse specialists meet most of the time during eclipse expeditions. Solar Eclipse meetings outside of an eclipse are rare, or are mainly focused on solar physics. Patrick and Joanne Poitevin, founders of the Solar Eclipse Mailing List (SEML), are the organizers of a series of conferences to bring together professionals and amateurs during years when there is no central solar eclipse.

International Solar Eclipse Conferences have been held in 2000, 2004, 2007, 2011 and 2014. The aim of these meetings is to bring together professionals and amateurs, addicts, enthusiasts, and chasers, as well as members of SEML, in a forum to share information, knowledge, and experience.

Two days of lectures are given in each of the disciplines: predictions, mathematics, solar physics, weather forecasting, eye safety, solar diameter measurement, edge and central solar eclipse observations, and ancient solar eclipse research. The most recent and upcoming solar eclipses are always featured topics of discussion, along with Venus and Mercury Transits.

The next International Solar Eclipse Conference (SEC2018) will be held from on 2018 July 27 to 29 in Tenerife. For more information and to get on the email list for SEC2018, contact Patrick Poitevin (patrick.poitevin@yahoo.com).

7.9 Algorithms, Ephemerides, and Parameters

The algorithms and software for the eclipse predictions were developed primarily from the *Explanatory Supplement* (Her Majesty's Nautical Almanac Office, 1974) with additional algorithms from Meeus et al. (1966), and Meeus (1989). The solar and lunar ephemerides were generated from the JPL DE405. All eclipse calculations were made using a value for the Moon's radius of k=0.2722810 for umbral contacts, and k=0.2725076 (adopted IAU value) for penumbral contacts. Center of mass coordinates for the Moon have been used without correction to the lunar limb profile except where noted. Extrapolating from 2015 April to 2017 August, a value for ΔT of 68.6 seconds was used to convert the predictions from Terrestrial Dynamical Time to Universal Time (UT1).

7.10 Total Solar Eclipse of 2024 April 8

The next total eclipse of the Sun visible from the USA after 2017 is on Monday 2024 April 8. The track of the Moon's umbral shadow begins in the Pacific Ocean and first reaches the coast of North America at Mazatlan, Mexico. Greatest Eclipse occurs near Torreon, Mexico (lat. = 25° 17.4' N, long. = 104° 8.4' W) at 18:17 UT1. The central line duration of totality will then be 4 minutes 28 seconds with the Sun at an altitude of 70°.

Crossing into the USA, the shadow path passes through parts of Texas, Oklahoma, Arkansas, Missouri, Kentucky, Illinois, Indiana, Ohio, Pennsylvania, New York, Vermont, New Hampshire, and Maine. The central line duration of totality slowly drops from 4 minutes 27 seconds in southern Texas to 3 minutes 21 seconds in eastern Maine.

Along the Ohio–Maine section of the track, Lake Erie and Lake Ontario fall within the path as does southern Ontario and Quebec. Beyond Maine, the track crosses New Brunswick, Prince Edward Island and Newfoundland. It ends in the North Atlantic 1400 km (870 miles) west of France.

For more information on the 2024 total solar eclipse, visit *EclipseWise.com* at:

eclipsewise.com/solar/SEprime/2001-2100/SE2024Apr08Tprime.html

Figure 7–1: Total Solar Eclipses of 2024 April 8

The total solar eclipse of 2024 April 8 is the next to cross the continental United Stated after 2017. © 2015 Michael Zeiler

7.11 Total Solar Eclipses Through the USA During the 21st Century

The 2017 eclipse is just the first of 8 total solar eclipses visible from parts of the contiguous United States during the 21st century. The eclipses occur during the years 2017, 2024, 2044, 2045, 2052, 2078, 2079, and 2099. The figure below contains a global map of each of these eclipses.

For more information on each of these eclipses, visit *EclipseWise.com* at:

eclipsewise.com/solar/SEnews/USA21.html

Figure 7–2: Total Solar Eclipses Through the USA During the 21st Century

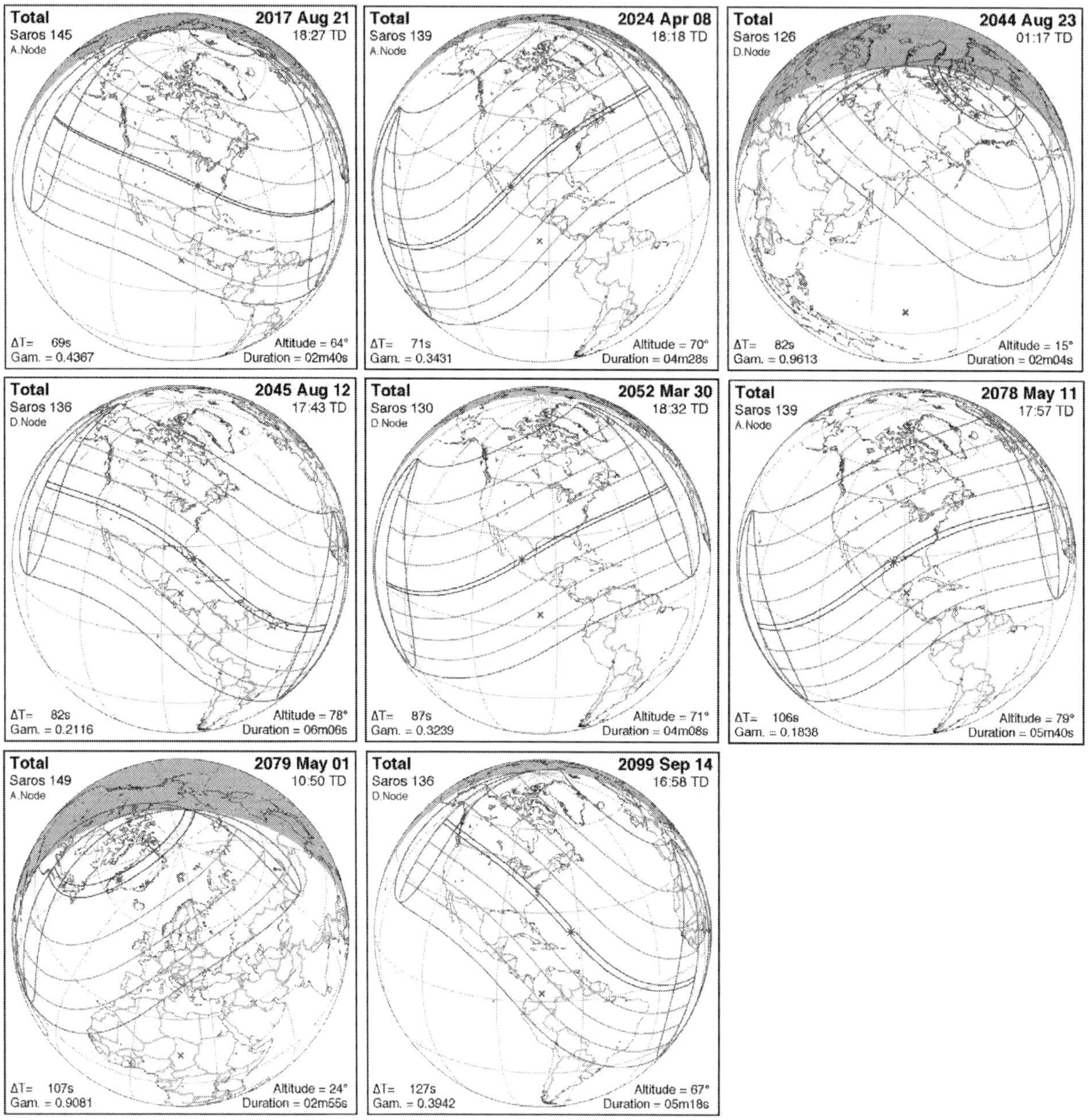

*There are 8 total solar eclipses visible from the continental United Stated during the 21st century. © 2015 by F. Espenak, courtesy of **Thousand Year Canon of Solar Eclipses: 1501 – 2500**, http://astropixels.com/pubs/TYCSE.html*

Bibliography

References

American Conference of Governmental Industrial Hygienists Worldwide (ACGIH): TLVs® and BEIs® Based on the Documentation of the Threshold Limit Values for Chemical Substances and Physical Agents & Biological Exposure Indices, ACGIH, Cincinnati, Ohio, 151–158 (2004).

Astronomical Almanac for 1986, Washington: US Government Printing Office, (1985).

Bretagnon, P., and J. L. Simon, *Planetary Programs and Tables from -4000 to +2800*, Willmann-Bell, Richmond, Virginia, 151 pp (1986).

Chou, B. R., "Safe solar filters". Sky & Telescope, 62(2), 119 pp (1981).

Chou, B. R., "Eye safety during solar eclipses—Myths and realities". In: Z. Mouradian and M. Stavinschi, eds., "Theoretical and Observational Problems Related to Solar Eclipses," Proc. NATO Advanced Research Workshop. Kluwer Academic Publishers, Dordrecht, Germany, 243–247 (1996).

Chou, B. R., and M. D. Krailo, "Eye injuries in Canada following the total solar eclipse of 26 February 1979". Can. J. Optom., 43, 40 (1981).

Del Priore, L. V., "Eye Damage from a Solar Eclipse." In: M. Littmann, F. Espenak and K. Willcox, *Totality, Eclipses of the Sun*, 3d Ed. Oxford University Press, New York, 156–157 (2008).

Espenak, F., *Fifty Year Canon of Solar Eclipses: 1986–2035*, NASA Ref. Pub. 1178, NASA Goddard Space Flight Center, Greenbelt, Maryland, 278 pp (1987).

Espenak, F., *Thousand Year Canon of Solar Eclipses: 1501 to 2500*, Astropixels Pub., Portal, Arizona, 294 pp (2014).

Espenak, F., *Road Atlas for the Total Solar Eclipse of 2017*, Astropixels Pub., Portal, Arizona, 50 pp (2015).

Espenak, F., and J. Meeus, *Five Millennium Canon of Solar Eclipses: -1999 to +3000 (2000 BCE to 3000 CE)*, NASA Tech. Pub. 2006-214141, NASA/GSFC, Greenbelt, Maryland, 648 pp (2006).

Her Majesty's Nautical Almanac Office, *Explanatory Supplement to the Astronomical Ephemeris and the American Ephemeris and Nautical Almanac*, prepared jointly by the Nautical Almanac Offices of the United Kingdom and the United States of America, London, 534 pp (1974).

Herald, D., "Correcting predictions of solar eclipse contact times for the effects of lunar limb irregularities". J. Brit. Ast. Assoc., 93, 241–246 (1983).

Marsh, J. C. D., "Observing the Sun in safety". J. Brit. Ast. Assoc., 92, 6 (1982).

Meeus, J., C. C. Grosjean, and W. Vanderleen, *Canon of Solar Eclipses*, Pergamon Press, N.Y., 779 pp. (1966)

Michaelides, M., R. Rajendram, J. Marshall, and S. Keightley, "Eclipse retinopathy". Eye, 15, 148–151 (2001).

Pasachoff, J. M., "Public Education in Developing Countries on the Occasions of Eclipses." In: A.H. Batten, Ed., Astronomy for Developing Countries, IAU special session at 24th General Assembly, 101–106 (2001).

Pasachoff, J. M., and M. Covington, *Cambridge Guide to Eclipse Photography*, Cambridge University Press, Cambridge and New York, 143 pp (1993).

Penner, R., and J. N. McNair, "Eclipse blindness—Report of an epidemic in the military population of Hawaii". Am. J. Ophthal., 61, 1452–1457 (1966).

Pitts, D. G., "Ocular Effects of Radiant Energy." In: D.G. Pitts and R.N. Kleinstein, Eds., Environmental Vision: Interactions of the Eye, Vision and the Environment, Butterworth-Heinemann, Toronto, Canada, 151 pp (1993).

Rand McNally, *The New International Atlas*, Chicago/New York/San Francisco, 560 pp (1991).

Reynolds, M. D., and R. A. Sweetsir, *Observe Eclipses*, Astronomical League, Washington, DC, 92 pp (1995).

Sherrod, P. C., *A Complete Manual of Amateur Astronomy*, Prentice-Hall, 319 pp (1981).

van den Bergh, *Periodicity and Variation of Solar (and Lunar) Eclipses*, Tjeenk Willink, and Haarlem, Netherlands (1955).

Further Reading on Eclipses

Brewer, B., *Eclipse*, Earth View, Seattle, Washington, 104 pp (1991).

Brunier, S., *Glorious Eclipses*, Cambridge University Press, New York, 192 pp (2001).

Covington, M., *Astrophotography for the Amateur*, Cambridge University Press, Cambridge, 346 pp.

Espenak, F., and J. Meeus., *Five Millennium Catalog of Solar Eclipses: –1999 to +3000 (2000 BCE to 3000 CE)*, NASA Tech. Pub. 2008-214170, NASA GSFC, Greenbelt, Maryland, 270 pp (2008).
Espenak, F., *Thousand Year Canon of Lunar Eclipses: 1501 to 2500*, Astropixels Pub., Portal, AZ, 298 pp (2014).
Golub, L., and J. M. Pasachoff, *The Solar Corona*, Cambridge University Press, Cambridge, MA, 390 pp (2010).
Golub, L., and J. M. Pasachoff, *Nearest Star: The Surprising Science of Our Sun*, Cambridge University Press, Cambridge, MA, 297 pp (2014).
Guillermier, P., and S. Koutchmy, *Total Eclipses, Science, Observations, Myths and Legends*, Springer, New York, 247 pp (1999).
Harrington, P. S., *Eclipse!*, John Wiley and Sons, New York, 280 pp (1997).
Harris, J., and R. Talcott, *Chasing the Shadow: An Observer's Guide to Solar Eclipses*, Kalmbach Publishing Company, Waukesha, Wisconsin, 160 pp (1994).
Littmann, M., F. Espenak, and K. Willcox, *Totality, Eclipses of the Sun*, 3d Ed. Oxford University Press, New York, 341 pp (2008).
Maunder, M., and P. Moore, *The Sun in Eclipse*, , Springer, New York, 211 pp (1998).
Mitchell, S. A., *Eclipses of the Sun*, Columbia University Press, New York, 425 pp (1923).
Meeus, J., *Elements of Solar Eclipses: 1951–2200*, Willmann-Bell, Inc., Richmond, Virginia, 112 pp. (1989)
Meeus, J., *Astronomical Algorithms*, Willmann-Bell, Inc., Richmond, Virginia, 477 pp (1998).
Mobberley, M., *Total Solar Eclipses and How to Observe Them*, Springer, New York, 202 pp (2007).
Mucke, H., and Meeus, J., *Canon of Solar Eclipses: –2003 to +2526*, Astronomisches Büro, Vienna, Austria, 908 pp (1983).
North, G., *Advanced Amateur Astronomy*, Edinburgh University Press, Edinburgh, Scotland, 441 pp (1991).
Ottewell, G., *The Understanding of Eclipses*, Astronomical Workshop, Greenville, South Carolina, 96 pp (1991).
Pasachoff, J.M., *Field Guide to the Stars and Planets*, 4th edition, Houghton Mifflin, Boston, Massachusetts, 578 pp (2000).
Pasachoff, J. M., *The Complete Idiot's Guide to the Sun*, Alpha Books, Indianapolis, Indiana, 360 pp (2004).
Pasachoff, J. M., "Observing solar eclipses in the developing world. In: Astronomy in the Developing World, Proc. IAU Special Session 5, J.B. Hearnshaw and P. Martinez, eds., Cambridge University Press, New York, 265–268 (2007).
Russo, K., *Total Addiction: The Life of an Eclipse Chaser*, Springer, New York, 193 pp (2012).
Steel, D., *Eclipse, The Celestial Phenomenon That Changed the Course of History*, Joseph Henry Press, Washington, DC, 492 pp (2001).
Todd, M. L., *Total Eclipses of the Sun*, Little, Brown, and Co., Boston, Massachusetts, 273 pp (1900).
von Oppolzer, T. R., *Canon of Eclipses*, Dover Publications, New York, 376 pp (1962).
Zirker, J. B., *Total Eclipses of the Sun*, Princeton University Press, Princeton, New Jersey, 228 pp (1995).

Further Reading on Eye Safety

Chou, B. R., "Solar filter safety". Sky & Telescope, 95(2), 119 (1998).
Pasachoff, J. M., "Public education and solar eclipses." In: L. Gouguenheim, D. McNally, and J.R. Percy, Eds., New Trends in Astronomy Teaching, IAU Colloquium 162, (London), Astronomical Society of the Pacific Conference Series, 202–204 (1998).

Further Reading on Meteorology

Griffiths, J. F., Ed., World Survey of Climatology, Vol. 10, Climates of Africa, Elsevier Pub., N.Y., 604 pp (1972).
National Climatic Data Center, International Station Meteorological Climate Summary; Vol. 4.0 (CD-ROM), NCDC, Asheville, North Carolina (1996).
Wallen, C. C., Ed.,: *World Survey of Climatology*, Vol. 6, Climates of Central and Southern Europe, Elsevier Publishing Company, New York, 258 pp (1977).
Warren, S. G., C. J. Hahn, J. London, R. M. Chervin, and R. L. Jenne,, Global *Distribution of Total Cloud Cover and Cloud Type Amounts Over Land.* NCAR Tech. Note NCAR/TN-273+STR and DOE Tech. Rept. No. DOE/ER/60085-H1, U.S. Department of Energy, Carbon Dioxide Research Division, Washington, DC, (NTIS number DE87-006903), 228 pp (1986).

Eclipse Marketplace

Great American Eclipse®

GreatAmericanEclipse.com is your resource for detailed maps, infographics, and animations for the August 21, 2017 total solar eclipse.

We are your partner in sharing the spectacle, wonder, and science of this sight of a lifetime with the American public.

Organizing a local event? Let us know and we will publicize it on our webpage dedicated to your state.

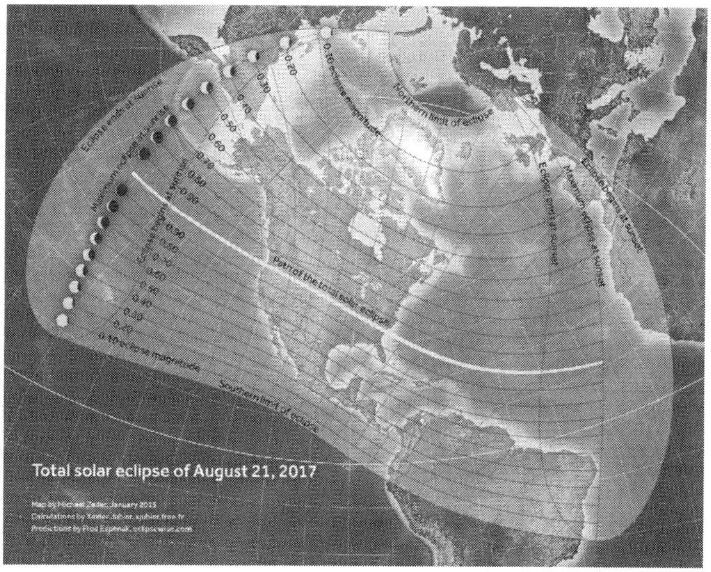

Visit **GreatAmericanEclipse.com/Store** for quality maps, apparel, eclipse viewers, and more.

Fundraising, reselling, or promoting? Astronomy clubs, telescope and science shops, and tour operators can receive quantity discounts for bulk and co-branded orders. Email us at greatamericaneclipse@gmail.com

Download our app for iOS and Android smartphones and tablets at **GreatAmericanEclipse.com/App** with

- interactive maps and animations
- strategies for successful eclipse viewing
- a primer on eclipse theory
- a gallery of spectacular eclipse photography
- a survey of the science done at eclipses
- facsimile reproductions of historical eclipse items

Our maps are based on eclipse predictions by Fred Espenak and we thank Fred for his decades of service to the eclipse community!

GreatAmericanEclipse

@AmericanEclipse

GreatAmericanEclipse

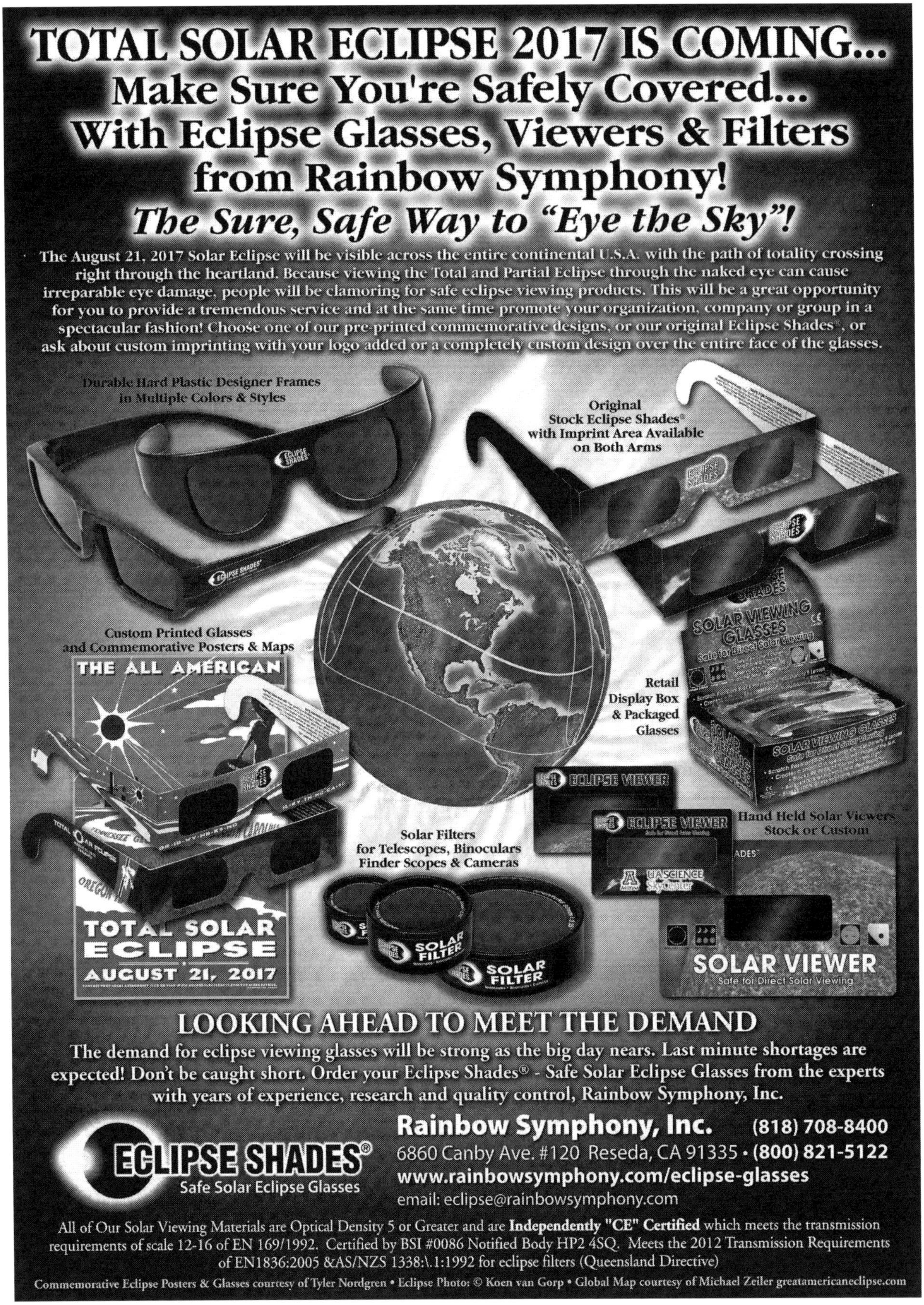
TOTAL SOLAR ECLIPSE 2017 IS COMING...
Make Sure You're Safely Covered...
With Eclipse Glasses, Viewers & Filters
from Rainbow Symphony!
The Sure, Safe Way to "Eye the Sky"!
The August 21, 2017 Solar Eclipse will be visible across the entire continental U.S.A. with the path of totality crossing right through the heartland. Because viewing the Total and Partial Eclipse through the naked eye can cause irreparable eye damage, people will be clamoring for safe eclipse viewing products. This will be a great opportunity for you to provide a tremendous service and at the same time promote your organization, company or group in a spectacular fashion! Choose one of our pre-printed commemorative designs, or our original Eclipse Shades®, or ask about custom imprinting with your logo added or a completely custom design over the entire face of the glasses.
Durable Hard Plastic Designer Frames in Multiple Colors & Styles
Original Stock Eclipse Shades® with Imprint Area Available on Both Arms
Custom Printed Glasses and Commemorative Posters & Maps
THE ALL AMERICAN
TOTAL SOLAR ECLIPSE AUGUST 21, 2017
Retail Display Box & Packaged Glasses
SOLAR VIEWING GLASSES
ECLIPSE VIEWER
Hand Held Solar Viewers Stock or Custom
Solar Filters for Telescopes, Binoculars Finder Scopes & Cameras
SOLAR FILTER
SOLAR VIEWER
Safe for Direct Solar Viewing
LOOKING AHEAD TO MEET THE DEMAND
The demand for eclipse viewing glasses will be strong as the big day nears. Last minute shortages are expected! Don't be caught short. Order your Eclipse Shades® - Safe Solar Eclipse Glasses from the experts with years of experience, research and quality control, Rainbow Symphony, Inc.
ECLIPSE SHADES®
Safe Solar Eclipse Glasses
Rainbow Symphony, Inc. (818) 708-8400
6860 Canby Ave. #120 Reseda, CA 91335 • (800) 821-5122
www.rainbowsymphony.com/eclipse-glasses
email: eclipse@rainbowsymphony.com
All of Our Solar Viewing Materials are Optical Density 5 or Greater and are Independently "CE" Certified which meets the transmission requirements of scale 12-16 of EN 169/1992. Certified by BSI #0086 Notified Body HP2 4SQ. Meets the 2012 Transmission Requirements of EN1836:2005 &AS/NZS 1338:\.1:1992 for eclipse filters (Queensland Directive)
Commemorative Eclipse Posters & Glasses courtesy of Tyler Nordgren • Eclipse Photo: © Koen van Gorp • Global Map courtesy of Michael Zeiler greatamericaneclipse.com

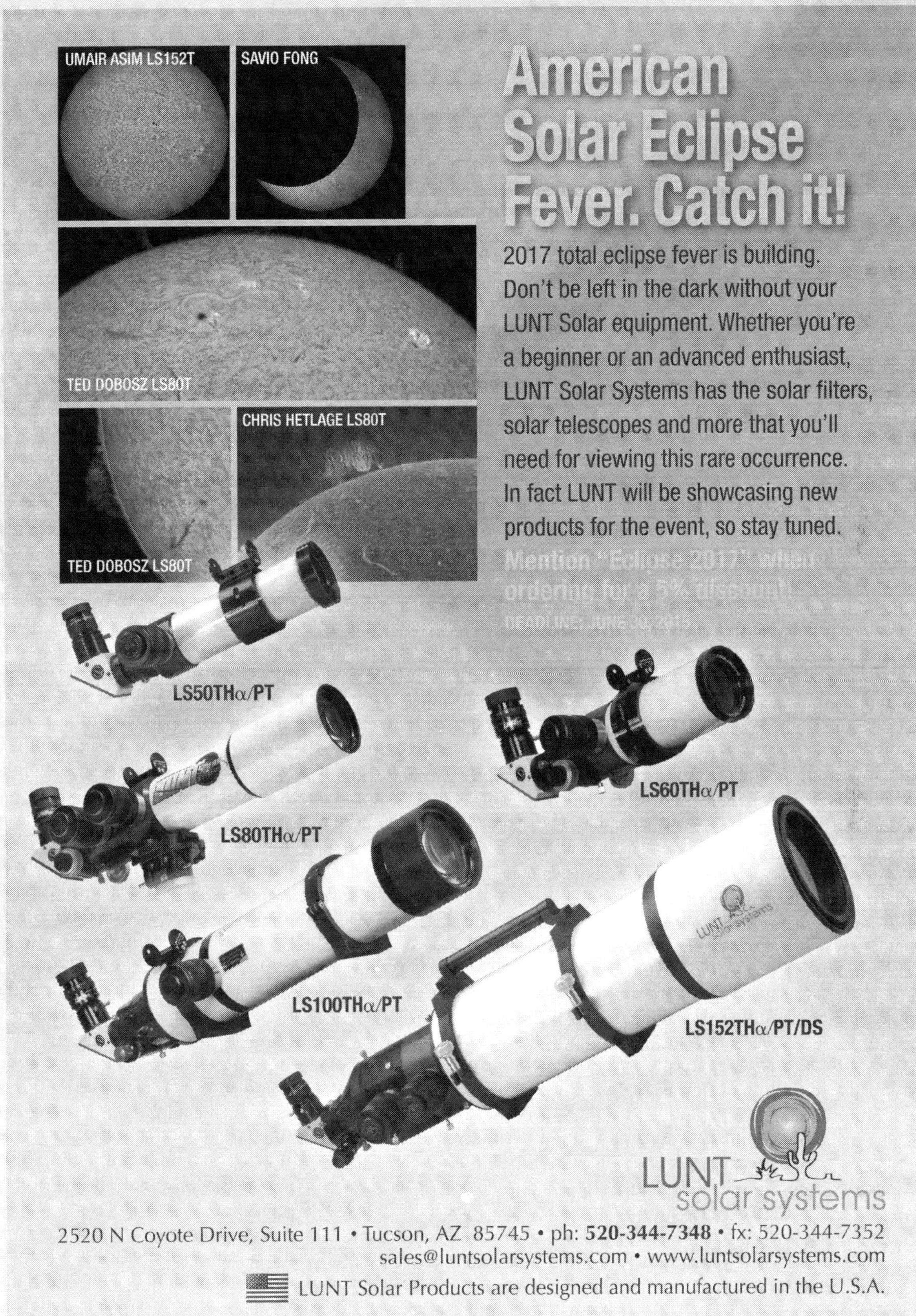
UMAIR ASIM LS152T
SAVIO FONG
TED DOBOSZ LS80T
CHRIS HETLAGE LS80T
TED DOBOSZ LS80T
American Solar Eclipse Fever. Catch it!
2017 total eclipse fever is building. Don't be left in the dark without your LUNT Solar equipment. Whether you're a beginner or an advanced enthusiast, LUNT Solar Systems has the solar filters, solar telescopes and more that you'll need for viewing this rare occurrence. In fact LUNT will be showcasing new products for the event, so stay tuned.
LS50THα/PT
LS60THα/PT
LS80THα/PT
LS100THα/PT
LS152THα/PT/DS
LUNT solar systems
2520 N Coyote Drive, Suite 111 • Tucson, AZ 85745 • ph: 520-344-7348 • fx: 520-344-7352
sales@luntsolarsystems.com • www.luntsolarsystems.com
LUNT Solar Products are designed and manufactured in the U.S.A.

SE
USA
French Quarter
TRAVELQUEST
international
TravelQuestTours.com
1 928 445-7754
Join us and expect the extraordinary.

Astronomy Tours
melitatrips.com
2016
Java Sea
TOTALITY
2017
U.S.A.
Reservations & More Info
1.877.707.7827

www.SpectrumTelescope.com
High Quality Solar Filters
We offer safe and effective
solar filters for use
with any optical device.
Visit our website today to view our full catalog!
Spectrum Telescope

MrEclipse.com
Eclipse & Astronomy Images
for Books, Magazines, Pubs, etc.
MrEclipse.com/main/photoindex.html

The Great American Eclipse
NASHVILLE, TN • August 17 – 22, 2017
http://InsightCruises.com/Sky-8
Join Sky & Telescope for The Great American Eclipse of August 2017. View the big event in Hopkinsville, KY near the point of greatest eclipse and enjoy Nashville's many attractions.
SKY
& TELESCOPE

Solar Eclipse Tours
2016- INDONESIA
3 tour options- 6/10/16 days including Balinese New Year
2017 – USA – OREGON
Tour with World Renowned Astronomer
Prof. Jay Pasachoff
A Classic Tours Collection
AClassicTour.com
(888) 605 8687 - (310) 373 1001
info@aclassictour.com
CA CST#2091259-40
Since 1980
'Successful Eclipse trip 2015 Svalbard'
Call Now! for Reservations

DON'T DO THIS
NEVER LOOK
STRAIGHT INTO THE SUN
without protecting your eyesight with
the proper solar filter
visit:
AstroSolar.com

Suit Up For Totality: 2017 August 21!
TOTAL SOLAR ECLIPSE
2017 August 21
Saros 145
www.cafepress.com/mreclipse

49601571R10088

Made in the USA
Charleston, SC
28 November 2015